Frequently Used Formulas

Chapter 2

1. $m = \dfrac{\Sigma X}{n}, \quad \overline{X} = \dfrac{\Sigma X}{n}$

2. $s = \sqrt{\dfrac{\Sigma(X - m)^2}{n - 1}}$

3. $s = \sqrt{\dfrac{\Sigma X^2 - \dfrac{(\Sigma X)^2}{n}}{n - 1}}$

4. $z_X = \dfrac{X - \mu}{\sigma}, \quad z_X = \dfrac{X - \mu}{s}, \quad z_X = \dfrac{X - m}{s}$

5. $X = \mu + z\sigma, \quad X = \mu + zs, \quad X = m + zs$

6. $PR_X = \dfrac{B + \frac{1}{2}E}{n}$ (100)

Chapter 4

1. $p = P(\text{event}) = \dfrac{F}{T}$

2. $q = 1 - p$

Chapter 7

1. $\mu_S = np$
2. $\sigma_S = \sqrt{npq}$
3. $S_? = \mu_S + z\sigma_S$

Chapter 8

1. $\mu_S = np$
2. $\sigma_S = \sqrt{npq}$
3. $S_c = \mu_S + z_c\sigma_S$
4. Power $= 1 - \beta$

Chapter 9

1. $\mu_{d\hat{p}} = p_1 - p_2$

2. $\hat{\sigma}_{d\hat{p}} = \sqrt{\dfrac{\hat{p}\hat{q}}{n_1} + \dfrac{\hat{p}\hat{q}}{n_2}}$ or $\sqrt{\hat{p}\hat{q}\left(\dfrac{1}{n_1} + \dfrac{1}{n_2}\right)}$

3. Sample difference $d\hat{p} = \hat{p}_1 - \hat{p}_2$
4. Critical difference $d\hat{p}_c = \mu_{d\hat{p}} + z_c\hat{\sigma}_{d\hat{p}}$

Chapter 10

One-Sample Tests

1. $\mu_m = \mu_{\text{pop}}$

2. $s = \sqrt{\dfrac{\Sigma X^2 - \dfrac{(\Sigma X)^2}{n}}{n - 1}}$

3. $s_m = \dfrac{s}{\sqrt{n}}$

4. $m_c = \mu_m + z_c s_m$

5. Experimental outcome, $m = \dfrac{\Sigma X}{n}$

Two-Sample Tests

6. $\mu_{dm} = \mu_1 - \mu_2$
   (if H_0 states that $\mu_1 = \mu_2$, then $\mu_1 - \mu_2 = 0$)

7. $s_{dm} = \sqrt{\dfrac{s_1^2}{n_1} + \dfrac{s_2^2}{n_2}}$

8. $dm_c = \mu_{dm} + z_c s_{dm}$

9. Experimental outcome, $dm = m_1 - m_2$

(Continued on back cover)

Understanding Statistics

THIRD EDITION

UNDERSTANDING STATISTICS

ARNOLD NAIMAN
Late Professor
Nassau Community College

ROBERT ROSENFELD
Nassau Community College

GENE ZIRKEL
Nassau Community College

McGRAW-HILL BOOK COMPANY
New York   St. Louis   San Francisco   Auckland
Bogotá   Hamburg   Johannesburg   London   Madrid
Mexico   Montreal   New Delhi   Panama   Paris
São Paulo   Singapore   Sydney   Tokyo   Toronto

UNDERSTANDING STATISTICS

Copyright © 1983, 1977, 1972 by McGraw-Hill, Inc.
All rights reserved.
Printed in the United States of America.
Except as permitted under the United States Copyright Act of 1976,
no part of this publication may be reproduced
or distributed in any form or by any means,
or stored in a data base or retrieval system,
without the prior written permission of the publisher.

567890 DOCDOC 89876

ISBN 0-07-045863-4

This book was set in Helvetica by Waldman Graphics, Inc.
The editors were John J. Corrigan,
Stephen Wagley, and Jo Satloff;
the designer was Nicholas Krenitsky;
the production supervisor was Charles Hess.
New drawings were done by J & R Services, Inc.
R. R. Donnelley & Sons Company was printer and binder.

Library of Congress Cataloging in Publication Data

Naiman, Arnold.
  Understanding statistics.

  Includes index.
  1. Statistics. I. Rosenfeld, Robert.
II. Zirkel, Gene. III. Title.
QA276.12.N34 1983   519.5   82-17183
ISBN 0-07-045863-4

In memory of our colleague
DR. ARNOLD NAIMAN

Contents

PREFACE ix

CHAPTERS
1 INTRODUCTION 1
2 COMMON STATISTICAL MEASURES 11
3 FREQUENCY TABLES AND GRAPHS 36
4 PROBABILITY 55
5 THE BINOMIAL DISTRIBUTION 67
6 THE NORMAL DISTRIBUTION 82
7 APPROXIMATION OF THE BINOMIAL DISTRIBUTION BY USE OF THE NORMAL DISTRIBUTION 98
8 HYPOTHESIS TESTING: BINOMIAL ONE-SAMPLE 111
9 HYPOTHESIS TESTING: BINOMIAL TWO-SAMPLE 141
10 HYPOTHESIS TESTING WITH SAMPLE MEANS: LARGE SAMPLES 155
11 HYPOTHESIS TESTING WITH SAMPLE MEANS: SMALL SAMPLES 171
12 CONFIDENCE INTERVALS 188
13 CHI-SQUARE TESTS 204
14 CORRELATION AND PREDICTION 229
15 TESTS INVOLVING VARIANCE 250
16 NONPARAMETRIC TESTS 269

APPENDIXES
A ARITHMETIC REVIEW 291
B PROBABILITY 294
C TABLES 302
ANSWERS TO SELECTED ODD-NUMBERED EXERCISES 326
INDEX 353

Preface

We wrote the first edition of *Understanding Statistics* with our late colleague Dr. Arnold Naiman for students with little mathematical background, such as those we had all worked with at Nassau Community College for a number of years. We wanted a text that would be elementary enough to reach these students and still be mathematically sound and suitable for a one-semester college-level course. We are pleased that the text has "worked" for so many people. Over a period of 10 years, through two editions, many students and professors who have used the book have given us their comments on it. We have learned from them that its major strengths are its readability, its often humorous approach, and its problem sets. In putting together this third edition, we have tried to maintain these strengths while responding to users who have recommended changes.

The objective of this book is to show readers how statistics is used, not to train them to be statisticians. Students using it will gain an appreciation of the proper use of statistics and statistical terms that confront them in textbooks, newspapers, magazines, and on TV and radio. Our major emphasis is on understanding sampling and hypothesis testing.

In this edition we have included a class survey at the beginning of the book so that students will have real data *about themselves* to use. We then pose problems based on these data in special sets of questions at the end of most exercise sets. We have found this option particularly useful during the early part of the course.

We feel that the best way to introduce inferential statistics is through probability theory. Therefore, after a brief discussion of descriptive statistics, probability is treated intuitively. This leads into the binomial distribution, and the normal distribution is then introduced as an approximation to the binomial distribution. Chapter 8 then brings everything together and discusses the method

Preface

of statistical hypothesis testing. One-sample binomial tests are used to introduce this important idea.

The next few chapters discuss other types of hypothesis testing. Two-sample binomial, one- and two-sample tests of sample means with both large samples (z scores) and small samples (t scores), chi-square tests, and tests about population variance (including a light introduction to analysis of variance) are included. A chapter on correlation and prediction and one on nonparametric tests conclude the text. Once students have mastered the basic material through Chapter 8, the instructor can select from the remaining chapters those topics appropriate to the needs of his or her students.

No formal proofs are presented. When feasible, theorems are motivated by an appeal to common sense. While this presentation is not mathematically rigorous, care is taken that the material is at all times mathematically accurate. Topics are introduced informally by questions and examples that lead naturally to the development of pertinent ideas. Notation is kept as simple as possible, and illustrations are used throughout for clarification.

Numerous examples and exercises are provided from various fields, including biology, medicine, business, psychology, education, and political science. Ranging from the frivolous to the serious, they have been chosen carefully to arouse student interest. They are not just lists of numerical exercises. In this edition we have added many new exercises including a nonroutine, thought-provoking question at the beginning of most problem sets.

A glossary of new words, symbols, and formulas is given at the end of each chapter, and answers to odd-numbered exercises are given at the end of the book.

Appendix A contains a selection of typical arithmetic problems that illustrate the mathematical skills needed for the material in the book. We strongly recommend that each student do these problems at the beginning of the course. Students should be able to handle signed numbers, but no manipulative skills from algebra are needed.

A number of changes have been made for this edition. In Chapter 2 we have added some elementary material on rates, especially birth rates and mortality rates. We rewrote Chapter 3 to focus on the histogram as a picture of a frequency table, and put less emphasis on details of intervals, boundaries, and so on. In Chapter 5 we now include a section on solving binomial problems with the aid of binomial probability tables, which we include in the appendix of tables.

In Chapter 13 we have expanded the section on 2 by 2 contingency tables, and we point out the relation between such problems and two-sample binomial problems and give a simplified way to compute X^2 for 2 by 2 tables. In Chapter 16 we have added the Mann-Whitney U test for comparing two samples. We have restored the appendix on probability written for the first edition by Dr. Naiman.

There have been a few notation changes. In Chapter 12 the notation on confidence intervals has been simplified, with less use of inequality symbols. We continue to mention both $\bar{X}$ and m for sample means, and although we use the symbol m most of the time, we do use $\bar{X}$ more than in the past. We have changed the symbol for the statistic in the runs test from U to R, and now reserve the U for the new material on the Mann-Whitney U test.

Since calculators and computers have become commonplace, we deleted most of the material on coding that is fast becoming obsolete. What is left is relegated to a few exercises that focus on the properties of the mean and the standard deviation. Similarly, the square root table was deleted, as was the random numbers table, since most teachers reported that they do not use them. Also in connection with the use of calculators, we have included some material in Chapter 1 on rounding off and working with approximate numbers.

Amid all these changes, we hope that you still find the essentials that have made our text popular over the years: simplicity, accuracy, and a blend of humorous examples with real-world problems. These are elements that have motivated students with little math background and interest.

We wish to thank our colleagues at Nassau Community College for their encouragement and helpful suggestions, particularly Professors Frank Avenoso, James Baldwin, Eli Berlinger, Alice Berridge, Mauro Cassano, Dennis Christy, Jerry Kornbluth, George Miller, Aaron Schein, Michael Steuer, and Abraham Weinstein. Thanks also to Roy McLeod of LaGuardia Community College.

For helpful detailed criticism of the manuscript we would like to thank Professors Daniel Brunk; Wilfrid Dixon, UCLA; James Edmondson, Santa Barbara City College; and Paul Kroll, formerly of William Paterson College of New Jersey. For this edition in particular we wish to thank Sister Mary Erwin Baker, Saint Mary's College; Professors Donald Evans, Polk Community College; David M. Crystal, Rochester Institute of Technology; John S. Mowbray, Shippensburg State College; Norman Neff, Trenton State College; Charles A. Oprian, Western Illinois University; Ronald E. Pierce, Eastern Kentucky University; Maxine D. Reed, Tennessee State Technical Institute at Memphis; William M. Self, Pittsburg State University; William Scott, Ocean County College; and Patricia L. Smith, Old Dominion University.

We are grateful to all who helped and whose suggestions made these improvements possible.

ROBERT ROSENFELD
GENE ZIRKEL

Introduction

1

When you applied to college, you probably filled out a form designed to establish your "financial need." Maybe you took the SAT or ACT exam. Then some stranger used these numbers to help decide what college you got into and how much money you had to pay. To a large degree you were being treated as a collection of numerical values, a collection of statistics.

You should understand how statistics are used because decisions that affect you personally are based on statistics. "Your grade-point average is only 1.47; sorry kid, we'll have to put you on probation. I know that you just had a bad time with your parents, but let's face it, everybody's got problems." That hurts.

There are probably times when you have mixed feelings about being treated as a source of statistics.

DEAR FRIEND:
 ØUR RELIABLE CØMPUTERIZED MATCHING SYSTEM HAS PRØCESSED YØUR VITAL STATISTICS AND WE ARE PLEASED TØ ENCLØSE HEREWITH THE NAMES, ADDRESSES (WITH ZIP CØDES) AND PHØNE NUMBERS (WITH AREA CØDES) ØF SIX IDEAL MATCHES. WE ARE HØPEFUL THAT YØU CAN ESTABLISH A LASTING RELATIØNSHIP WITH AT LEAST ØNE ØF THEM.
 WE REMIND YØU THAT YØUR FEE IS NØT REFUNDABLE.
  SINCERELY,
  CØMPUTERIZED DATING SERVICE

Products are sold to you all the time with numbers thrown at you: (1) "I used Grit toothpaste, and now I have 20 percent fewer cavities." (Fewer than what?) (2) "Hey, kids! Start your day with Daystart Cereal! It has twice as much iron as a delicious slice of toast and more vitamin C than two slices of bacon!" (So who said that bread was a good source of iron in the first place, or that bacon has a lot of vitamin C?)

Doctors prescribe medicine and treatment for you, basing their judgment on statistical information. (1) Use of this pill will cause deleterious side effects in 1.4 percent of its users. (Is the risk worth taking?) (2) There is a 40 percent chance that an adult suffering from a herniated spinal disk will recover spontaneously. (Should we go ahead with the operation?)

TWO USES OF STATISTICS

Consider the following example. Maybe you have participated in this kind of survey if you live near a big shopping mall.

> A random sample of adults was taken in a large shopping plaza in the city of Niles. Of those questioned, 15 percent used NoCav brand toothpaste. Subsequently a concentrated advertising campaign was undertaken to sell NoCav to the public. A second survey taken 3 weeks after this campaign showed that 19 percent of those questioned used NoCav toothpaste.

Are we correct in assuming that the rise from 15 percent in the first sample to 19 percent in the second sample is due to advertising? If we have doubts that the advertising caused a substantial increase in the use of NoCav toothpaste, what questions should we ask concerning the data presented? What about the data that were omitted from the presentation?

In this example we see numbers used in two different ways. The number 15 percent is used to describe the fraction of people in the first sample who used NoCav. As such it summarizes with conciseness and clarity the unreported fact that of 140 persons interviewed, 21 used NoCav. This is an example of descriptive statistics. **Descriptive statistics** is the use of numbers to summarize information which is known about some situation. In contrast to this use of numbers, if we use this sample to imply that approximately 15 percent of *all* the adults in Niles used NoCav, then we are using the number to infer something about a larger population for which we do not have complete information. This is an example of statistical inference. **Statistical inference** is the use of numbers to give numerical information about larger groups than those from which the original raw data were taken.

In characterizing a large amount of data by a few descriptive statistics we gain clarity and compactness, but we lose detail. The following statistics, by summarizing information, describe in some way the populations from which they were taken.

1. The average IQ at Nostrum College is 109.

2. The marks on the last exam ranged from 51 to 98.

3. Nielsen reports that 25 percent of those who were interviewed watched the President's news conference last Sunday night.

The following are examples of statistical inference. We might infer from appropriate samples that

1. Between 20 and 25 percent of American college students are married.

2. Cholesterol level and heart trouble are related.

3. 25 percent of all television viewers watched the President's news conference last Sunday.

Here is a more detailed example of statistical inference.

> Suppose there were a disease in which three-fourths of the patients recovered without treatment within 3 months of contracting the disease. Suppose also that a doctor claims to have discovered a new drug to cure this disease. We shall administer the drug to 100 patients. Even if the drug were useless, we would still expect about 75 (three-fourths) of these people to recover. Due to chance variations more or less than 75 may recover.

One of the problems of statistical inference for the example given above is to decide how many must recover before we are willing to accept the drug as a cure. Certainly if all 100 recovered, we would be enthusiastic about the drug's potential. But how about 95 or 90 or 80 recoveries? Where should we draw the line?

The job of deciding where to draw the line is an important one for the statistician. It is one of the main skills we hope you will develop from this book. Can we confidently say that the new drug saves lives, or is it likely that this result occurred by chance? Even if all 100 recovered, it is possible (though very unlikely) that they would have recovered anyway. Perhaps, just by luck, this particular group of 100 patients was unusually resistant to the disease.

It is important for the statistician to pick the sample in an impartial way. If by chance we happened to test the drug on only mild cases, our results would be misleading. We would hope that the sample is truly a mirror of the population we want to learn about (in this case all victims of this disease).

Sample surveys, polls, and statistical tests have become a part of our way of life. Every day, people present figures to prove or disprove some claim: Does a certain food additive cause cancer? Does smoking marijuana lead to heroin use? In this book we will study some of the tests that statisticians use when making claims. We hope to show you how such tests should be done properly and how to interpret the "proof" of such claims.

SOME STATISTICAL TERMS

If we are listing the ages of students in a certain school, then each age is called a **raw score.** In general, a raw score is any number as it originally appears in an experiment. A collection of such scores about one particular thing is frequently referred to as a **distribution** of scores. If we consider the grades that your class gets on the first test in this course, then you will be very interested not only in the entire distribution of the class's grades, but also in one particular raw score, namely, your own grade on the test.

Often we collect raw scores about several different things. We may, for instance, collect information on heights, weights, and ages of the people who belong to a particular organization. All the information we have is called our **data.** By the way, "data" is a plural noun. For example, you should say, "The data *show* that X is more popular than Y." It is incorrect to say "The data *shows* that . . . ." The singular of data is datum.

Understanding statistics

The word **population** is used to refer to *all* the persons, objects, scores, or measurements under consideration. The word **sample** refers to any portion of the population. A population may be large or small.

Suppose a scientist is trying to determine the average weight of all 1-year-old male white rabbits which are raised in laboratories using a certain diet. It is impossible for her to weigh every rabbit in the population because the population never exists completely at any one time. If she selects 50 rabbits and determines their average weight, these 50 would be referred to as a sample from the population.

We used the word **random** in examples at the beginning of this chapter. No word is more important to the theory of inferential statistics than this word. An item is chosen "at random" from a population if in the selection every item in the population has the same probability or chance of being selected; the *process* of selection does not favor any particular item either intentionally or inadvertently. A sample of items in which each item is chosen this way is called a **random sample.** Entire textbooks have been written describing procedures for selecting random samples, and the process can become quite technical. In this text we prefer to leave the idea to your intuition. It will be sufficient to think of a random sample as one that has been picked "fairly"—without prejudicing the chances of any member of the population to be chosen. For example, if we want to pick a random sample of 20 people from some population, then every possible grouping of 20 people should have an equal chance of being selected as the sample. The practice of putting paper slips into a large drum, mixing them well, and then picking one without looking is a simple model of random selection.

Statistical testing is frequently based upon the assumption that the sample was picked randomly. If it turns out that the sample was not random, the results may not be useful. Hidden, unsuspected bias can completely destroy the usefulness of statistical information and statistical inferences made from such information. For example, if random phone calls were made at 1 P.M. to sample the population of all voters, many people with full-time day jobs would be missed.

Note: The word *random* describes the *process* by which the sample was chosen. This does *not guarantee* that the sample will turn out to be representative.

ARITHMETIC, CALCULATORS, AND ROUNDING OFF

During this course you will have to do a lot of problems, and most likely you will use a calculator for some of the arithmetic. One of the characteristics of calculators is that they routinely display a lot of digits, often more than make sense in a particular problem. This means that you will want to round off your results, to some convenient *approximate* value. This section gives you some informal rules for reasonable rounding off.

BASIC RULE FOR ROUNDING OFF AT A GIVEN DIGIT

> Look at the digit following the one that is to be rounded. If it is 4 or less, simply drop it; but if it is 5 or more, then add one to the digit that is being rounded.

Introduction

EXAMPLE 1-1 Round these numbers off to two decimal places: 16.837, 8.00319, 9.105, and 10.1349.

SOLUTION

original data	next digit	rule	result
16.837	7	add 1 to 3	16.84
8.00319	3	drop	8.00
9.105	5	add 1 to 0	9.11
10.1349	4	drop	10.13

Notice that in rounding the last number we did *not* round off twice. A common *mistake* is to first change 10.1349 to 10.135 and then to 10.14. This is not correct. We look *only* at the next digit when rounding off. ∎

EXAMPLE 1-2 Round off these numbers to the nearest hundred: 5826, 9084, 163.7, and 4041.

SOLUTION

original data	next digit	rule	result
5826	2	drop	5800
9084	8	add 1 to 0	9100
163.7	6	add 1 to 1	200
4041	4	drop	4000

Note two things here. The first answer is 5800 and not 58. We need the zeros to indicate that the number is large, in the thousands. Note also that in the last answer the first zero is accurate and the last two are not. When it is important to indicate this, we place a bar over the last accurate digit. In this case we would write $4\bar{0}00$.

In the same vein we only write final zeros after a decimal point if needed. The numbers 6.0, 6, and 6.00 are all different if they are approximate values. ∎

CALCULATING WITH APPROXIMATE NUMBERS

If we find the average of the eight numbers 6, 9, 9, 0, 7, 7, 5, and 4, we obtain 47/8 = 5.875. Now if these eight numbers represent the number of people in eight families, then they are **exact numbers.** No family had *about* 7 people. Therefore we can say that the average is exactly 5.875.

However, if the data represent eight measurements, then they are **approximate numbers.** For example, the numbers could be the weight in tons of feed that a large ranch possessed at the end of each of the last eight months, or the number of quarts of strawberries eight people picked on a picnic. In the case of measurements the data are not exact, but approximate. They were rounded off to the nearest whole number. For us to claim to know the average correct to the nearest thousandth is silly. Since our data were only given correct to the nearest whole number, we can only approximate the average to the nearest whole number. Thus to say that the average is "about 6" is more sensible than to give the false impression that we know this figure correct to three decimal places.

Our basic rule of thumb is this:

The final result of calculations with approximate numbers should be in agreement with the data you started with, so round them off appropriately.

There are other more complicated rules, and in fact the whole science of numerical analysis deals with the accuracy and precision of numbers after calculations. However, this simple rule of thumb will suffice for our purposes.

INTERMEDIATE RESULTS

When reporting intermediate results, before the final answer, it is not unusual to keep one or more extra digits. Thus to find the average of the approximate numbers 19.62, 18.3, 17.064, and 16.21, we first find the sum, 71.194. Dividing this sum by 4 yields an average of 17.7985. Since the weakest item in our data, 18.3, was correct to only one decimal place, we round off the average to 17.8. It would not be uncommon to record the sum of the numbers, which is an intermediate result, to two decimal places as 71.18.

Different calculators may yield different numbers of digits in their answers, and if you use pencil and paper, you will probably use less digits than a calculator. This can sometimes lead to a slightly different final answer.

Which answer is wrong? Neither one. Remember we are dealing with approximate numbers. Do not be concerned if your answer differs by 1 in the least accurate digit from the answer obtained by your neighbor, by your instructor, or from an answer key.

WARNING

You should be careful *not to round off too early in your calculations.* Always carry at least one extra digit in your intermediate results until you obtain your final answer.

Because many people do their arithmetic on calculators or even on computers, we must be careful. For example, if the sum of 17 approximate numbers is 68.23, their average, 68.23/17, may be displayed on a calculator as 4.013 529 411 764. Of course, you know that this figure is wrong and that the answer is better expressed as 4.01. Years ago, when we all used pencil and paper, there was little danger that students would carry out a long division to 13 digits. Today, however, it is important to realize that calculators often give senseless and misleading information when dealing with approximate numbers. It is up to us to use these machines intelligently. The following quotation from *Statistical Method in Biological Assay* by D. J. Finney underscores our point.

Bioassays are seldom sufficiently precise to warrant quotation of results to more than 4 significant digits: a statement that a test preparation is estimated to have a potency of 35.71685 units per mg is both stupid and confusing.

One calculator says that the square root of 34.26 is 5.853 204 250 664. If 34.26 is an exact value, this is fine, but if 34.26 represents an approximate number, then, depending on the original data, 5.8 might be a better answer. Remember, it is your responsibility to round off calculator results reasonably.

Throughout this book we will usually assume that all data in any one problem

Introduction

are given to the same accuracy or precision. This is in keeping with ordinary usage. Statisticians do not measure some people to the nearest half inch and others to the nearest inch. Thus if some data are given as 17, 21, 19, and 30, we presume that 30 is also correct to the nearest unit. Similarly, 18.02, 191, 19.61, and 10 imply that the 10 and the 191 are correct to two decimal places. It would be better to write them as 10.00 and 191.00 to indicate this.

STUDY AIDS

VOCABULARY

1. Descriptive statistics
2. Statistical inference
3. Raw score
4. Distribution
5. Data
6. Population
7. Sample
8. Random
9. Random sample
10. Exact number
11. Approximate number

EXERCISES

1-1 Michelangelo tosses a fair coin 4 times and obtains 4 heads. Is this sample of the results of a coin-tossing experiment a *random* sample?

1-2 What is the point of the NoCav example in this chapter? What questions would you want to ask to help you decide if the advertising was effective? What are some other possible explanations for the increase in users of NoCav?

1-3 "There are three kinds of lies: lies, damned lies, and statistics."—Benjamin Disraeli, Prime Minister of England (1804–1881). Why do people both admire and fear statistics? What are some of the advantages to the use of statistics? Are there any disadvantages?

1-4 95 percent of the people who use heroin started out using marijuana regularly. Therefore, using marijuana regularly leads to using heroin. Comment.

1-5 98 percent of the people who use marijuana first drank milk on a regular basis. Therefore, drinking milk on a regular basis leads to using marijuana. Comment.

1-6 Classify each of the following as either statistical inference or descriptive statistics.
(a) Walter Krankrite predicts the results of an election after looking at the votes in 15 of 100 districts.
(b) Dr. Bea Kareful, an ecologist, says that the flesh of fish in a certain lake contains an average of 400 units of mercury.
(c) At Webelo Normal High School last year the average SAT score was 528.
(d) The safety councils of Pessam and Mystic counties predict 600 automobile accidents for the next July 4 weekend.
(e) Last year 72 percent of the workers in Scrooge and Marley's accounting firm missed at least 1 day of work.

1-7 For each of the following statements describe the *population* or *populations* that should have been sampled to get this information. If necessary, clarify the question until it is clear what population is meant.
(a) 30 percent of all suicides are widows.
(b) Malignant tumors were found in 80 percent of the rats injected with 10 ml of chemical X.

(c) English majors at Hudson University have higher grade-point averages than chemistry majors.
(d) Too much cholesterol is bad for your heart.
(e) Girls learn to speak before boys do.

1-8 Have someone in the class secretly mix in a large bag any amount of dried yellow split peas with any amount of dried green split peas. Without counting or even seeing *all* the peas, discuss any methods that could be used to estimate what fraction of the peas in the bag are green. Test your methods. Were they successful? In this experiment, what is the population? What is the sample?

1-9 Find some uses of statistics (sample, average, percentile, etc.) in texts that you use in other courses. Can you classify them as either descriptive or inferential?

1-10 Find some uses of statistics in current magazines and newspapers. Classify them as descriptive or inferential.

1-11 An advertisement states that three-fourths of doctors interviewed recommended Brand X. What is your reaction?

1-12 Answer (a), (b), or (c).
(a) Find some reference to the 1936 survey by *Literary Digest* which predicted that Alf Landon would easily win the United States presidential election (e.g., Huff, *How to Lie with Statistics*).
(b) Find some reference to the polls on the June 18, 1970, British election. (Check newspapers of that week.)
(c) Find some references which discuss the randomness of the December 1, 1969, draft lottery. (Check newspapers of that week, or see the book, *A Sampler on Sampling* by Bill Williams.)

1-13 Round off as indicated.
(a) 16.43 (tenths)
(b) 50,631 (hundreds)
(c) 40,538 (tens)
(d) 18.062 (tenths)
(e) 40,100 (thousands)
(f) 19.8963 (hundredths)

1-14 Estimate these square roots if the original data were given as indicated.
(a) $\sqrt{3.120}$ inches (tenth of an inch)
(b) $\sqrt{0.0196}$ tons (hundredths of a ton)
(c) $\sqrt{800}$ degrees (nearest 10 degrees)
(d) $\sqrt{89}$ volts (nearest volt)
(e) $\sqrt{26{,}000{,}000}$ people (nearest 1000 people)

1-15 Average these expenditures: $16,000, $120,000, $40\overline{0},000.

1-16 Find the sum of 0.00160, 0.00058, and 0.002098.

1-17 (a) Calculate $z = \dfrac{90.3 - 46.12}{20.3}$ if the original data were given to tenths of an inch.

(b) Calculate $z = \dfrac{1031 - 982.8}{2.41}$ if the original data were given to the nearest fathom.

1-18 We measure the left thumbs of eight people and obtain the following measurements: 2.30, 1.92, 2.10, 2.41, 1.88, 1.70, 2.00, and 1.80 inches. Using your rule of thumb, of course, find the average length of left thumbs.

1-19 (a) Multiply 0.18422 × 1.9 and round to the nearest tenth.

(b) Round to 0.1842, multiply by 1.9, and round to the nearest tenth.
(c) Comment on your answers to (a) and (b).
1-20 Find examples of exact numbers and numbers which have been rounded off in a newspaper, magazine, or textbook.

CLASS SURVEY

Learning the basic ideas of statistics is more interesting when you can work with information that you collected yourself, and the simplest such data to obtain are your class's own personal data. The following brief survey will provide enough raw data from your own class to let you consider some interesting questions. Perhaps you will want to add one or two questions of your own. At the end of most sets of exercises we will include a question or two based on the class survey.

SURVEY

Write X if you don't know some answer.

1. Your sex
2. Your age
3. Your height in inches
4. Your father's height
5. Your mother's height
6. Fifth digit of your social security number
7. Last digit of your social security number
8. Your hair color
9. Your eye color
10. Have you ever broken a bone in your body?
11. Are you left-handed, right-handed, or ambidextrous?
12. Do you smoke cigarettes regularly?

Here is a convenient way to collect data so that each student gets a full copy. First, have each student record his or her own responses on a sheet of paper. Second, pass around a master sheet (perhaps a spirit master or a paper that can be photocopied) onto which students copy their responses. You should end up with one piece of paper which looks like the one below. Copies of this piece of paper can be given to each student.

person	\multicolumn{12}{c}{answers to survey questions}
	1	2	3	4	5	6	7	8	9	10	11	12
1	M	20	68	70	63	0	5	brown	brown	no	R	no
2	F	20	62	72	X	6	8	brown	blue	no	A	no
3												
.
.
.

CLASS SURVEY EXERCISES

1 Find the number of males and the number of females and the total number of people in the survey.

2 Find the *percentage* of males and the *percentage* of females in the survey. What is the sum of these two percentages? Why?

3 How many of the females smoke cigarettes regularly? What *percentage* of the females is that? Do you think there is any relationship between sex and cigarette smoking? Explain.

4 Look over all the data. Report one observation that seems interesting to you. Present your finding in a clear fashion.

FIELD PROJECT

1 Suppose you had a random sample of students at this school. If the sample were representative by age and by sex, then we would be confident that the average age of our sample would be close to the average age of the entire population and that the proportion of males in our sample would be near the proportion of males in the population. Your assignment, if you decide to accept it, is to devise a method of obtaining a sample of 100 students so that the age and sex will be random. Outline this method in a clear, detailed, and specific paragraph. Include the exact questions you will ask. Comment on some of the strengths and some of the weaknesses of your method.

2 After your method has been approved by your instructor, gather these data. Include in your report the data, the average age of your sample, the number of males and the number of females in your sample, and comment on anything that occurred that was not expected. Do you think that the average age of all the students is close to the average you computed for your sample? Do you think that the proportion of males on campus is close to the proportion in your sample?

Common Statistical Measures

2 MEASURES OF CENTRAL TENDENCY

Tommy Tufluque just got his first D. He complained to the head of the mathematics department that Professor Noays grades too low. The grades on the first test were as follows:

100 100 100 63 62 60 12 12 6 2 0

Tommy indicated that the class average was 47, which he felt was rather low. Professor Noays stated that nevertheless there were more 100s than any other grade. The department head said that the middle grade was 60, which was not unusual.

    Each of these three people was looking for one number to represent the general trend of these test grades. Such a number is called an **average** or a **measure of central tendency.** Mr. Tufluque used the **mean** or arithmetic average, which is obtained by adding the grades and dividing by the number of grades. Professor Noays used the **mode,** which is the most frequent number. The department head used the **median,** which is the middle number when the group of numbers is written in numerical order.

    These are three commonly used averages. Which of them is the best? That depends on the particular situation. Consider these nine numbers: 71, 71, 71, 71, 73, 74, 74, 75, and 95. If they represent style numbers of dresses sold today in the Chic Dress Boutique, you can see that the style number 71 was the most popular. It was the mode. This would be important in reordering stock. If they represent grades from a psychology final exam, then perhaps you would want the mean, 75, for use in certain statistical testing. If they represent the annual salary, in hundreds of dollars, for the employees of Smith's Emporium, then you might take the median, $7300, as the average salary. Note that the mean salary of $7500 is larger than seven of the nine salaries.

Each average has certain properties. Depending on the context, these properties may or may not be useful. For example, the median is less affected by extremely large or extremely small values, while the mean is affected by every score. In this book we will generally use the mean because it lends itself to much statistical testing.

In the above example we found the median of a distribution with an *odd* number of raw scores. Finding the number in the middle was no problem. For example, the median of 3, 7, 5, 6, 8 is _____ ? We hope that you did not say 5, since the definition of median indicates the middle term *when the numbers are arranged according to size.* Thus the median of 3, 5, 6, 7, 8 is 6. If the distribution includes an *even* number of raw scores, then there are two scores in the middle and the median is defined as their arithmetic mean. Example: The median of 3, 3, 5, 6, 8, 13 is found by adding the two middle numbers, 5 and 6, together and dividing by 2. Thus, the median is 5.5. Note that half the scores are less than 5.5 and the other half are greater.

SYMBOLS AND FORMULAS

We will use n to indicate the number of numbers or raw scores in a given distribution. For the distribution 3, 1, 8, 9, $n = 4$.

For the mean of a sample we use the letter m. For the mean of the population we use the Greek letter for m, which is μ (read: mu). Measures of a population, that is, measures that take into account *every* member of the population, are called **parameters**; μ is a parameter. Measures based on sample data, that is, measures which take into account only *some* members of the population, are called **statistics**; m is a statistic.

Many statisticians use Greek letters for parameters and English letters for statistics. Thus, μ is used for the population mean, and m is used for the sample mean. It is important for you to learn how to write the Greek letters and to know their names.

We will often use capital letters such as X or Y to stand for the list of numbers in a distribution. For example, X: 1, 5, 3, 2. (At times we will use X to name just one of these numbers, such as $X = 3$. Some texts use X_i for this purpose.) It will be clear from the context in which way X is being used.

We will use the Greek letter for capital S, Σ (read: sigma; it looks like a sideways M) to stand for the command "sum." Thus, if a distribution labeled X consists of the four numbers 1, 5, 3, and 2, then ΣX is $1 + 5 + 3 + 2 = 11$. The mean of this sample is $m = 2.75$. Another popular symbol for the mean of a sample is $\overline{X}$ (read: X bar).

A formula for the mean of a sample is

$$m = \frac{\Sigma X}{n} \quad \text{or} \quad \overline{X} = \frac{\Sigma X}{n}$$

In the above example we have

$$m = \frac{\Sigma X}{n} = \frac{1 + 5 + 3 + 2}{4} = \frac{11}{4} = 2.75$$

A second way of interpreting the symbols ΣX is as follows: If in a column labeled X we have the numbers 1, 5, 3, and 2, then ΣX means the sum of

Common statistical measures

numbers in this column. Similarly, X^2 will indicate a column of numbers obtained by squaring each number in column X. The column X^2 will consist of 1, 25, 9, and 4, and $\Sigma X^2 = 39$. The symbol $(\Sigma X)^2$ represents $(11)^2$ or 121. Notice that $(\Sigma X)^2$ is different from ΣX^2.

X	X^2
1	1
5	25
3	9
2	4
$\Sigma X = 11$	$\Sigma X^2 = 39$

$(\Sigma X)^2 = 121$

In the same way, $X - 1$ will be the heading for the column obtained by subtracting 1 from each number in column X. The column will consist of 0, 4, 2, and 1. Therefore, $\Sigma(X - 1) = 7$.

X	$X - 1$
1	0
5	4
3	2
2	1
$\Sigma X = 11$	$\Sigma(X - 1) = 7$

Note that $\Sigma(X - 1) = 7$, while $\Sigma X - 1 = 11 - 1 = 10$.

Sometimes we want to differentiate between two different populations in the same example. If we label the first population X and the second Y, then the number of elements in the first population would be n_X (read: n sub X) and in the second population n_Y (read: n sub Y). The mean of the first would be μ_X and the mean of the second μ_Y. We have

$$\mu_X = \frac{\Sigma X}{n_X} \quad \text{and} \quad \mu_Y = \frac{\Sigma Y}{n_Y}$$

For samples, we write

$$m_X = \frac{\Sigma X}{n_X}$$

For example, the president of the local Planned Parenthood Association has four boys, ages 18, 11, 15, and 9, and three girls, ages 18, 2, and 10. If we let X represent the distribution of boys' ages, and Y the distribution of girls' ages, we have

$n_X = 4 \qquad n_Y = 3$

X	Y
18	18
11	2
15	10
9	
$\Sigma X = 53$	$\Sigma Y = 30$

$$\mu_X = \frac{\Sigma X}{n_X} = \frac{53}{4} = 13.25$$

$$\mu_Y = \frac{\Sigma Y}{n_Y} = \frac{30}{3} = 10$$

EXERCISES

2-1 "The average number of children in a family simply cannot be 2.3. Whoever heard of 3/10 of a child?" Comment on this often heard criticism of averages.

2-2 Let μ be the mean of a distribution labeled X. What reason is there for denoting a sample mean by the symbol m? by the symbol $\bar{X}$?

2-3 Given Y: 2, 3, 4, 5, 6, 7, 8, calculate each of the following quantities.

(a) $\Sigma Y =$ (b) $\Sigma Y^2 =$ (c) $(\Sigma Y)^2 =$

(d) $\dfrac{\Sigma Y}{n} =$ (e) $\bar{Y} =$ (f) $\Sigma(Y - 2) =$

(g) $\Sigma(Y - 5) =$ (h) $\Sigma Y - 5 =$ (i) $\dfrac{\Sigma(Y - 5)^2}{n - 1} =$

(j) $\dfrac{\Sigma Y^2 - \dfrac{(\Sigma Y)^2}{n}}{n - 1} =$ (k) Which is bigger, $(\Sigma Y)^2$ or ΣY^2?

2-4 Repeat the previous exercise with Y: 3, 4, 5, 6, 7, 8, 9.

2-5 Given a sample of six values, X: 4, 4, 3, 0, -1, 2, calculate each of the following quantities.

(a) ΣX (b) m (c) $\bar{X}$

(d) $\Sigma(X - m)$ (e) $(\Sigma X)^2$ (f) ΣX^2

(g) $\dfrac{\Sigma X^2 - \dfrac{(\Sigma X)^2}{n}}{n - 1}$

2-6 Repeat the previous exercise with X: 3, 3, 2, -1, -2, 1.

2-7 Given a population of five values, X: 2, 7, 6, 11, 0, calculate each of these quantities.

(a) $\dfrac{\Sigma X^2 - \dfrac{(\Sigma X)^2}{n}}{n}$ (b) $\dfrac{\Sigma(X - \mu)^2}{n}$

2-8 Repeat the previous exercise with X: 4, 14, 12, 22, 0.

2-9 Make up a set of three numbers for which $(\Sigma X)^2 = \Sigma X^2$.

2-10 A family had kept track of the age at death of its members over several generations. The ages are 72, 68, 0, 67, 45, 7, 70, 68, 72, 66, 70. Compute the mean, median, and mode ages, and decide which you think is most meaningful.

2-11 Salaries in a mathematics department were as follows: four people at $15,000, six at $16,000, two at $21,000, and one at $28,000. Compute the mean, median, and mode salaries. Which seems most meaningful?

Common statistical measures

2-12 Find the mean, median, and mode of the following grade-point averages: 2.9, 3.1, 3.4, and 3.8.

2-13 Illustrate by an example the sentence from this chapter which states "the median is less affected by extremely large or extremely small values, while the mean is affected by every score."

2-14 Write three different distributions where each distribution contains five numbers and has a mean equal to 70. How many such distributions of numbers is it possible to find?

2-15 A student trying Exercise 2-14 took for the first four numbers in one distribution 0, 1, 2, 3. Can he still complete the distribution so that it will have a mean of 70? Could he have started off with *any* four numbers?

2-16 Find the mean of the following arithmetic test grades: 70, 75, 80, 81, 82, 83, 85, 85, 86, 86, 86, 89, 90, 90, 91, 92, 94, and 95.

2-17 The U.S. government reports each year the median age at first marriage of brides and grooms. Here are the results for the decade of the 70s.

year	median age of bride	median age of groom
1970	20.6	22.5
1971	20.5	22.5
1972	20.5	22.4
1973	20.6	22.5
1974	20.6	22.5
1975	20.8	22.7
1976	21.0	22.9
1977	21.1	23.0
1978	21.4	23.2
1979	21.6	23.4

(a) Do you see any trends?

(b) The results would be different if they reported mean age instead of median. Do you think that they would have been higher or lower?

2-18 Some properties of the mean Knowing some properties of the mean can be useful in calculation. If the same value is added to every number in a distribution, then the mean goes up by that value.

Example	X	$Y = X + 6$
	2	8
	3	9
	4	10
	7	13
	$\overline{X} = 4$	$\overline{Y} = \overline{X} + 6 = 10$

The same property holds for subtraction, multiplication, and division.

Use the above idea to answer the following questions.

(a) A mistake was made in grading the arithmetic papers in Exercise 2-16, and each student is entitled to five more credits than given. Correct the value for the mean found in the answer to Exercise 2-16.

(b) If the mean of 15, 18, 23, 24 equals 20, what is the mean of 1.5, 1.8, 2.3, 2.4? Why?

Understanding statistics

2-19 Grouped data A sample distribution X consists of 25 threes and 6 fives.

X	Frequency
3	25
5	6

(a) What is n? (b) What is ΣX? (c) What is ΣX^2?
(d) If the 31 values are arranged in numerical order, which value would be the middle one? (It would be the 16th one. Why?) This middle value is the median.

2-20 A sample distribution X consists of 1000 fives, 500 sixes, and 500 eights. Find the mean, median, and mode.

2-21 Over the past month, the cost of a share of stock in the industrial contractor, Pollution Industrial Group, has been 587, 588, 588, 590, 593, 597, 597, 600, 601, 599, 598, 597, 599, 600, 603, 605, 605, 604, 607, 605, and 607. Find the mean cost of a share of stock.

2-22 The mode was defined to be the most frequent number appearing in a distribution. Some distributions may have more than one mode. Find the modes in each of the following.
(a) 5, 3, 7, 3, 8, 5, 7, 1, 3, 6, 2, 8, 7
(b) 2, 0, 3, 3, 0, 5, 2, 6, 0, 7, −1, 2, 3
(c) 1, 5, 9, 7

2-23 A billboard ad for Smithsonian magazine in 1981 stated that the average income of its subscribers was $42,500. Which average would you guess they used? How do you suppose they got this information?

MEASURES OF VARIABILITY

Suppose you are planning to go on a Caribbean cruise during your spring vacation. A travel agent tells you that there are three possible cruises, and that the mean ages of the passengers on each ship are 20, 29, and 41. Which cruise will you select? Which one will your mother select?

Did you pick the ship with a mean age of 20? 29? Or 41? After you have made your choice, look at Table 2-1 for a detailed listing of the passengers' ages. Having seen the passenger lists, would you like to change your selection? You can see that the mean does not accurately reflect the distribution of ages in ships 1 and 2. We need a measure that will indicate whether the numbers in a distribution are close together or far apart. Such a measure is called a measure of **variability, scatter,** or **spread.** Ideally such a measure should be large if the raw scores are spread out and small if they are close together.

One simple measure of variability is the **range.** The range is the difference between the largest number in the distribution and the smallest number. Thus in ship 1 the range is 62 − 2 = 60 years, in ship 2 the range is 52 − 19 = 33 years, and in ship 3 the range is 43 − 39 = 4 years.

For most "everyday" problems, which you might have to solve on an intuitive basis, the range serves very well as the measure of variability. For many more "technical" problems, especially the type we will be doing later in this book, there is another measure of variability that is useful. It is called the **standard deviation.**

Common statistical measures

Table 2-1   Ages of the Passengers on Each Cruise Ship

ship 1	ship 2	ship 3
2	19	39
3	20	39
4	20	39
5	21	39
8	22	41
9	23	41
9	23	41
10	24	41
40	25	43
44	49	43
44	50	43
62	52	43
sum = 240	sum = 348	sum = 492
$\mu = \dfrac{240}{12} = 20$	$\mu = \dfrac{348}{12} = 29$	$\mu = \dfrac{492}{12} = 41$

To illustrate the concept of standard deviation, let us consider two small populations. Two students, David and Laura, have the same mean grade in algebra, 70. David's grades were 67, 70, 72, and 71. Laura's grades were 100, 62, and 48. Laura's grades are spread out while David's are close together. One way this can be seen clearly is to look at the **deviations from the mean.** The deviation of a score from the mean is found by subtracting the mean from that score (Table 2-2).

Table 2-2

David's grades X	mean μ	deviations from the mean $X - \mu$	Laura's grades Y	mean μ	deviations from the mean $Y - \mu$
67	70	−3	100	70	+30
70	70	0	62	70	−8
72	70	+2	48	70	−22
71	70	+1			
		$\Sigma(X - \mu) = 0$			$\Sigma(Y - \mu) = 0$

A positive sign on a deviation tells us that a grade is above the mean, while a negative sign indicates that it is below the mean. A zero deviation indicates that a particular grade equals the mean. Note that the deviations in David's grades are closer to zero than those in Laura's grades. This is a result of the fact that David's grades are less scattered than Laura's.

If you compute the mean of the deviations for David's grades you will find that it is zero: $(-3 + 0 + 2 + 1)/4 = 0/4 = 0$. If you compute the mean of the deviations of Laura's grades you will see that it also is zero. In fact it is true for any distribution that the sum of the deviations is zero, and therefore the mean of the deviations is zero.

Recall that we are trying to introduce a new measure of variability, called the standard deviation. We want the standard deviation to be representative of the deviations. You might think to use the mean of the deviations as the representative deviation, but we have just mentioned that it is always zero, no matter how much variability there is in the distribution. So statisticians have developed a procedure that is not immediately obvious. First the *squares* of the deviations are obtained. None of these can be negative. Then their mean is found. For example, the deviations in David's grades were −3, 0, 2, and 1, and so the squared deviations are 9, 0, 4, and 1. The mean of the squared deviations is (9 + 0 + 4 + 1)/4 = 14/4 = 3.5. This number, the mean of the squared deviations, can be used as a measure of variability. It is called the **variance** of David's grades. Chapter 15 deals with several situations where the variance is the easiest measure of variability to use.

You will notice, though, that in this process we have squared all the original deviations, so that the variance is representative of the *squares* of the deviations. To get a number representative of the original deviations, we take the square root of the variance. This final number is called the *standard deviation*. In the case of David's grades, since the variance was 3.5, the standard deviation was $\sqrt{3.5} = 1.9$. Since the deviations from the mean were between 0 and 3 units, 1.9 is reasonable as a representation of the deviations. The standard deviation is always in the same units as the original raw scores. In this case the standard deviation is 1.9 grade points. If we have an example where the raw scores are in feet, then the standard deviation is also in feet.

Let us now calculate the standard deviation of Laura's grades (Table 2-3).

Table 2-3

Laura's grades X	deviation from the mean of 70 $X - \mu$	squared deviation $(X - \mu)^2$
100	30	900
62	−8	64
48	−22	484
$\Sigma X = 210$	$\Sigma(X - \mu) = 0$	$\Sigma(X - \mu)^2 = 1448$

Variance = mean of squared deviations

$$= \frac{\Sigma(X - \mu)^2}{n} = \frac{1448}{3} = 482.67$$

Standard deviation = square root of variance

$$= \sqrt{\frac{\Sigma(X - \mu)^2}{n}} = \sqrt{482.67} = 22$$

Again notice that the original deviations were 8, 22, and 30. So 22 is reasonable as a representative deviation.

FORMULAS

We use the Greek letter for lowercase s, which is σ (read: sigma), to represent the standard deviation of a population. Therefore σ^2 represents the variance.

Thus the formula for the variance is

$$\sigma^2 = \frac{\Sigma(X - \mu)^2}{n}$$

Common statistical measures

and the formula for the standard deviation is

$$\sigma = \sqrt{\frac{\Sigma(X - \mu)^2}{n}}$$

In most statistical applications we do not know all the data for the population. We usually have only a sample of these data. A common problem for the statistician is **to estimate the standard deviation** of the population from the sample data. The formula given above for σ gives the standard deviation of the population. It is used when you have *all the population data*. However, *when you wish to estimate σ using only sample data, the formula must be adjusted.* This estimate is denoted by s.

The formula for s is

$$s = \sqrt{\frac{\Sigma(X - m)^2}{n - 1}}$$

Using $n - 1$ instead of n gives a larger value. This compensates for the fact that estimates which use n in the formula tend to be too small because there is usually less variability in a sample than there is in the whole population. Since s is an estimate of σ based on sample data, s is a statistic, while σ is a parameter. When no confusion will result, statisticians sometimes refer to s as the standard deviation, even though it is only an estimate. The estimate for the variance is given by s^2.

The formula for s^2 is

$$s^2 = \frac{\Sigma(X - m)^2}{n - 1}$$

A COMPUTATIONAL FORMULA FOR s

It turns out that in practice the above formula for s is often awkward to use because of all the subtractions. A second, more convenient formula is

$$s = \sqrt{\frac{\Sigma X^2 - \frac{(\Sigma X)^2}{n}}{n - 1}}$$

This formula produces the same answer as the previous one. Let us illustrate this by computing s both ways for the sample data 1, 8, 0, 3, 9.

Formula 1

Using $s = \sqrt{\dfrac{\Sigma(X - m)^2}{n - 1}}$

we need m, $X - m$, and $\Sigma(X - m)^2$:

$$m = \frac{\Sigma X}{n} = \frac{21}{5} = 4.2$$

Formula 2

Using $s = \sqrt{\dfrac{\Sigma X^2 - \dfrac{(\Sigma X)^2}{n}}{n - 1}}$

we need ΣX and ΣX^2:

X	$X - 4.2$	$(X - 4.2)^2$
1	−3.2	10.24
8	3.8	14.44
0	−4.2	17.64
3	−1.2	1.44
9	4.8	23.04
$\Sigma X = 21$		$\Sigma(X - 4.2)^2 = 66.8$

$$s = \sqrt{\frac{66.8}{4}} = \sqrt{16.7} = 4.1$$

X	X^2
1	1
8	64
0	0
3	9
9	81
$\Sigma X = 21$	$\Sigma X^2 = 155$

$$s = \sqrt{\frac{155 - \frac{(21)^2}{5}}{4}}$$

$$= \sqrt{\frac{155 - \frac{441}{5}}{4}}$$

$$= \sqrt{\frac{155 - 88.2}{4}}$$

$$= \sqrt{\frac{66.8}{4}} = \sqrt{16.7}$$

$$= 4.1$$

The corresponding computational formula for the variance is

$$s^2 = \frac{\Sigma X^2 - \frac{(\Sigma X)^2}{n}}{n - 1}$$

EXERCISES

2-24 Last year the mean high temperature in two cities, Squaresville and Octothorpe, was 70 °F. Explain how the two cities might be very different from one another in daily temperatures.

2-25 For the following two distributions find s first by the definition

$$s = \sqrt{\frac{\Sigma(X - m)^2}{n - 1}}$$

and then by the computational formula

$$s = \sqrt{\frac{\Sigma X^2 - \frac{(\Sigma X)^2}{n}}{n - 1}}$$

State which method is easier for each part.
(a) 4, 6, and 8
(b) 3, 8, 9, 17, and 20

2-26 List all the raw scores in part (a) of Exercise 2-25 which are 1 or more standard deviations below the mean. List all the raw scores in part (b) of Exercise 2-25 which are more than 1 standard deviation away from the mean.

2-27 A random sample of prisoners at Singsong State Prison were given an honesty test. The scores went from −31 to +9, giving a range equal to 40. The average score was −17, and the standard deviation for the sample was

Common statistical measures

7. Warden Warren Wardon estimates that the average score for the entire prison population is also about −17, but that the range is larger than 40 and the standard deviation is larger than 7. Explain why Warden Wardon is probably correct.

In Exercises 2-28 to 2-32 use the following ideas.

Some properties of s If you add the same value to all the numbers in a distribution, the standard deviation is not changed at all. The same holds for subtraction.

Example X: 1, 2, 3 gives $s = 1$. $X + 6$ gives 7, 8, 9, which still has $s = 1$.

If you multiply all the numbers in a distribution by the same positive value, the standard deviation is multiplied by that value. The same holds for division.

Example X: 1, 2, 3 gives $s = 1$. $X \times 6$ gives 6, 12, 18 which has $s = 6$.

2-28(a) Using either method discussed in this chapter, find m, s^2, and s for distribution 2, 5, 6, and 7.
(b) In part (a) we found m, s^2, and s for the distribution 2, 5, 6, and 7. Use the ideas above to find m, s^2, and s for the distribution 12, 15, 16, and 17.
(c) For the distribution in part (a), list all the values that are less than 1 standard deviation away from the mean. *1st formula*
2-29(a) Using either method, find m, s^2, and s for the distribution 3, 4, 7, 9, and 11. *2nd formula in problem 25*
(b) Find m, s^2, and s for the distribution 30, 40, 70, 90 and 110.
(c) For the distribution in part (a), list all the values that are more than 3 standard deviations away from the mean.
2-30 As the result of a strike, the International Brotherhood of Pogo Stick Workers obtained a contract in which the mean salary was 11,000 marks per year with a standard deviation of 800 marks.
(a) This year each worker will receive an increase of 500 marks per year. What will the new mean and standard deviation be?
(b) Next year each worker will receive a 10 percent increase. What will the new mean and standard deviation be?
2-31 Without actually calculating, compare the mean and the standard deviation of the following ages:
X: 5, 2, 7, and 3
Y: 65, 62, 67, and 63
2-32 Without actually calculating, compare the mean and the standard deviation of the following temperatures:
Kodiak, Alaska, 10, 8, 0, and −1
Coldfoot, Alaska, −10, −8, 0, and 1
2-33 Here are the bills and the tips that a waiter collected in one night.

bill, dollars	tip, dollars
12.46	1.75
20.16	3.00
6.25	.75
22.00	3.25
15.88	2.50
38.50	5.50

(a) What is the average tip?
(b) What are the variance and the standard deviation of these tips?
(c) Determine what percentage of the bill each tip represents. What is the average percentage?
(d) What are the variance and the standard deviation of these percentages?

2-34 Make a list of five numbers whose variance is:
(a) Larger than its standard deviation.
(b) Smaller than its standard deviation.
(c) Zero.

2-35 Consider the following grades on an arithmetic test: 70, 75, 80, 81, 82, 83, 85, 85, 86, 86, 89, 90, 90, 91, 92, 94, and 95. In this distribution $\bar{X} = 85.53$ and $s = 6.7$.
(a) How many scores are within 1 standard deviation of the mean?
(b) What *percentage* of the scores are within 1 standard deviation of the mean?
(c) How many scores are within 2 standard deviations of the mean?
(d) What *percentage* of the scores are within 2 standard deviations of the mean?

2-36 The number of UFOs reported to the National UFO Reporting Center each month for the preceding 12 months was 30, 3, 27, 0, 15, 40, 37, 1, 1, 20, 10, and 5. Compute the mean and s for this distribution. Answer questions (a) to (d) listed in Exercise 2-35 for this distribution.

2-37 An IQ test was given to two groups of fourth-grade students. One group was from a grade school in Nassau County; the other group was from a school for psychotic children. Both groups have a mean score of 100, but the standard deviation for the normal students was 14, and the standard deviation for the psychotic children was 23. Interpret the statistics.

2-38 A mine foreman is comparing the products of two manufacturers of blasting materials. Both companies' materials explode a mean time of 40 minutes after they are set, but the standard deviation for Brand A materials is 4 minutes, and the standard deviation for Brand B materials is 14 minutes. Which brand should he choose?

2-39 Two brands of yardsticks both have mean lengths of 36 inches. The standard deviation for the Euclidean brand is .002 inch, while the standard deviation for the Pythagorean brand is .001 inch. Which brand is better?

2-40 After a class test in statistics, you go to your teacher's office to find out your grade. The teacher is not there but on the desk you see the following results:
Class I: mean = 80, standard deviation = 5
Class II: mean = 80, standard deviation = 10
(a) Which class would you rather be in? Why?
(b) If Mary is in class I and she is within 3 standard deviations of the mean, find the interval containing her grade.
(c) If Bill is in class II and he is more than 2 standard deviations below the mean, find the interval containing his grade.

2-41 Two experiments were done on different brands of artificial hearts. The first compared Brand A with Brand B. The second compared Brand X with Brand Y. They all wear out eventually and must be replaced. We show here the results of tests on 10 hearts of each brand. Explain why the results are conclusive in the first experiment, but not in the second.

Common statistical measures

	experiment I		experiment II	
	brand A	brand B	brand X	brand Y
average lifetime, days	1000	1400	1000	1400
standard deviation	22	22	300	320

2-42 John follows the prices of spaghetti at the local supermarkets. Each Monday he goes to 20 markets and checks the price of a 1-pound box of House Brand spaghetti. Here are his results for the past 10 weeks.

week	average price, cents	standard deviation, cents
1	49.3	2.01
2	50.0	1.79
3	48.7	2.00
4	52.1	2.61
5	51.0	1.98
6	50.0	1.90
7	49.7	1.82
8	51.0	1.00
9	51.2	2.45
10	51.4	2.30

(a) What is the mean price for the 10-week period?
(b) What is the mean standard deviation for the 10-week period?

2-43 Every week the U.S. government publishes certain health statistics. One such report gives the number of reported cases of various diseases so far that year. (The count starts January 1.) Here is an illustration for the first 29 weeks of 1980 and the first 29 weeks of 1981, for reported cases of measles. The figures are reported by region of the country.

region	first 29 weeks of 1980	first 29 weeks of 1981
New England	665	72
Mid Atlantic	3612	756
Central	4754	961
South Atlantic	1839	332
Mountain	415	32
Pacific	971	315
	12,256	2468

(a) For the first 29 weeks of 1980 find the mean and the standard deviation for the number of reported cases of measles in the various regions. Do the same for 1981.
(b) For 1980 find the percentage of reported cases in each region.
(c) What is your general impression of what this table means?

SOME MEASURES OF AN INDIVIDUAL IN A POPULATION

z SCORES

The results of the last exam in this class were good: the class mean was 83, the median was 87, the range was 24, and the standard deviation was 5.

Understanding statistics

All the above information is nice, but what you really want to know is: How did *you* do on the examination?

We have already mentioned one measure of an individual score, the *raw score*. Another very important measure of an individual's rank in a population is called the **z score**. The z score measures how many standard deviations a raw score is from the mean. In the example above, $\mu = 83$ and $\sigma = 5$. Therefore, the z score corresponding to 88 (written z_{88}) is $+1$, since 88 is 1 standard deviation (5 units) above the mean; $z_{73} = -2$, since 73 is 2 standard deviations (10 units) below the mean.

A formula for finding the z score corresponding to a particular raw score X is

$$z_{\text{corresponding to a raw score}} = \frac{\text{raw score} - \text{mean}}{\text{standard deviation}}$$

Represented in symbols, the formula for the z score is

$$z_X = \frac{X - \mu}{\sigma}$$

When there is no confusion, we often drop the subscript and simply write

$$z = \frac{X - \mu}{\sigma}$$

EXAMPLE 2-1 Mike is in the above class. His grade was 69. Find the z score for his grade.

SOLUTION We simply substitute into the formula $X = 69$, $\mu = 83$, and $\sigma = 5$:

$$z_{69} = \frac{69 - 83}{5} = \frac{-14}{5} = -2.8$$

This tells us that Mike's score was below average by almost 3 standard deviations. In general, you will see that this is quite a lot below average.

If in the above class Beverly had a z score of $+2$, what was her score on the test? Since the standard deviation was 5 and the mean 83, 2 standard deviations above the mean would be $83 + 2(5) = 93$.

A formula for the raw score corresponding to a particular z score is

$$\text{Raw score}_{\text{corresponding to a } z \text{ score}} = \text{mean} + z \text{ score} \cdot \text{standard deviation}$$

In symbols, this is

$$X_z = \mu + z\sigma$$

or simply

$$X = \mu + z\sigma$$

If David had a z score of $-.7$, then his raw score is

$$83 + (-.7)(5) = 83 - 3.5 = 79.5$$

Common statistical measures

Notice that we do the multiplication first.

If we are using s as an estimate of σ, these two formulas become

$$z_X = \frac{X - \mu}{s} \quad \text{and} \quad X = \mu + zs$$

EXAMPLE 2-2 Suppose someone claimed that the mean depth of successful oil-well drillings is 2500 feet. If you estimate the standard deviation of these depths from some sample data and you get $s = 100$ feet, find the z score corresponding to a depth of 2250 feet.

SOLUTION

$$z_{2250} = \frac{X - \mu}{s} = \frac{2250 - 2500}{100} = -\frac{250}{100} = -2.5$$

Similarly, to find the depth that corresponds to a z score of 1.65 we have

$$X = \mu + zs = 2500 + 1.65(100) = 2665 \text{ feet}$$

RELATIVE STANDING VIA z SCORE

Let us consider two test grades you might receive. Suppose you received an 85 in English and a 65 in physics. Clearly you would rather receive a raw score of 85 than a raw score of 65, but a second consideration is how well you did relative to the other students in the class. Suppose we tell you that the mean in English was 70 and the mean in physics was 50. Thus, in both classes you scored 15 points above the mean. Does this mean that, relatively speaking, you did the same in both classes? The answer is no; the number of points above or below the mean is insufficient information to give you a rating relative to your position in the class, as you can see from the class scores given in Table 2-4.

Table 2-4

English	physics
100	65 (your score)
99	57
98	55
85 (your score)	53
73	50
67	49
60 (Alice's score)	47
53	44
45	44 (Alice's score)
20	36
mean = 70	mean = 50
$s = 26.4$	$s = 8.1$

We can see from this table that although you scored 15 points above the mean in both classes, when compared to the other students you did better in physics than in English, since your physics grade was the top in the class, while three students scored higher than you did on the English test.

In order to see how well you did compared to the rest of the class, you can use z scores. In English your z score is

$$z = \frac{85 - 70}{26.4} = \frac{15}{26.4} = .6$$

In physics your z score is

$$z = \frac{65 - 50}{8.1} = \frac{15}{8.1} = 1.9$$

Thus you see that even though you scored 15 points above the class average in both subjects, your physics score was *relatively* better.

EXAMPLE 2-3 Using the information in Table 2-4, if Alice scored 60 in English and 44 in physics, which was a better grade relative to the class?

SOLUTION In English, Alice's z score is

$$z = \frac{60 - 70}{26.4} = \frac{-10}{26.4} = -.4$$

In physics, Alice's z score is

$$z = \frac{44 - 50}{8.1} = \frac{-6}{8.1} = -.7$$

Since $-.4$ is greater than $-.7$, the 60 in English had the higher z score than the 44 in physics, and Alice did better, relatively speaking, in the English class.

PERCENTILE RANK

Another measure of an individual's position in a population is the **percentile rank.** This is primarily used for large populations. In a small population one would simply use ordinary rankings, such as "fifth out of nine." Essentially, the percentile rank of a raw score tells us the percentage of the distribution that is *below* that raw score. Consider a person whose raw score has percentile rank 75. Approximately 75 percent, or three-fourths, of the population scored below this individual.

In a large distribution, if a raw score of 72 has a percentile rank of 80, then 80 percent of the scores are below 72 and 20 percent are above. Now consider the following example. In a distribution of weights of babies, 70 percent of the babies weighed less than Steven, 10 percent weighed the same as Steven, and 20 percent weighed more than Steven. Since 70 percent weighed less than Steven and 20 percent weighed more than Steven, his percentile rank should be between 70 and 80. We will use 75, the value that is halfway between 70 and 80. You can find the percentile rank of a raw score directly by finding the percent of scores that are below the given score and adding one-half of the percent of scores that are the same as that raw score. In our example, 70 percent weighed less than Steven and 10 percent weighed the same as Steven, so the percentile rank of Steven's weight = 70 + ½(10) = 75.

Common statistical measures

Suppose in a class of 50 Andrea's score on an aptitude test was 603. She finds that 6 grades are above hers, 3 (including Andrea's) are the same as hers, and 41 are below her grade. Thus 41/50 = .82 = 82 percent scored below Andrea's grade and 3/50 = 6 percent are at Andrea's grade. The percentile rank of Andrea's grade is 82 + ½(6) = 85.

We can express the procedure for finding the percentile rank with a formula. Let B equal the number of raw scores below a particular score X. Let E be the number of scores equal to X, including X itself. Let n be the total number of raw scores.

Then the percentile rank PR of X is

$$PR_X = \frac{B + \frac{1}{2}E}{n}(100)$$

placing her in the middle — 3 people w/same grade which is below middle top

In the preceding example, the percentile rank of Andrea's grade is

$$PR_{603} = \frac{41 + \frac{1}{2}(3)}{50}(100) = \frac{41 + 1.5}{50}(100) = \frac{42.5}{50}(100) = 85$$

A grade which has a percentile rank of 85 is said to be "at the 85th percentile." We will write P_{85} (read: P sub 85) to indicate the raw score that is at the 85th percentile. Thus, in the example that we just did, Andrea's grade would be at the 85th percentile, and would be symbolized by P_{85}. The idea here is that the percentile *rank* is a value from 0 to 100, while a percentile can be any raw score. Thus we can write both $PR_{603} = 85$ and $P_{85} = 603$.

Generally, we round off percentile ranks to the nearest whole number. For example, in a distribution of 150 numbers, suppose that 97 of the numbers are below 700 and that 11 numbers are equal to 700. We wish to find the percentile rank of 700:

$$PR_{700} = \frac{B + \frac{1}{2}E}{n}(100)$$

$$= \frac{97 + \frac{1}{2}(11)}{150}(100) = .683(100) = 68.3$$

So we say that the percentile rank of 700 is 68, or $PR_{700} = 68$. We also say that 700 is at the 68th percentile, or $P_{68} = 700$. When you read, for example, that your SAT score is at the 82nd percentile, this means that about 82 percent of those taking the examination scored lower than you.

EXAMPLE 2-4 Refer to the data in Table 2-4. Find the percentile ranks for your grades in English and physics. Find the percentile ranks for Alice's grades.

SOLUTION The percentile rank for your 85 in English is given by

$$PR_{85} = \frac{6 + .5(1)}{10}(100) = 65$$

and the percentile rank for your 65 in physics is given by

$$PR_{65} = \frac{9 + .5(1)}{10}(100) = 95$$

Understanding statistics

The percentile rank for Alice's 60 in English is given by

$$PR_{60} = \frac{3 + .5(1)}{10}(100) = 35$$

and the percentile rank for Alice's 44 in physics is given by

$$PR_{44} = \frac{1 + .5(2)}{10}(100) = 20$$

We summarize these results in Table 2-5.

Table 2-5

	grade	z score	percentile rank
English, you	85	.57	65
physics, you	65	1.86	95
English, Alice	60	−.38	35
physics, Alice	44	−.74	20

EXERCISES

2-44 Complete the following table. (The first row is already done.)

	symbol	pronunciation	meaning
	n	en	number of items in a distribution
(a)	m		
(b)		X bar	
(c)		mu	
(d)			the command to add
(e)	σ		
(f)		z sub 98	
(g)			a score at the 78th percentile
(h)			an estimate of the standard deviation
(i)	PR_{198}		

2-45 Give an example where possible of a negative value for each of the following: mean, median, mode, standard deviation, variance, range, z score, percentile rank, 25th percentile.

2-46 A distribution of temperatures has a mean of 98.6° and a standard deviation of 0.5°. Find the z scores corresponding to the following temperatures.
(a) 99.1° (b) 97.6° (c) 98.6° (d) 100° (e) 98°

2-47 A distribution of air pressures has a mean of 32 pounds per square inch, and a standard deviation of 1.2 pounds per square inch. Find the air pressures corresponding to the following z scores.
(a) 2 (b) 0 (c) −3 (d) 1.35 (e) −.06

Common statistical measures

2-48 A sensitivity test was given to 1000 people. Here are the z scores for five of them: Adelbert $-.02$, Bastion 1.27, Carmulon .001, Joe -2.03, Elfremde .48.
(a) Which of these 5 people scored above the mean for the test?
(b) Which of these 5 people scored below the mean for the test?
(c) Rank these 5 people from the highest scorer to the lowest. B E C A J

2-49 In Exercise 2-48 suppose that the mean test score for the 1000 people who took the sensitivity test was 10 and the standard deviation was 5.
(a) Find the raw score for each of the 5 people named.
(b) In general, must negative z scores always correspond to negative raw scores?
(c) In general, must positive z scores always correspond to positive raw scores? *Hint:* Consider temperatures at the North Pole.
(d) Another 6 people scored as follows on the sensitivity test: Lou 10, Nes 9, Bates 5, Inez 15, Pat 12, Sue 11. Find their z scores.

2-50 Astronauts discover that the mean height of a Martian is 3.6 Martian inches with a standard deviation of .2 Martian inch. *Note:* 12 Martian inches equals 1 Martian foot.
(a) Given the heights of the following Martians, find their z scores: Xgol 3.8, Zib 2.6, Mni 2.6, Rfd 4.
(b) How tall is President Mil, whose z score is $-.5$?
(c) Mr. Zar is 3.9 Martian inches tall. Ms. Zar's height has a z score of 1.6. Who is taller?

2-51 True or false? If *everyone* in a population has the same z score, then they must all have a z score equal to 0.

2-52 In a large distribution of ages at Golden Vista Nursing Home, the percentile rank of age 72 is 50. True or false?
(a) The median age in this population is about 72.
(b) The mean age in this population is about 72.

2-53 In a given year the mean length of American-made cars was 171 inches and the standard deviation was 5 inches.
(a) Find the z scores for cars with lengths 169 inches, 171 inches, and 180 inches.
(b) Three models of one manufacturer had z scores of -1, 0, and .3. Find the lengths of these models.
(c) All Colonel Motors cars measured within 2 standard deviations of the mean. (1) James claims his CM car was 185 inches long. Find the z score for 185 inches. Why is James's claim not possible? (2) What is the maximum possible length of James's car? (3) What is the minimum possible length of James's car?

2-54 Below are given student Pentak's scores on some standard exams. Also given are other statistics for the exams.

test	mean	standard deviation	Pentak's score
math	47.2	10.4	83
verbal	64.6	8.3	71
geography	74.5	11.7	72

(a) Transform each of Pentak's test scores to a z score.
(b) On which test did Pentak stand relatively highest? Relatively lowest?

Understanding statistics

2-55 A doctor collects the heights, weights, and blood pressures of a large group of people, called a *control group,* and then computes the three means and the three standard deviations. After taking your height, weight, and blood pressure, the doctor computes your z scores with regard to the control group. They are: height, $z = 2.1$; weight, $z = -1.3$; and blood pressure, $z = .003$. Interpret these results.

2-56 (a) Sam Safety took the National Safe Driver Test. Of the 120,000 people who took the test there were 100,000 who scored lower than Sam and 2400 who scored the same as Sam. Find the percentile rank for Sam's score.
(b) John Hasty took the National Safe Driver Test. Of 120,000 people who took the test there were 18,000 who scored lower than John and 1000 who scored the same as John. Find the percentile rank for John's score.

2-57 According to the statistics of the Quick and Easy Data Company, some percentiles for family incomes in Nowso County are as follows:

$P_{25} = \$4000 \qquad P_{50} = \$6800 \qquad P_{75} = \$10,200 \qquad P_{90} = \$14,500$

State approximately the percentage of the families that earn:
(a) Less than $4000.
(b) Less than $6800.
(c) Less than $10,200.
(d) Less than $14,500.
(e) More than $14,500.
(f) Between $4000 and $10,200.

2-58 There are about 500,000 families in Nowso County. Using the figures from Exercise 2-57, state approximately *how many* families earn:
(a) Less than $4000.
(b) Less than $6800.
(c) Less than $10,200.
(d) Less than $14,500.
(e) More than $14,500.
(f) Between $4000 and $10,200.

2-59 A test of susceptibility to photographic stimuli was given to 500 subscribers to *Sportfellow* magazine. Some results are tabled below.

	\multicolumn{6}{c}{test score}
	68	84	100	116	132	148
z score	−2	−1	0	1	2	3
percentile rank	2	16	30	50	98	99

By inspection of the table is it possible to tell:
(a) What is the value of n? Why?
(b) What was the mean test score? Why?
(c) What was the median test score? Why?
(d) What percentage of the readers scored below 68? between 84 and 116?
(e) What is the standard deviation for the distribution?
(f) What test score would be 1.5 standard deviations above the mean?
(g) What test score would be transformed to a z score of -1.2?

2-60 In a study of waste disposal in Nosewer County, it was discovered that the mean amount of garbage was 30 pounds per day per family and the median was 35 pounds per day per family. Which one of the following is true?
(a) Exactly half the families produced 30 pounds or more of garbage.
(b) More than half the families produced 30 pounds or more of garbage.
(c) Less than half produced 30 pounds or more of garbage.

2-61 On an important exam Phil's z score was negative. He claimed that his score had a percentile rank of 60. Can this happen?

2-62 If exactly 50 percent of the scores in a population are below 70, then which of these are correct?
(a) $PR_{50} = 70$ (b) $P_{50} = 70$ (c) $PR_{70} = 50$ (d) $P_{70} = 50$

Common statistical measures

2-63 Danny Kazort is a demolition expert. He has been able to knock down 75 apartment houses at an average time of 3.5 weeks per house, with a standard deviation of 4 days. One tricky job took 5 weeks. What z score would that job have?

2-64 Bess scored 87 on her typing exam and 78 on her physics exam, yet the z score for the 87 was 0 and the z score for the 78 was 2. Explain how this can be and what it means. With relation to her fellow students, is she a better typing student or a better physics student?

RATES

In Chapter 1 we said that one major use of statistics is "descriptive," the summarizing of large amounts of data. So far in this chapter we have discussed several quantities useful for such a description. For example, the *mean* summarizes the location of the "center" of the data; the *variance* summarizes the variability of the data. Another descriptive device often used in statistical reports is called a **rate**. Rates are especially useful in the area called **vital statistics**, which is concerned largely with population problems such as birth (natality), death (mortality), and various social phenomena. No doubt you have heard statements like, "the birth rate is dropping," "the divorce rate is going up every year," "the death rate due to lung cancer among women is rising rapidly."

A rate is basically a fraction, though it is usually expressed in a convenient decimal form. Let us consider birth rates as our first example. According to U.S. government data, during 1980 there were about 3,598,000 (3.598 million) live births in the United States. Also, during 1980 the population of the United States was about 222,300,000 (222.3 million) people.[†] We say that the birth rate for the year 1980 was 3.598 million births per 222.3 million population. As a fraction this is

$$\frac{3{,}598{,}000}{222{,}300{,}000}$$

To express this as a decimal, we divide top by bottom and find:

$$\frac{3{,}598{,}000}{222{,}300{,}000} = 0.016$$

This represents the birth rate expressed as births per *one* member of the population. This is usually not a sensible or convenient way to think of the rate. It is common to want rates in terms of each 1000 (or 100,000) members of the population. So we multiply the decimal by 1000 (or 100,000). This gives, for example, .016 × 1000 = 16.

We say that the 1980 birth rate was 16 births per thousand population. This means that on the average for every 1000 people in the United States 16 babies were added to the population. We can do the calculation in one step:

$$\text{Annual birth rate} = \frac{\text{number of births during year}}{\text{size of population during year}} (1000)$$

$$= \frac{3{,}598{,}000}{222{,}300{,}000} (1000) = 16$$

[†]Technically this is the Census Bureau's estimate of the population at midyear, July 1, 1980.

Understanding statistics

We use rates when it is natural to refer to a base population. Suppose, for example, we are studying how people in various cultures cope with stress, and we believe it would be useful to study suicide in, say, the United States and Finland. Suppose we find that in 1979 there were about 1190 suicides in Finland and about 27,500 suicides in the United States. Does this tell us that people in the United States are more likely to commit suicide than people in Finland? No, because we have not used the information that there are many more people in the United States to begin with. We incorporate the base populations by using them as the denominators in the suicide rate fractions:

$$\text{Annual suicide rate} = \frac{\text{number of suicides}}{\text{size of population}} (100{,}000)$$

Now we get:

$$\text{Finland, 1979 suicide rate} = \frac{1190}{4{,}760{,}000} (100{,}000)$$
$$= 25 \text{ suicides per } 100{,}000 \text{ people}$$

$$\text{United States, 1979 suicide rate} = \frac{27{,}500}{220{,}000{,}000} (100{,}000)$$
$$= 12.5 \text{ suicides per } 100{,}000 \text{ people}$$

We see that the Finnish suicide *rate* is double that of the United States. People are more likely to commit suicide in Finland than in the United States.

Often rates are computed in the same location but at different times to see if there are trends in the rate. For example, here are some birth rates for different years in the United States (expressed per 1000 population).

year	1970	1971	1972	1973	1974	1975	1976	1977	1978	1979	1980
birth rate	18.4	17.2	15.6	14.9	14.9	14.8	14.8	15.4	15.3	15.5	16.3

It appears that the trend was decreasing, "bottomed out" about 1975, and then began to increase. Time trends in rates are often shown in graphs, as illustrated in Figure 2-1.

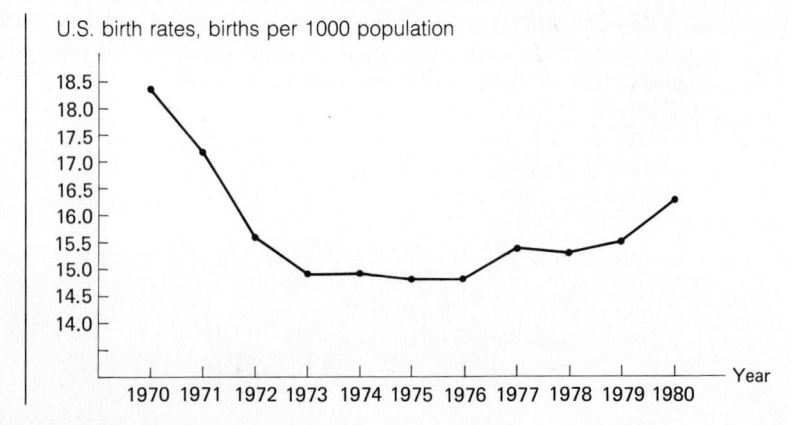

Figure 2-1

U.S. birth rates, births per 1000 population

Common statistical measures

In summary, to understand a rate fully you must know:

1. What period of time it covers (one year, one month, etc.)
2. What it is counting (births, deaths, etc.). This is the numerator.
3. What is the base population (U.S. citizens, males, etc.). This is the denominator.
4. What are the units (per 1000 population, per 100,000 population).

Here are some examples.

1. The 1980 U.S. marriage rate was 10.9 marriages per 1000 population. This means that in 1980 for every 1000 people there were about 10.9 marriages. So about 22 people out of every 1000 got married. Note that the 1000 includes children as well as people who were already married.
2. The 1980 U.S. rate of infant death per 1000 live births was 12.5. This means that in 1980 about 13 babies out of every 1000 born died before reaching age 1.
3. The 1979 death rate in the United States for the age group 15–24 years was 118.2 per 100,000. This means that in 1979 about 118 out of 100,000 people in this age group died. This is called an *age-specific* death rate.

STUDY AIDS

VOCABULARY

1. Mean
2. Mode
3. Median
4. Parameter
5. Statistic
6. Variability
7. Range
8. Standard deviation of a population
9. Variance
10. Estimate of standard deviation
11. z score
12. Percentile rank
13. Percentile
14. Rate

SYMBOLS

1. X, Y
2. n, n_X, n_Y
3. Σ
4. $\mu, m, \overline{X}$
5. σ, s
6. z, z_X
7. P_{83}
8. PR_{85}
9. B
10. E

FORMULAS

1. $\mu = \dfrac{\Sigma X}{n}, \quad m = \dfrac{\Sigma X}{n}, \quad \overline{X} = \dfrac{\Sigma X}{n}$

2. $s = \sqrt{\dfrac{\Sigma(X - m)^2}{n - 1}}, \quad s^2 = \dfrac{\Sigma(X - m)^2}{n - 1}$

3. $s = \sqrt{\dfrac{\Sigma X^2 - \dfrac{(\Sigma X)^2}{n}}{n-1}}$, $s^2 = \dfrac{\Sigma X^2 - \dfrac{(\Sigma X)^2}{n}}{n-1}$

4. $z_X = \dfrac{X - \mu}{\sigma}$, $z_X = \dfrac{X - \mu}{s}$, $z_X = \dfrac{X - m}{s}$

5. $X = \mu + z\sigma$, $X = zs$, $X = m + zs$

6. $PR_X = \dfrac{B + \frac{1}{2}E}{n}(100)$

EXERCISES

2-65 Coward Hossel stated, "In 1924 Babe Ruth's batting average was three-seventy-eight." What kind of an average is a batting average?

2-66 A government report stated that in 1979 the marriage rate in the United States was 63.6 per 1000 in the category "unmarried women 15 years and over," but was 107.9 in the category "unmarried women 15–44 years." What do these two rates mean? Why are they so different?

2-67 For each of these rates give: (1) The time period. (2) The base population. (3) What is being counted.
(a) The 1978 U.S. birth rate.
(b) The death rate for people 65 and over in the United States in 1983.
(c) The divorce rate in the United States for July 1981.
(d) The apple consumption rate in the Garden of Eden in the year 1.

2-68 Express these data as rates.
(a) The population of Conception City in June 1982 was 400,000. During June there were 400 births. Find: (1) The monthly birth rate per 1000 population. (2) The annual estimated birth rate per 1000 population (by multiplying the monthly rate by 12).
(b) In 1979 the U.S. population was about 220 million. During that year there were about 2,359,000 marriages, 1,170,000 divorces, 953,100 deaths due to cardiovascular disease, 403,100 deaths due to cancer, and 53,990 deaths due to motor vehicle accidents. Express the marriage and divorce rates as rates per 1000 population. Express the death rates as rates per 100,000 population.
(c) In Amusement Land 3500 people ride the bumper cars each day and there are about 68,000 collisions each day. Express the daily accident rate per person.

2-69 In 1975 there were 2,152,662 marriages in the United States, giving a rate of 10.1 per 1000 population. In 1976 there were 2,154,807 marriages, giving a rate of 10.0 per 1000 population. How can the rate go down if the number of marriages went up?

2-70 Why are these reports incomplete or uninterpretable?
(a) The U.S. marriage rate is 45 per 1000 population.
(b) The annual motor vehicle death rate is 24.2.
(c) The death rate is 2.4 for disease A and 24.0 for disease B.

2-71 *Epidemiology* is the study of who in a population gets sick and who does not. Epidemiologists are usually called in when there is a sudden outbreak of a disease. Their job is to help pinpoint the cause of the disease. A typical example would be to try to locate which food was the source of an outbreak

Common statistical measures

of food poisoning. A useful rate in such cases is the *attack rate* for and against various foods:

$$\text{Attack rate "for" a food} = \frac{\text{number ill who ate the food}}{\text{number of people who ate the food}}$$

$$\text{Attack rate "against" a food} = \frac{\text{number ill who did not eat the food}}{\text{number of people who did not eat the food}}$$

The table below shows the information the epidemiologist collected about a picnic where there was an outbreak of food poisoning. Look over the data and decide which food is the likely source of infection. Calculate the various attack rates to confirm your intuition.

food	persons who ate these ill	not ill	total	persons who did not eat these ill	not ill	total
baked ham	29	17	46	17	12	29
spinach	26	17	43	20	12	32
mashed potatoes	23	14	37	23	15	38
cabbage salad	18	10	28	28	19	47
jello	16	7	23	30	22	52
rolls	21	16	37	25	13	38
bread	18	9	27	28	20	48
milk	2	2	4	44	27	71
coffee	19	12	31	27	17	44
water	13	11	24	33	18	51
cakes	27	13	40	19	16	35
vanilla ice cream	43	11	54	3	18	21
chocolate ice cream	25	22	49	21	7	28
fruit salad	4	2	6	42	27	69

CLASS SURVEY

1 (a) Find the mean, median, and modal ages of this class.
(b) Do you think the mean age of this class would be a good estimate of the mean age of the school? Why or why not?
2 (a) Find the range and the standard deviation of the heights of the smokers.
(b) Find the range and the standard deviation of the heights of the nonsmokers.
(c) Which group has more variability?
(d) Find the z score and the percentile rank for your height in whichever group you belong.

Frequency Tables and Graphs

3

ORGANIZING THE DATA: FREQUENCY TABLES

Usually, when a collection of statistical data is gathered, it must be organized in some way before much sense can be made of it. Probably the most common way to organize data is to combine the individual raw scores into fewer categories and then to summarize the grouping in a short table. Here is an illustration.

EXAMPLE 3-1 George Stephen, a famous mathematician, gets bored waiting for the 8 A.M. bus each morning. He decides to record to the nearest minute the length of time the bus is late each day. His raw data for the past 30 days look like this.

day	1	2	3	4	5	6	7	8	9	10	11	12	13	14	15
minutes late	9	7	3	4	2	5	3	7	2	6	5	3	10	1	10

day	16	17	18	19	20	21	22	23	24	25	26	27	28	29	30
minutes late	3	3	2	5	1	4	6	4	3	5	6	3	5	3	4

(a) To organize the data he first lists the values from smallest to largest:

Minutes late: 1 1 2 2 2 3 3 3 3 3 3 3 4 4 4 4 5 5 5 5 5 6
6 6 7 7 9 10 10

Frequency tables and graphs

(b) Then he condenses the data into a **frequency table.** To do this he must first decide the size of each category. If he decides that each category will be 1 minute, he gets the following frequency table (Table 3-1). The **frequency** of an outcome is the number of times it occurs.

Table 3-1

minutes late	frequency
10	2
9	1
8	0
7	2
6	3
5	5
4	4
3	8
2	3
1	2

The frequency table allows him to answer certain common questions easily. For example:

(a) What percentage of the latenesses were more than 5 minutes?

Answer Add the frequencies for all latenesses of more than 5 minutes and divide by 30:

$$\frac{2 + 1 + 0 + 2 + 3}{30} = \frac{8}{30} = .27$$

About 27 percent of the latenesses were more than 5 minutes.

(b) Which lateness occurred most often? That is, find the mode of the distribution of latenesses.

Answer See which has the highest frequency. The most common was 3 minutes. It occurred 8 times.

■

HISTOGRAMS AND BAR GRAPHS

Often it is a good idea to present a graphical display of the data. This gives shape to the data and may make certain trends or patterns in the data very clear. The simplest graph is the **bar graph.** We can translate a frequency table directly into a bar graph by labeling the horizontal axis according to our grouping categories and then drawing bars according to the corresponding frequencies. Each line of the frequency table becomes one bar of the graph. For the data in Table 3-1 we get the bar graph in Figure 3-1. Now it stands out very clearly, for example, that a 3-minute wait occurred more than any other.

Closely related to the bar graph is the **histogram.** This is a graph where we let the bars touch and then erase the inside vertical lines. If there is a bar

Understanding statistics

Figure 3-1

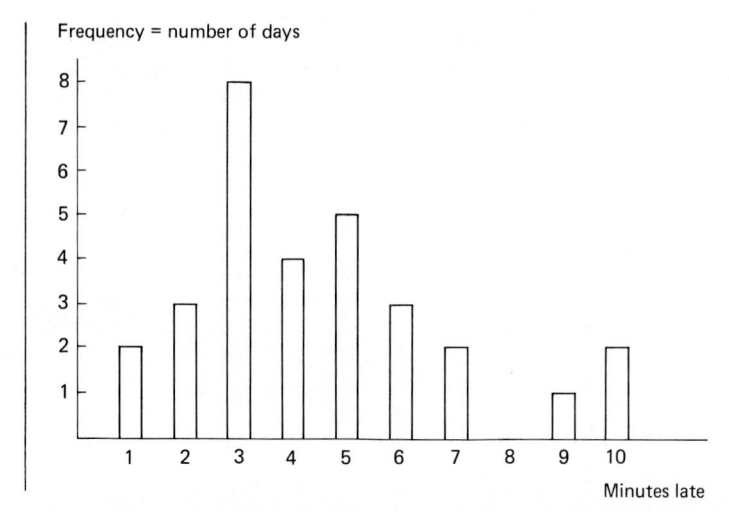

Frequency = number of days

[Bar graph showing frequency of minutes late: 1 min=2, 2 min=3, 3 min=8, 4 min=4, 5 min=5, 6 min=3, 7 min=2, 8 min=0, 9 min=1, 10 min=2]

Minutes late

missing, then we imagine that it is there and that its height is zero. For the bar graph of Figure 3-1 we get the histogram in Figure 3-2.

This type of graph makes sense when the horizontal axis describes some *increasing quantity,* such as time in our example, where the categories can be put in a natural numerical order. It does not make much sense when the groups do not have any natural order. For example, if we were counting the number of blond-, brown-, red-, and black-haired people who made appointments one week at a certain hair stylist, we might get a bar graph like Figure 3-3.

It is not helpful to join these bars to make a histogram. There is no logical numerical order to the categories, they are just names of colors. It is best to just use the bar graph.

It is useful to learn some specific vocabulary for describing histograms.

1. **Boundary** is the value on the horizontal axis where two bars of a histogram meet. For example, George measured bus waits *to the nearest minute.* So

Figure 3-2

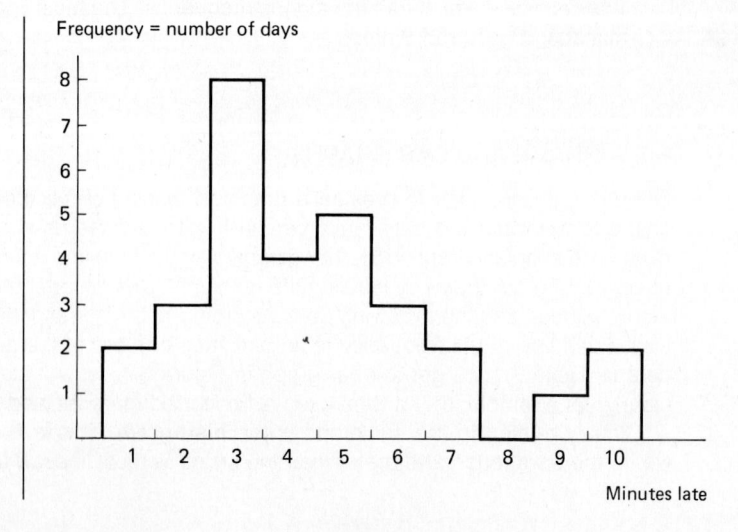

Frequency = number of days

[Histogram version of the same data]

Minutes late

Frequency tables and graphs

Figure 3-3

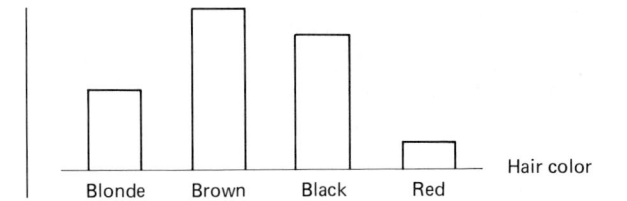

[Bar graph showing hair color categories: Blonde, Brown, Black, Red]

the waits which he recorded as 3 minutes really go from 2.5 to 3.5 minutes. So 2.5 and 3.5 are the boundaries for that category.

Usually a histogram shows the labels for the boundaries. In Figure 3-4, we redraw the histogram of Figure 3-2 showing the boundaries.

Figure 3-4

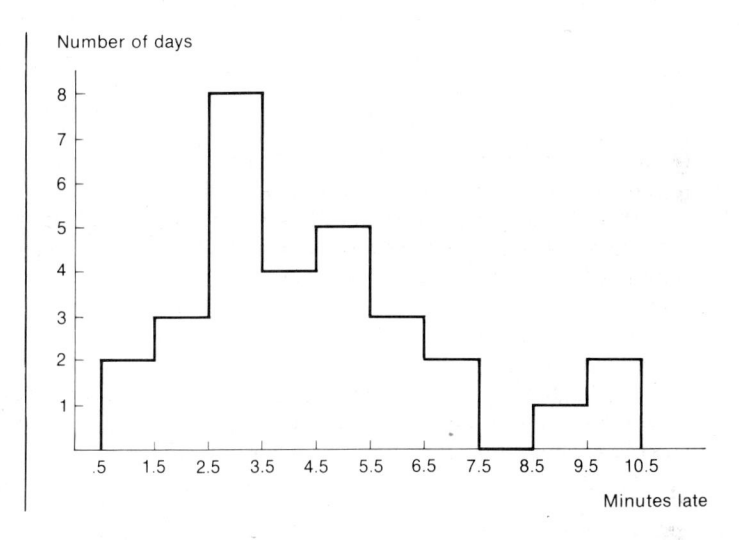

[Histogram with Number of days on vertical axis (1-8) and Minutes late on horizontal axis (.5, 1.5, 2.5, 3.5, 4.5, 5.5, 6.5, 7.5, 8.5, 9.5, 10.5)]

2. The piece of the horizontal axis between two consecutive boundaries is called an **interval**. For example, the piece from .5 to 1.5 is called the interval from .5 to 1.5.

3. When you subtract the lower boundary of an interval from the upper boundary, the result is called the **width of the interval**. For example, the width of the left most interval is $1.5 - .5 = 1$. Notice that all the intervals in any one histogram are of the same width. The width of the intervals in a histogram is decided by the person who is making the graph. The width can be small or large, depending on the purpose of the graph. For example, George Stephen could have measured how long he waited for the bus to the nearest .5 minute instead of the nearest minute; then he could have chosen intervals of width .5. On the other hand, he might only be interested in waits to the nearest 2 minutes and might set up intervals with width 2.

EXAMPLE 3-2 An experiment consists of tossing a silver dollar 20 times and counting the number of heads. This experiment was carried out 30 times. The results are shown in Table 3-2. The bar graph is shown in Figure 3-5. Draw the histogram for this data.

Understanding statistics

Table 3-2

number of heads in 20 tosses	number of experiments
15	2
14	1
13	3
12	0
11	4
10	5
9	7
8	3
7	2
6	2
5	0
4	0
3	1

total number of experiments = 30

Figure 3-5

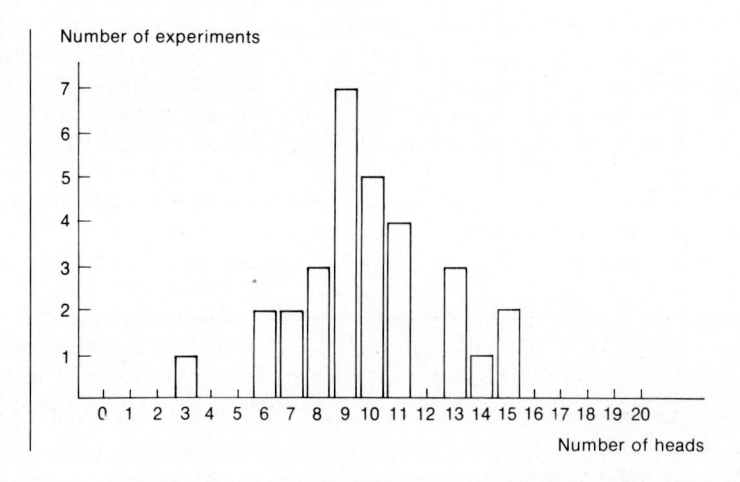

SOLUTION We make the histogram by widening the bars and finding the boundaries of the intervals. Figure 3-6 shows what we would get.

What is the effect of such a graph? It gives the impression that the number of heads is a continuously increasing quantity. It appears, for example, that in the 7 experiments represented by the interval 8.5 to 9.5 the number of heads could have been anything from 8.5 to 9.5, when in fact the number of heads could only have been exactly 9. On the other hand, the graph still gives an accurate picture of the outcomes of the experiments and their frequencies. For example, you can still clearly see that 9 heads was the outcome that occurred more often than any other outcome. And you can still see that no experiment resulted in 12 heads. In other words, the basic *shape* of the distribution is preserved. We will see in Chapters 5 and 7 that it will be very useful to graph even some noncontinuous quantities using histograms because we will be asking questions about areas, which will be easier to answer if we have one

Frequency tables and graphs

Figure 3-6

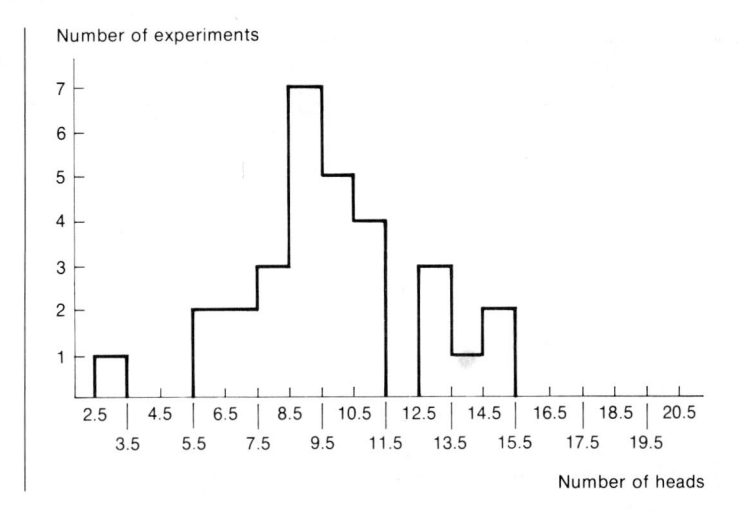

Number of experiments

[Histogram with x-axis "Number of heads" showing boundaries 2.5, 3.5, 4.5, 5.5, 6.5, 7.5, 8.5, 9.5, 10.5, 11.5, 12.5, 13.5, 14.5, 15.5, 16.5, 17.5, 18.5, 19.5, 20.5]

continuous graph rather than separate bars. As long as you keep in mind the kinds of data you have, continuous or not, no confusion will result.

Let us go back to the coin-tossing experiment and discuss drawing a histogram for those data. Two decisions have to be made first: (1) What width should the intervals be? (2) Where should we start the lowest interval? There are no fixed rules for answering these questions. It is up to the person drawing the graph. You might try graphing the same data in histograms of different width intervals to see what the overall effect is. Generally speaking, if the intervals are too wide, you will have too much data lumped together, and the trends in the data will be hard to spot. At the other extreme, if the intervals are too narrow, the graph will be too spread out and is likely to become very spotty, and once again any trends will be hard to spot. Suppose that for the coin-tossing data we decide to use intervals of width 3, and suppose that we start the lowest interval with a boundary of 2.5. We could tabulate the data as shown in Table 3-3. The histogram for these data would be as shown in Figure 3-7.

Table 3-3

number of heads in 20 tosses	number of experiments	boundaries	frequency (number of experiments)
15	2	14.5 to 17.5	2
14	1		
13	3	11.5 to 14.5	4
12	0		
11	4		
10	5	8.5 to 11.5	16
9	7		
8	3		
7	2	5.5 to 8.5	7
6	2		
5	0		
4	0	2.5 to 5.5	1
3	1		

Understanding statistics

Figure 3-7

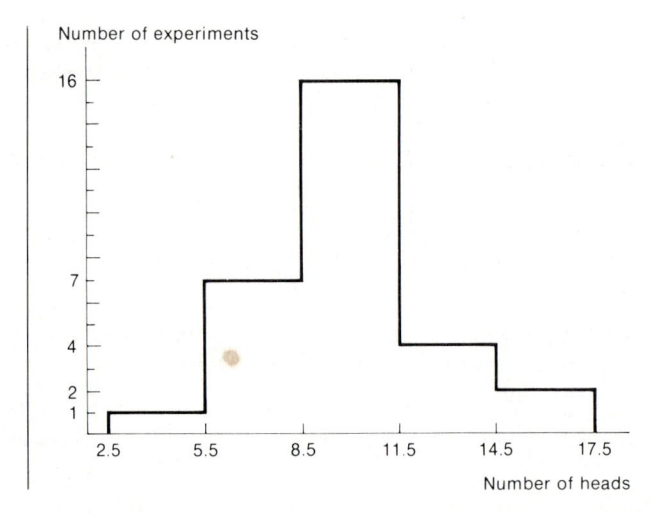

[Histogram: Number of experiments vs Number of heads. Bars with heights 1, 7, 16, 4, 2 over intervals with boundaries at 2.5, 5.5, 8.5, 11.5, 14.5, 17.5]

You can see that this histogram is similar to that in Figure 3-6, but it is easier to understand. It shows clearly the idea that in this experiment we often get *around* 10 heads, and that outcomes become more and more rare the further away they are from 10 heads.

READING HISTOGRAMS

What information is contained in a histogram? We note that the percentage of area of the graph which is over any particular interval is equal to the percentage of the outcomes that are in that interval. Let us see why. In Figure 3-8 we redraw the histogram of Figure 3-7 and we add some vertical and horizontal lines to show how each of the 30 outcomes is represented by an equal amount of area. This is the key to histograms: *each outcome is represented by an equal amount of area.*

Figure 3-8

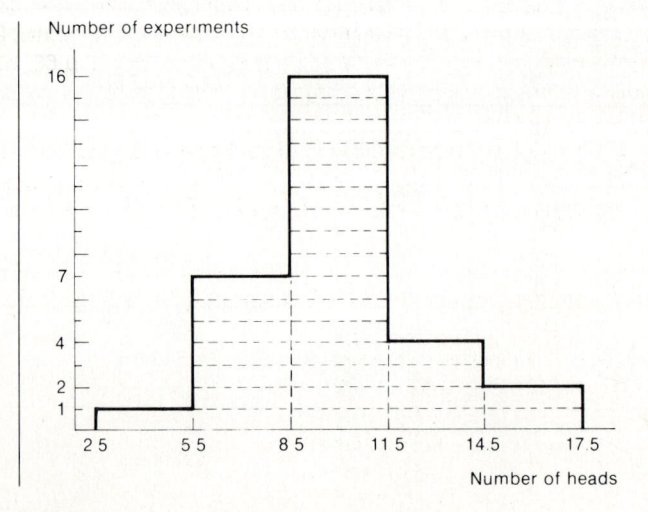

[Histogram: Number of experiments vs Number of heads, same as Figure 3-7 but subdivided with dashed lines showing each outcome as equal area units.]

Frequency tables and graphs

You see for example that there are 16 equal sections in the third interval. This corresponds to the frequency of 16 in Table 3-3. Now let us answer some specific questions to illustrate the principles involved.

1. In what percentage of the experiments was the result from 3 to 5 heads?

   Answer From the last two columns of Table 3-3 we see that the frequency for these outcomes was 1. The total number of outcomes was 30. The percentage of outcomes in this interval therefore is 1/30 = .0333 = 3.33 percent.

2. What percentage of the *area* of the histogram corresponds to experiments where the outcome was from 3 to 5 heads?

   Answer We see that over this interval is 1 unit of area. The total number of units of area in the graph is 30. Therefore the percentage of area in the graph over this interval is 1/30 = .0333 = 3.33 percent.

You notice then that as far as percentages of outcomes are concerned, the graph and the table contain the same information. In the next chapter you will see that percentages of outcomes are very important in discussing basic ideas of probability. This, in turn, means that we will be able to discuss questions of probability by looking at graphs and measuring percentages of area over various intervals. It will turn out that this is indeed a major tool in statistics because many apparently different questions lead us to graphs of the same shape. So once we know how to measure percentages of area over the intervals of that graph, we will be in good shape to answer those statistical questions. All that remains in this chapter is to show how we can relate what is known about percentages of area over various intervals and the actual overall *shape* of the graph.

PERCENTILE RANK, z SCORES, AND GRAPHS

In different distributions, the same z score may be associated with different percentile ranks. This is because the distributions have *different shapes*. For example, in a particular distribution of ages of college students, 21 years might correspond to $z = 1$ and have a percentile rank equal to 60, while in a distribution of incomes, $12,000 might also correspond to $z = 1$, but have a percentile rank equal to 75.

If we know which percentile ranks correspond to the z scores at the boundaries of the intervals, we can draw the histogram. We demonstrate this relationship between z scores, percentile ranks, and area in the following examples.

EXAMPLE 3-3 Draw the histogram for the following data.

z score	percentile rank
2	100
1	70
0	30
−1	20
−2	0

Understanding statistics

44

SOLUTION   Since the z scores of 0 and 1 have percentile ranks of 30 and 70, respectively, 40 percent of the distribution must lie between $z = 0$ and $z = 1$. In order to draw the histogram, we need to know the area corresponding to *each* interval. As we have just shown, you find the area corresponding to an interval by subtracting the percentile ranks of its boundaries.

boundaries	percentage of area
1 to 2	100 − 70 = 30
0 to 1	70 − 30 = 40
−1 to 0	30 − 20 = 10
−2 to −1	20 − 0 = 20

We can now draw the histogram (Figure 3-9).

Figure 3-9

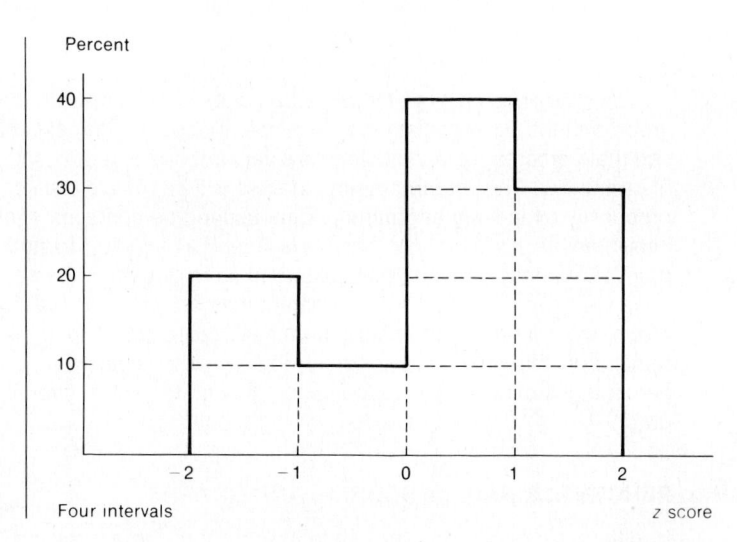

[Histogram showing percent vs z score over four intervals: −2 to −1 at 20%, −1 to 0 at 10%, 0 to 1 at 40%, 1 to 2 at 30%]

Four intervals   z score

If we had nine z scores instead of five, we could draw a more accurate histogram, as indicated in the next example.

EXAMPLE 3-4   Draw a histogram similar to the one shown in Example 3-3, given these nine z scores and their corresponding percentile ranks.

z score	percentile rank
2	100
1.5	90
1	70
0.5	55
0	30
−0.5	25
−1	20
−1.5	5
−2	0

Frequency tables and graphs

SOLUTION We find the percentage of area for each interval as we did in Example 3-3.

boundaries	percentage of area
1.5 to 2	10
1 to 1.5	20
0.5 to 1	15
0 to 0.5	25
−0.5 to 0	5
−1 to −0.5	5
−1.5 to −1	15
−2 to −1.5	5

The histogram for these data is shown in Figure 3-10.

Figure 3-10

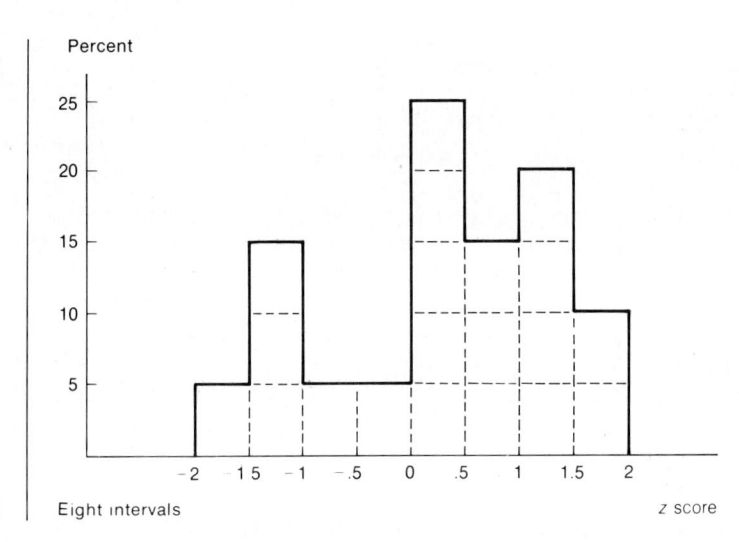

Eight intervals

If we had 100 z scores, we would have a histogram with 99 intervals. This histogram would be difficult to draw and is often approximated by a smooth curve, as shown in Figure 3-11.

Figure 3-11

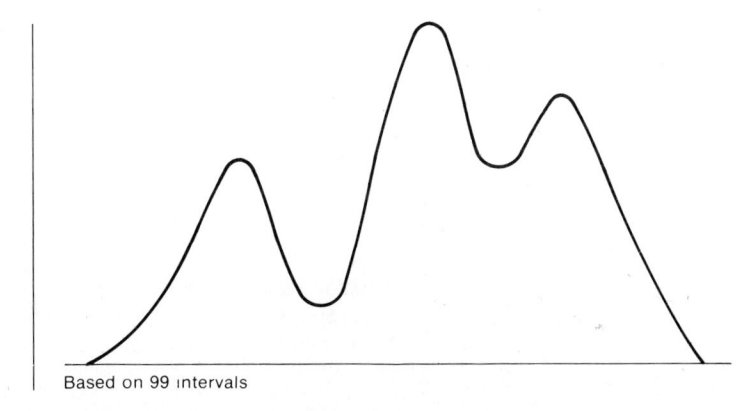

Based on 99 intervals

Understanding statistics

STUDY AIDS

VOCABULARY

1. Frequency table
2. Bar graph
3. Histogram
4. Boundary
5. Interval
6. Width of interval
7. Frequency

EXERCISES

3-1 The following histogram was the result of a study of the time it took to do a major engine repair job at a school of auto repair.

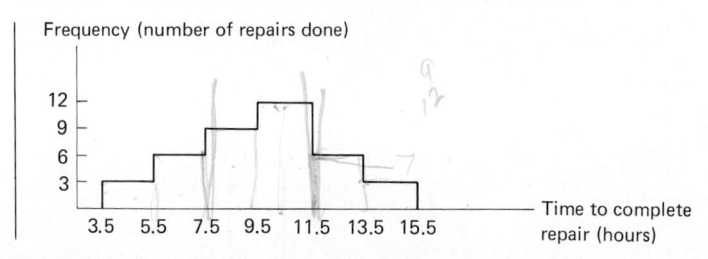

(a) How many repair jobs were included in this study?
(b) How many of the repair jobs took more than 11.5 hours? What percentage of the jobs took more than 11.5 hours?
(c) What percentage of the jobs took between 7.5 and 11.5 hours?

3-2 For the histogram shown below answer these questions.

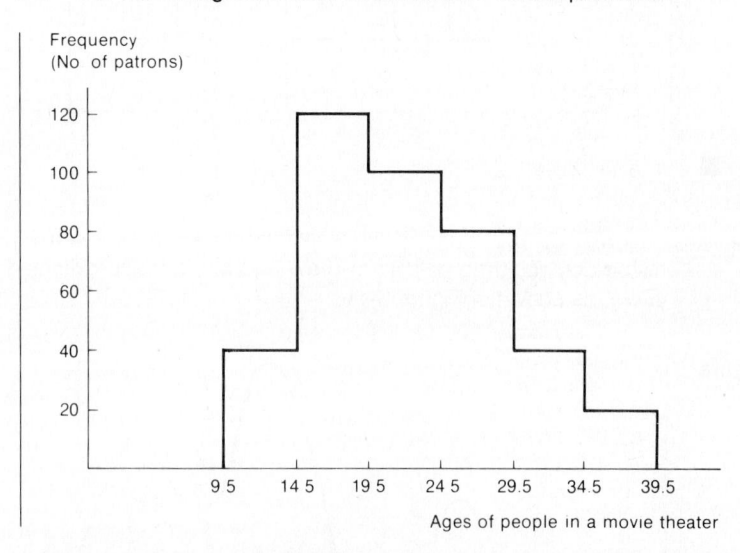

(a) Find the number of people represented in each interval, and find the total number of people represented in the graph.
(b) Find the percentage of people in each interval.
(c) Find the *percentage of area* of the graph which is in the space over each interval, and check that the total of these is 100 percent.

3-3 Here are the ages at inauguration of the U.S. presidents from Washington to Reagan.

name (party)	age at inauguration
1. Washington (F)	57
2. J. Adams (F)	61
3. Jefferson (DR)	57
4. Madison (DR)	57
5. Monroe (DR)	58
6. J. Q. Adams (DR)	57
7. Jackson (D)	61
8. Van Buren (D)	54
9. W. H. Harrison (W)	68
10. Tyler (W)	51
11. Polk (D)	49
12. Taylor (W)	64
13. Fillmore (W)	50
14. Pierce (D)	48
15. Buchanan (D)	65
16. Lincoln (R)	52
17. A. Johnson (U)	56
18. Grant (R)	46
19. Hayes (R)	54
20. Garfield (R)	49
21. Arthur (R)	50
22. Cleveland (D)	47
23. B. Harrison (R)	55
24. Cleveland (D)	55
25. McKinley (R)	54
26. T. Roosevelt (R)	42
27. Taft (R)	51
28. Wilson (D)	56
29. Harding (R)	55
30. Coolidge (R)	51
31. Hoover (R)	54
32. F. D. Roosevelt (D)	51
33. Truman (D)	60
34. Eisenhower (R)	62
35. Kennedy (D)	43
36. L. B. Johnson (D)	55
37. Nixon (R)	56
38. Ford (R)	61
39. Carter (D)	52
40. Reagan (R)	68

(a) Construct a frequency table for the ages and draw the corresponding histogram. Use 5-year categories starting with age 40. (First category is 40–44.)
(b) Compute the mean and median ages.
(c) What percentage of the presidents were in their 50s when inaugurated?

Understanding statistics

3-4 Here are the percentages of the U.S. population in different age categories for different years.

	age				
year	under 5	5–19	20–44	45–64	65 and over
1860	15.4	35.8	35.7	10.4	2.7
1870	14.3	35.4	35.4	11.9	3.0
1880	13.8	34.3	35.9	12.6	3.4
1890	12.2	33.9	36.9	13.1	3.9
1900	12.1	32.3	37.8	13.7	4.1
1910	11.6	30.4	39.1	14.6	4.3
1920	11.0	29.8	38.4	16.1	4.7
1930	9.3	29.5	38.3	17.5	5.4
1940	8.0	26.4	38.9	19.8	6.9
1950	10.7	23.2	37.7	20.3	8.1
1960	11.3	27.1	32.4	20.0	9.2
1970	8.4	29.4	31.7	20.6	9.9

(a) Using these categories, construct a bar graph for the data from 1870. Do the same for 1920 and 1970. Describe in a few sentences what is going on.
(b) Notice that the categories are not of equal width. This makes it difficult to draw a histogram. Why do you suppose these categories were picked?

3-5 Draw a histogram for the job performance ratings given below:

interval	boundaries	frequency
10	9.5 to 10.5	7
9	8.5 to 9.5	16
8	7.5 to 8.5	15
7	6.5 to 7.5	36
6	5.5 to 6.5	99
5	4.5 to 5.5	150
4	3.5 to 4.5	216
3	2.5 to 3.5	304
2	1.5 to 2.5	401
1	.5 to 1.5	197
0	−.5 to .5	253

3-6 Compute the frequency in each interval by using the data given in Exercise 3.5, and draw the histogram.

interval	boundaries	frequency
9 to 11	8.5 to 11.5	
6 to 8	5.5 to 8.5	
3 to 5	2.5 to 5.5	
0 to 2	−.5 to 2.5	

3-7 Here are the salary standings a few seasons ago for the 26 major-league baseball teams. Construct a frequency table and draw a histogram for these data, starting at a lower boundary of $50,000 and using intervals of width (a) $25,000 and (b) $40,000. Which grouping do you prefer? Why?

Frequency tables and graphs

club	average salary
N.Y. Yankees	$242,937
Philadelphia	221,274
Pittsburgh	199,185
California	191,014
Boston	184,686
Los Angeles	183,124
Houston	176,720
St. Louis	173,480
Cincinnati	162,655
Chicago Cubs	160,209
Milwaukee	159,086
Montreal	158,196
Texas	148,792
San Francisco	148,265
Atlanta	147,989
San Diego	138,978
Cleveland	127,505
N.Y. Mets	126,448
Baltimore	116,156
Kansas City	100,453
Detroit	86,998
Seattle	82,244
Minnesota	80,538
Chicago White Sox	72,415
Toronto	67,218
Oakland	54,994

3-8 The table shows the breakdown of the expenses of one investment company for 3 consecutive years. We wish to compare the breakdowns to see if there is any pattern emerging.

All amounts are in *thousands of dollars*

expenses	year 1	year 2	year 3
management fee	1147	3086	7478
shareholder service	639	1541	2702
printing	45	167	506
postage	38	119	419
professional fees	78	65	325
rent	148	156	307
registration fees	75	195	261
equipment maintenance	72	102	151
custodian fees	38	101	138
directors fees	13	21	40
state and local taxes	11	8	7
Total	2304	5561	12,334

(a) For the first year, draw a bar graph corresponding to the dollar amounts spent in each category.
(b) For the first year, convert the dollar amounts to "fraction of total expense."

(For example, management fee = $\frac{1147}{2304}$ = .50.) Then draw a bar graph using the fractions to determine the heights of the bars. Compare the bar graphs from parts (a) and (b).

(c) For years 2 and 3 convert the amounts to fractions of total expense as in part (b). Then draw bar graphs for years 2 and 3.

(d) Compare the graphs for years 1, 2, and 3. Do you see any interesting patterns in the company's expenses?

3-9 Here are some U.S. government figures on income distribution from a few years ago. The frequencies are counting what the government calls STATS units for the purpose of estimating tax income. These are interdependent financial units (such as families, single people living alone, groups of unrelated people living together). They give you a rough idea of the incomes of households.

wage and salary income	percentage of households
none	25.1
$ 1– 999	5.0
1000– 1499	1.9
1500– 1999	2.2
2000– 2499	1.7
2500– 2999	1.7
3000– 3499	1.5
3500– 3999	1.9
4000– 4999	3.0
5000– 5999	3.6
6000– 6999	4.0
7000– 7999	3.9
8000– 9999	7.4
10,000– 14,999	16.9
15,000– 24,999	15.8
25,000 and over	4.5

(a) Regroup the table using these categories:

```
    0–      4999
 5000–      9999
10,000–  14,999
15,000–  24,999
25,000 and over
```

(b) Draw a bar graph corresponding to part (a). Note that the bars do not all cover categories of the same width.

(c) What percentage of the households had wage and salary incomes less than $5000? between $10,000 and $25,000?

(d) Approximately what was the *median* income from wages and salaries?

3-10 An experiment is to roll a die 6 times. This experiment was repeated 20 times, and the number of times that a 2 appeared in each experiment was recorded. The results are tabulated by making one tally mark after each 6 tosses of the die.

Frequency tables and graphs

number of times a 2 appeared	tally	frequency
6		0
5	/	1
4	/	1
3	///	3
2	//	2
1	///// //	7
0	///// /	6
		20

(a) Draw a histogram for the data in the table.
(b) What is the mean number of times that a 2 appeared on the die?

3-11 (a) Toss a coin 5 times, and count the number of times that tails appear. Repeat this experiment 15 times. Record the data in the table below, and draw a histogram for the data. Make one tally mark after each 5 tosses of the coin.

number of tails	tally	frequency
5		
4		
3		
2		
1		
0		
		15

(b) What is the mean number of tails that occurred in 5 tosses?

3-12 What visual impression can be created by lengthening or shortening the scale of numbers on the vertical axis of a bar graph? What happens if you do not start with zero? Suppose we are comparing sales of two competing brands of portable radio-cassette players. Here are the data.

brand	sales for last year
Nipco	30,000 units
Sanyaha	25,000 units
Monimo	15,000 units

(a) Draw bar graphs for these results using three different vertical scales as shown.

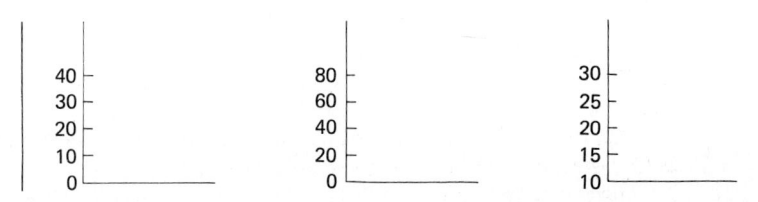

(b) Describe the different overall effects of the three graphs. Which graph do you prefer? Why? Which would you prefer if you worked for Monimo? Why?

Understanding statistics

3-13 The histogram illustrated below is symmetrical. The data represent the heights of children in Uncle Don's Nursery School. Some of the z scores and percentile ranks are related as follows.

z score	percentile rank
3	100
2	98
1	84
0	50

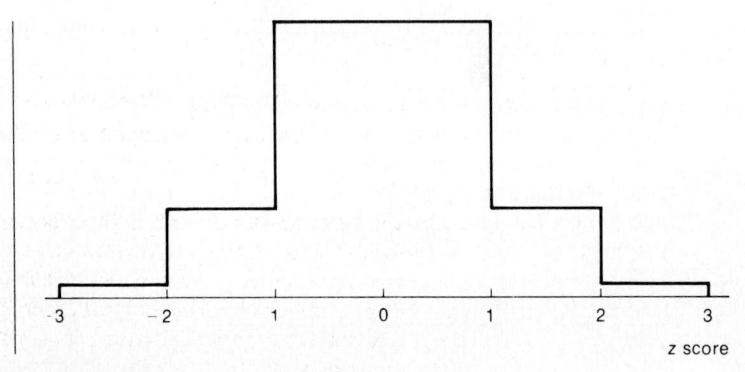

What percentage of the area of the graph is:
(a) To the left of 0?
(b) To the left of 1?
(c) To the left of -1?
(d) To the right of 2?
(e) To the right of -2?
(f) Between 0 and 3?
(g) Between 0 and 2?
(h) Between -2 and 0?
(i) Between -1 and 1?
(j) Between -2 and 2?

3-14 The following data describe the grade distribution in Scuba Diving 2.

z score	percentile rank
3	100
2	80
1	50
0	40
-1	20
-2	10
-3	0

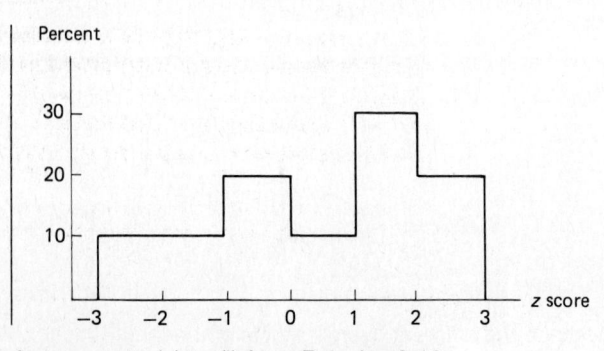

Answer parts (a) to (j) from Exercise 3-13.

Frequency tables and graphs

3-15 A distribution of salaries of supermarket clerks is as follows.

z score	percentile rank
3	100
2	85
1	45
0	35
−1	30
−2	10
−3	0

(a) For the given data, what percentage of the salaries lie between $z = 1$ and $z = 2$?

(b) What percentage of the area of a histogram based on these data will be between $z = 1$ and $z = 2$?

(c) Complete the following table.

boundaries	percentage of salaries	area of a histogram
2 to 3	15	
1 to 2		
.		
.		
.		

(d) Draw the histogram.

3-16 Construct a table of boundaries and percentages of area and draw the histogram for the following data derived from test scores in a prenursing course.

z score	percentile rank
3	100
2.5	75
2	65
1.5	50
1	45
0.5	40
0	35
−0.5	33
−1	30
−1.5	25
−2	10
−2.5	5
−3	0

3-17 The data given here represent amounts of money bet one night at the races. Construct a table of boundaries and percentages of area and draw a histogram.

z score	percentile rank
3	100
2	90
1	60
0	40
−1	0

CLASS SURVEY

1 Draw 3 histograms for the *male heights*. Graph 1 has intervals of width *1* inch; graph 2 has intervals of width *2* inches; graph 3 has intervals of width *3* inches. Start all graphs at the same height.
2 Which of the three graphs do you prefer for showing the height? Why?
3 Draw a histogram for the female heights using intervals equal to those you selected in answering part 2.
4 Compare the two histograms in parts 2 and 3 and comment on any differences.

Probability

4

The probability of drawing a heart out of an ordinary deck of playing cards is 1/4.
The probability of tossing a head on a fair coin is 1/2.
The weather bureau announces that the probability of rain tomorrow is 40 percent.

What meaning does the word "probability" have? If I toss a fair coin 2 times, will I get exactly 1 head? If I toss it 10 times, will I get exactly 5 heads? If I toss it once, will I get one-half of a head? Does a 40 percent possibility of rain indicate that it will rain 40 percent of the time tomorrow? Or that it will almost rain tomorrow?

Probability is usually interpreted in reference to a *large number* of trials. The probability that a fair coin will come up heads equals 1/2. This means that in a *large number* of trials we expect to get about 1/2 heads.

The probability of rain tomorrow is 40 percent. This means that out of a large number of days with weather conditions like today's, about 40 percent of the time it will rain on the following day.

Consider the following example.

At a party Peter Parapsych claimed to have powers of extrasensory perception. To test this claim we placed 6 pieces of round, hard candy in a paper bag; 2 of the candies were strawberry and the other 4 were orange. The candies were well mixed and Peter, who was blindfolded, was asked to reach in and select a strawberry. Peter did select a strawberry. Does this mean he has ESP? What were his chances of getting a strawberry candy just by luck?

In this experiment there were 6 *equally likely* possible outcomes, since he could have selected any of the 6 candies. Two of these outcomes, the strawberry candies, were favorable to Peter's claim. Thus we say that the probability of his selecting a strawberry candy *just by luck* was 2 chances in 6; that is, the probability equals 2/6 or 1/3. This does not imply that he will pick strawberry exactly once every 3 trials, but, rather, that in a large number of trials, about 1/3 of the picks will be strawberry just by luck. We wouldn't even begin to believe his claim unless he could pick the proper candy much more often than 1/3 of the times he tried.

DEFINITION OF PROBABILITY

When all the possible outcomes of an experiment are equally likely, then we define the **probability of an event** as follows.

> The probability of an event is the number of *favorable* outcomes of the experiment divided by the *total* number of possible outcomes.

A FORMULA FOR PROBABILITY

> The formula for the probability of an event is
>
> $$P(\text{event}) = \frac{F}{T}$$
>
> where F is the number of favorable outcomes, and T is the total number of outcomes.

Put another way, this is: The fraction of the population that makes up the event is the probability of the event. In the above example, P(selecting a strawberry candy) = 2/6 = 1/3.

The idea of probability is related to the idea of *randomness*. For example, when you play a game of cards, to ensure fairness you shuffle the cards. Why? To ensure that no card or cards have a *special place* in the deck; to ensure that *each card* has just the *same chance* of being in any given part of the deck. The statistician says you shuffle the deck to make certain that the cards are arranged *randomly* or *in random order*. When an item (or number) is picked <u>at random</u> from some population, then that item has the same likelihood of being picked as any other item in the population; that is, it had no special, privileged position in the population and neither did any other item in the population.

EXAMPLES OF PROBABILITY CALCULATIONS

EXAMPLE 4-1

If we toss a fair coin, there are two possible outcomes, heads or tails. These two outcomes are equally likely. The probability of tossing heads is written

■ $P(\text{heads}) = 1/2 = .5$

EXAMPLE 4-2 A carnival spinning wheel has the numbers 1 to 20 on it, all equally marked off. When it is spun, it will stop, at random, on one of the numbers.

(a) The probability that it will stop on the number 14 is 1/20 (or .05) because:
 (1) There are 20 numbers altogether, and
 (2) One of the numbers is a 14.

(b) The probability it will stop on an even number is 10/20 (or .5) because:
 (1) There are 20 numbers altogether, and
 (2) Ten of them are even.

(c) The probability that it will stop on a number 15 or higher is 6/20 (or .3) because:
 (1) There are 20 numbers altogether, and
 (2) Six of the numbers are 15 or higher (15, 16, 17, 18, 19, 20).

■

EXAMPLE 4-3 If we roll 1 die, there are 6 possible outcomes. If we let X represent the value of the outcome, then

P(outcome is 3) $= P(X = 3) = 1/6$

P(outcome is even) $= P(X$ is even$) = 3/6 = 1/2$

■ P(outcome is more than 4) $= P(X > 4) = 2/6 = 1/3$

EXAMPLE 4-4 In the town of Juvena, 40 percent of the population is below 25 years old (the young folks); 60 percent of the population is 25 years old or older (the old folks). If a person is picked at random from the population, the probability that the person will be young is .40, which is the fraction of young folks in the entire population.

■

The probability fraction F/T is always between 0 and 1. If an outcome is impossible, its probability is 0. Thus, the probability of rolling a 13 with 2 ordinary dice is 0. If an outcome is sure to happen, its probability is 1. Thus, if you draw 5 cards from a deck of playing cards, the probability that at least 2 are the same suit is 1.

If the probability of your passing this course is .9, then the probability of your not passing is .1. In general, if p is the probability of an event happening and q is the probability of the event not happening, then

$p + q = 1$ or $q = 1 - p$

Consider a spinning wheel which has the numbers 1 to 36 marked off on it in equally spaced divisions. Let X stand for the number at which the wheel stops. If we let $p = P(X < 13)$, then $q = P(X \geq 13)$. Since $p = 12/36 = 1/3$, then $q = 1 - 1/3 = 2/3$. Note that we could have computed q directly: $q = 24/36 = 2/3$.

Understanding statistics

EXAMPLE 4-5   A fair coin is tossed twice. What is the probability of getting 2 tails?

SOLUTION   If on the first toss we get heads, the second outcome can be heads or tails. Similarly, if the first toss is tails, then the second toss can be heads or tails. Since each toss has 2 possible outcomes, we can see that there are 2 × 2 or 4 equally likely possible outcomes for the 2 tosses.

outcome	first toss, second toss
1	heads, heads
2	heads, tails
3	tails, heads
4	tails, tails

Only the last of the 4 possible outcomes was favorable.
Therefore, $P(2$ tails$) = 1/4$.

EXAMPLE 4-6   A penny is tossed once and a die is rolled once. List the possible outcomes.

SOLUTION   Since the penny can come up 2 ways and the die 6 ways, there are 2 × 6 or 12 possible outcomes, namely, H1, H2, H3, H4, H5, H6, T1, T2, T3, T4, T5, T6.

EXAMPLE 4-7   Find the probabilities of the following outcomes for Example 4-6.

(a) Tossing a head *and* rolling an even number
(b) Tossing a head *or* rolling an even number
(c) Tossing a head and rolling a 5
(d) Tossing a head or rolling a 5
(e) Rolling either a 4 or a 6
(f) Rolling both a 4 and a 6
(g) Tossing a head or rolling a 7
(h) Tossing a head and rolling a 7

SOLUTION   (a) There are 3 favorable outcomes: H2, H4, H6. The answer is 3/12 or 1/4.

(b) There are 9 favorable outcomes: H1, H2, H3, H4, H5, H6, T2, T4, T6. The answer is 9/12 = 3/4.

(c) There is 1 favorable outcome: H5. The answer is 1/12.

(d) There are 7 favorable outcomes: H1, H2, H3, H4, H5, H6, T5. The answer is 7/12.

(e) There are 4 favorable outcomes: H4, H6, T4, T6. The answer is 4/12 = 1/3.

(f) There are no favorable outcomes. The answer is 0.

(g) There are 6 favorable outcomes: H1, H2, H3, H4, H5, H6. The answer is 6/12 = 1/2.

(h) There are no favorable outcomes. The answer is 0.

Probability

EXAMPLE 4-8 Find the probability of rolling a sum of 7 with 2 fair dice.

SOLUTION There are 6 faces on each die, numbered 1 to 6. If the first die lands with a 1 showing, the second die could land with a 1, 2, 3, 4, 5, or 6 showing. If the first die lands with a 2 showing, the second die could still land with a 1, 2, 3, 4, 5, or 6 showing. Thus, given any way the first die lands, there are still 6 possible ways the second die can land. Our listing of all possible outcomes consists of 6 × 6 or 36 outcomes, all of which are equally likely.

die 1, die 2	die 1, die 2	die 1, die 2	die 1, die 2	die 1, die 2	die 1, die 2
1, 1	2, 1	3, 1	4, 1	5, 1	6, 1
1, 2	2, 2	3, 2	4, 2	5, 2	6, 2
1, 3	2, 3	3, 3	4, 3	5, 3	6, 3
1, 4	2, 4	3, 4	4, 4	5, 4	6, 4
1, 5	2, 5	3, 5	4, 5	5, 5	6, 5
1, 6	2, 6	3, 6	4, 6	5, 6	6, 6

Of the 36 possible outcomes, there are 6 in which the sum of the dice is 7, and so

■ P(rolling a sum of 7) = 6/36 = 1/6

EXAMPLE 4-9 Pamela Purloiner, a clerk in a jewelry store, has a passion for diamonds. In a tray containing 4 diamonds, she has replaced 2 gems with imitations. Her employer reaches into the tray and picks up 2 stones at random. What is the probability that her employer has gotten *both real* diamonds?

SOLUTION We must determine the number of possible outcomes. To distinguish between the stones, let us label the diamonds D_1 and D_2, and the imitations I_1 and I_2. If the employer picks up a real diamond first, then any 1 of the 3 remaining stones can be picked second. Similarly, taking any one of the 4 stones first leaves a choice of 3 stones for the second selection. Therefore, there are 4 × 3 or 12 equally likely outcomes.

	1	2	3	4	5	6	7	8	9	10	11	12
1st stone	D_1	D_1	D_1	D_2	D_2	D_2	I_1	I_1	I_1	I_2	I_2	I_2
2d stone	D_2	I_1	I_2	D_1	I_1	I_2	D_1	D_2	I_2	D_1	D_2	I_1

There are only 2 outcomes of the 12 that would be favorable, that is, selecting *both real diamonds*, and so the probability of taking the 2 diamonds is 2/12 or

■ 1/6.

PROBABILITY AND HISTOGRAMS

Since the fraction of the total population that corresponds to an event is the probability of that event, in a histogram the fraction of the area which represents the event is equal to the probability of that event.

Understanding statistics

EXAMPLE 4-10 A statistician has been studying fuel mileage on the Datsun 1595-cc engine. The graph is shown in Figure 4-1.

mileage	percentile rank
15	0
18	2
21	16
24	50
27	84
30	98
33	100

Figure 4-1

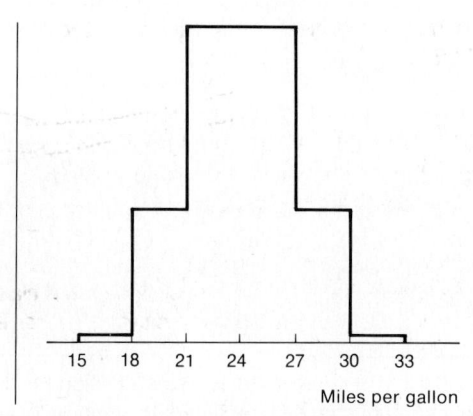

Miles per gallon

What is the probability that a Datsun 1595-cc engine picked at random will get less than 27 miles per gallon?

SOLUTION We need to know the fraction of engines in this category. This is equal to the percentage of the area of the graph to the left of 27. The percentile rank of 27 is 84, and so to the left of 27 is 84 percent of the graph. Therefore,

■ P(less than 27 miles per gallon) $= .84$

Note: See Appendix B for a further development of probability.

STUDY AIDS

VOCABULARY

1. Probability

SYMBOLS

1. P(event)   2. p 3. q

FORMULAS

1. $p = \dfrac{F}{T}$ 2. $q = 1 - p$

EXERCISES

4-1 If we toss two coins in the air, there are 3 possible outcomes: 2 heads, 1 head, or no heads. Why is the probability of obtaining 2 heads *not* equal to 1/3?

4-2 From an ordinary playing deck of 52 cards, 1 card is selected at random. Find the probability:
(a) That a three of diamonds is selected.
(b) That a heart is selected.
(c) That a jack is selected.

4-3 What is the probability that 1 day selected at random from this month is a Monday?

4-4 Find the following probabilities for the random selection of 1 card from an ordinary deck of 52:
(a) $P(\text{queen})$
(b) $P(\text{heart})$
(c) $P(\text{queen or heart})$
(d) $P(\text{queen and heart})$
(e) $P(\text{king or queen})$
(f) $P(\text{king and queen})$
(g) $P(\text{heart or diamond})$
(h) $P(\text{red or black})$

4-5 Are the following both true or both false?
(a) Let X be the middle digit in a 9-digit social security number. If a social security number is picked at random, then $P(X \text{ is even}) = 1/2$.
(b) Insecurity Trust and Savings Bank employs 12 different classifications of workers, including tellers. Count de Munnee works there. Hence, the probability that the Count is a teller is 1/12.

4-6 An experiment consists of tossing a penny and a dime.
(a) List the possible outcomes.
Find the probability of the following:
(b) $P(\text{heads on the penny})$
(c) $P(\text{heads on both})$
(d) $P(\text{exactly 1 head})$
(e) $P(\text{at least 1 head})$

4-7 A tile floor is made up of squares as illustrated. The tiles are numbered, and the odd-numbered tiles are black, while the even-numbered tiles are green.

1	2	3	4	5
6	7	8	9	10
11	12	13	14	15

A pin is dropped from a great height by a blindfolded person, and is just as likely to fall in one square as in any other. If the pin falls in a crack between two tiles, it does not count, and the pin is dropped again. The outcome will yield both a color and a number. Find:
(a) $P(\text{outcome is black})$.
(b) $P(\text{outcome is green})$.
(c) $P(\text{number is less than 5})$.
(d) $P(\text{number is greater than 5})$.
(e) $P(\text{outcome is green and less than 5})$.

4-8 In a certain high school, 20 percent of the students are seniors, and 48 percent of the students are girls. Assuming that the students enter the cafeteria

randomly, we stand at the door of the cafeteria and watch for the next pupil to enter. What is:
(a) $P(X$ is a senior)?   (b) $P(X$ is a boy)?

4-9   There are 100 males and 100 females belonging to an organization; 80 of the males and 10 of the females are scientists. What is the probability that *a person* picked at random from the membership list is:
(a) Female?
(b) A scientist?
(c) A female scientist?
(d) What is the probability that a scientist picked at random in the organization is female?
(e) What is the probability that a female picked at random from the organization is a scientist?

4-10   *Rates and Probability*   We can often interpret rates as probabilities. For example, in Chapter 2 we discussed death rates. Suppose the annual death rate for a particular disease is 2 per 100,000 in your community. One could say that the probability is about 2/100,000 that an individual picked at random in your community will die of this disease in the next year.

The highway patrol in Speed City predicts 20 traffic fatalities over the upcoming St. Swithin's Day weekend. There will be about 5000 travelers.
(a) What is the predicted death rate per 1000 travelers?
(b) What is the probability that a traveler picked at random will be killed?

4-11   On a true-false test you have no idea of the answers to three questions, and so you decide to guess.
(a) List all the possible outcomes, using C for correct and W to stand for wrong.
(b) If X = number of correct guesses, find: (1) $P(X = 3)$, (2) $P(X \geq 2)$, (3) $P(X = 0)$.

4-12   A spinner is labeled as indicated in the diagram.

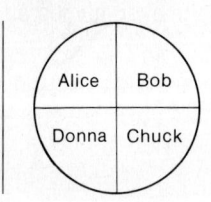

If we spin it once, find:
(a) P(outcome is Alice).
(b) P(outcome is a boy).
(c) If we play the game twice, there are 16 possible outcomes. List them. What is:
(d) P(first spin = Alice, and second = Bob)?
(e) P(first spin = Alice, and second = Alice)?
(f) P(outcome includes Alice exactly once)?
(g) P(outcome includes Alice at least once)?

4-13   In the game of Monopoly, Joe Capitalist needs to roll an 8 on the 2 dice to land on Park Place. Refer to the list of possible outcomes for 2 dice in Example 4-8.
(a) What is the probability that Joe will succeed in rolling an 8?

(b) Betty Badluck will go to jail if she rolls doubles one more time. What is the probability of that?

4-14 A group of people are going to play Russian roulette continuously until only one is "left alive." A *statistical model* of this experiment can be constructed as follows. Each person will roll a die once. If the outcome is a 1, he or she is "dead." The die is passed around among the "living" until only one is left. Find:
(a) P(first person "dies" on first roll).
(b) P(second person "dies" on first roll).
(c) Using the list of possible outcomes for the 2-dice experiment in Example 4-8, find P(a particular person "lives" through first two rolls). That is, find P(the first roll is not a 1, and the second roll is not a 1).

4-15 Sometimes it is easy to picture all the possible outcomes of an experiment by drawing a *tree diagram*. In Example 4-5, where we tossed a coin twice, the tree diagram looks like this.

Tree Diagram

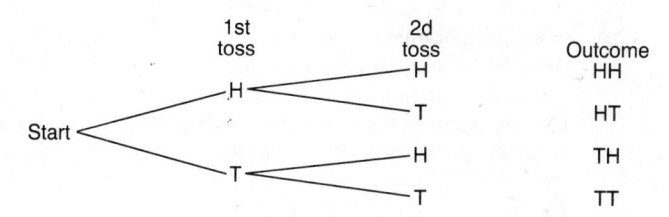

```
              1st         2d
              toss        toss        Outcome
                          H           HH
                  H
                          T           HT
       Start
                          H           TH
                  T
                          T           TT
```

Each possible path from the left starting point to a right end point represents one possible outcome. It is easy to see from this diagram what the 4 possible outcomes are.
(a) Draw a tree diagram for Example 4-6 about tossing a penny and a die, and list the outcomes corresponding to each branch. Your answer should be the same as that given in the example.
(b) Four pennies are tossed. Draw a tree diagram for this experiment. Find the probability of tossing exactly 2 heads; more than 2 heads.
(c) If the experiment in part (b) consisted of tossing 4 dice, would you want to solve it by use of a tree diagram? Why or why not?
(d) Construct a tree diagram for the outcomes of the Russian roulette problem in part (c) of Exercise 4-14. Why are there only 31 branches and not 36?

4-16 A young girl has been picking mushrooms. She accidentally picks 2 toadstools which to her appear identical to the 3 mushrooms she picked. She will eat 2 of the 5 picked.
(a) Using M_1, M_2, M_3 for the mushrooms and T_1, T_2 for the toadstools, list all 20 possible outcomes.
What is the probability that she:
(b) Eats 2 toadstools?
(c) Eats at least 1 toadstool?
(d) Eats no toadstools?

4-17 A child's toy slot machine has 3 wheels, and each wheel has 1 banana, 1 cherry, and 1 lemon.
(a) List the possible outcomes.
(b) Find P(3 lemons).
(c) Find P(3 the same).
(d) Find P(at least 2 lemons).
(e) If we play the slot machine 54 times, about how many times would you

expect to see 2 or more lemons? Suppose this happened only once. What would your reaction be?

4-18 Go the cafeteria at lunchtime. Record 50 consecutive purchases at one register. Draw a graph of the distribution of purchases.
(a) What fraction of the purchases was $2.00 or more?
(b) Based on your answer to part (a), what is the probability that a single purchase picked at random will be $2.00 or more?
(c) Repeat the problem on a different day to see if there is much change in the answers to parts (a) and (b).
(d) Discuss whether these samples are random or not.
(e) Find the mean and variance for your 50 purchases.
(f) What percentage of the purchases are more than 1 standard deviation from the mean?

4-19 Prunella Fructus wants to select 2 items at random from a basket containing 2 figs, 2 dates, and 1 lime.
(a) If she selects 1 item and does not replace it before she selects the second item, find the probability that she selects 2 different items.
(b) Repeat part (a) if she replaces the first item before selecting the second.
(c) Suppose she picks 2 items without replacement. If you see the first item selected and it is a fig, what is the probability that she selected 2 different items?
(d) Repeat part (c) if the first item is a lime.

4-20 365 capsules, each containing the date of 1 day of the year, are mixed, and 1 capsule is picked at random. What is the probability that the one picked will be:
(a) Some day in January?
(b) March 2?
(c) Either in March or in April?
(d) Not in December?

4-21 A person is thinking of a random letter of the alphabet. In front of a room of 260 people, he asks each person to concentrate very hard and to write down what the letter is. It turns out that 9 people have the correct letter. Have we discovered 9 people with ESP?

4-22 Here is a chart showing the racial breakdown of employees at a certain factory.

race	percentage
Asian	6
Black	46
White	43
Other	5

What is the probability that 1 employee picked at random will be:
(a) Asian?
(b) Not white?
(c) Either black or Asian?
(d) Not black?

4-23 When a baby is born, it is either male or female. If it is true that there is an equal chance that a newborn baby will be male or female, then the graph below shows the theoretical distribution of the relative frequencies of males in a family of 4 children.

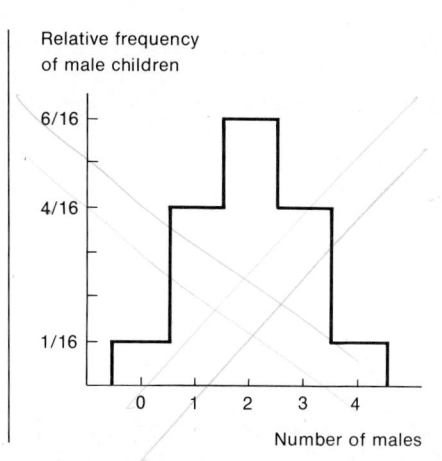

Using the histogram above, what is the probability that a family with 4 children will have:
(a) No male children?
(b) 2 male children?
(c) All male children?
(d) 2 or more male children?
(e) Either 1 male child or 3 male children?

4-24 The table and graph of relative frequencies for family incomes in Padsville are shown below.

income, dollars	percentile rank
0	0
10,000	35
20,000	50
30,000	67
40,000	100

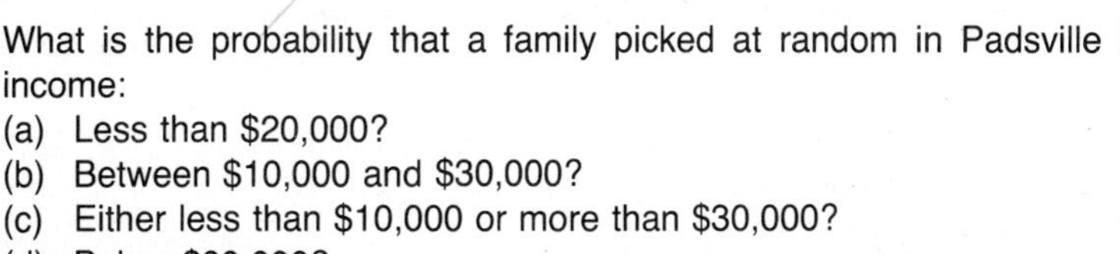

What is the probability that a family picked at random in Padsville has an income:
(a) Less than $20,000?
(b) Between $10,000 and $30,000?
(c) Either less than $10,000 or more than $30,000?
(d) Below $30,000?

4-25 Using the Datsun data from Example 4-10, find the probability that an engine picked at random gets:
(a) Less than 21 miles per gallon.
(b) Between 21 and 27 miles per gallon.
(c) More than 24 miles per gallon.

4-26 If you have a container with 500 red poker chips, 400 blue poker chips, and 300 green poker chips, and you draw 50 chips at random, then which outcome is more probable:

(10 red, 20 blue, and 20 green) or (20 red, 20 blue, and 10 green)?

CLASS SURVEY

What is the probability that someone picked at random in this class:
(a) Is a smoker?
(b) Is a nonsmoker?
(c) Is left-handed?
(d) Is right-handed?
(e) Has had one or more broken bones?

SAMPLE TEST FOR CHAPTERS 1, 2, 3, AND 4

1 *Inferential* statistics makes inferences about (a) a population or (b) a sample.
2 An important question in statistics is whether a sample is a _____ sample or not.
3 Which average would you use to measure the central tendency of a distribution of salaries? of test grades? of the time it takes 10 horses to run a measured mile? of the current time as read on 10 different clocks?
4 Which are not measures of variability: range, z score, variance, standard deviation, percentile rank, raw scores?
5 s^2 is (a parameter) or (a statistic), which estimates the value of σ^2.
6 What is the probability that a fair coin tossed 3 times will come up heads each time? that 3 fair coins tossed together once will all come up heads?
7 In a history class $P(X$ is female$) = .43$. What percentage of the class is female? male?
8 In a histogram 62 percent of the area is to the left of a grade of 75. What percentage of the class got below 75? What is the value of $P(X \geq 75)$?
9 A biased coin has $P($heads$) = 70$ percent. If it is tossed 50 times, how many heads will you expect to appear?
10 A sample of people were asked how many brothers and sisters they had. Their responses were: 3, 2, 1, 0, 2, 3, 2, 1, 6, 1.
(a) Find the mean, median, and mode of this distribution.
(b) Find the range, variance, and standard deviation.
(c) Find the largest z score and the smallest z score.
(d) If somebody had a z score equal to 0.54, how many brothers and sisters does the person have?
11 (a) Draw a histogram for these heights: 72, 50, 59, 60, 63, 71, 70, 35, 65, 63, 61, 75, 60, and 67 inches.
(b) Find the percentile rank for a height of 61 inches.
12 A sociologist, Andrew Grievely, wants to determine the average number of children in the households of Apawling, AZ. He goes to all the schools in the community and takes a random sample of school children. He asks each child how many children are in his or her household. He uses all of this data to estimate the average. Why will this be a biased sample for his study?

The Binomial Distribution

5

If we are investigating some population, then any particular characteristic of that population which can take on different values can be called a **variable**. Some variables can take on *many* values, such as the following.

1. Test scores from 0 to 100. An individual score can be any one of many values from 0 to 100.
2. The 6 outcomes from rolling 1 die. An individual outcome can be any 1 of 6 values.
3. The 4 marital states: single, married, widowed, divorced. An individual can be in any 1 of 4 states.

Other variables have only 2 values or outcomes, such as the following.

1. Answers in a true-false test.
2. Answers to yes-no questions.
3. Results of a contest in terms of win-lose (but not in terms of win, lose, draw).
4. Grades in a pass-fail grading system.
5. Outcomes of tossing a coin (heads-tails).
6. Sex of newborn infants (male-female).
7. Outcomes of a dice game in terms of rolling a 7 or not rolling a 7.

(1) If a variable has only 2 possible outcomes *and* (2) if the probabilities of these outcomes do not change for each trial regardless of what has happened

Understanding statistics

EXAMPLE 5-1

The Knights of Columbus run an annual fund-raising bazaar for a local hospital. A game of chance at the bazaar is set up like this: A spinner has 3 equal areas colored red, green, and blue. A patron bets on 1 color. The wheel is spun, and if it stops with the pointer on the color the patron selected, he wins. Marty decides to play the game twice, betting on red both times. The result of his betting is a binomial variable (win-lose) with 2 trials (since he will bet 2 times).

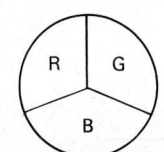

What is the probability that Marty will win:

(a) Both times?   (b) Just one time?   (c) Not at all?

We will solve this problem two ways; first, by looking at all the equally likely outcomes as before; and second, by a new approach we will call the "binomial" approach.

SOLUTION

Method 1 Since there are 3 possible outcomes on each game, there are 3 × 3 = 9 possible outcomes for two games. These equally likely outcomes are listed below.

outcome	1st game, 2d game	number of reds
1	red, red	2 reds
2	red, green	
3	red, blue	1 red
4	blue, red	
5	green, red	
6	blue, green	
7	green, blue	no reds
8	blue, blue	
9	green, green	

(a) The probability that Marty wins twice is $P(2 \text{ wins}) = 1/9$, since there are 9 equally likely outcomes and only 1 of them consists of 2 reds.

(b) The probability that Marty wins exactly 1 time is $P(1 \text{ win}) = 4/9$, because 4 of the outcomes consist of 1 red.

(c) Similarly the probability that Marty does not win at all is $P(0 \text{ wins}) = 4/9$.

We can present these probabilities in a histogram, as shown in Figure 5-1.

The binomial distribution

Figure 5-1

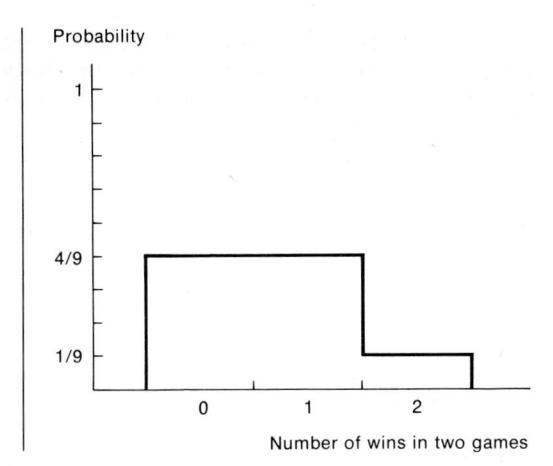

[Histogram showing Probability vs Number of wins in two games: at 0 wins probability is 1/9, at 1 win probability is 4/9, at 2 wins probability is 4/9]

SOLUTION *Method 2* There is an alternative way to compute these probabilities (1/9, 4/9, 4/9), which is often easier. We can analyze the game directly in terms of win and lose, instead of in terms of color. As far as Marty is concerned, when he plays twice, he will win twice, once, or not at all. There is only one way he can win twice (he must win on both games); there are two ways he can win once (either win on the first game, *or* win on the second game); and there is only one way he can lose twice (he must lose on both games). We summarize this in Table 5-1.

Table 5-1

number of wins	ways the wins can occur	number of different ways the wins can occur
2	WW	1
1	WL, LW	2
0	LL	1

Recall that in a binomial problem there are 2 possible outcomes on each trial. One of these is called a "success" and the other is called a "failure." We are now treating this problem as a binomial problem where "success" is win, "failure" is lose, and a "trial" is one spin. It is standard in this type of problem to let n stand for the number of trials, to let S stand for the number of successes in n trials, and to let $\binom{n}{S}$ stand for the number of different ways S can occur. The symbol $\binom{n}{S}$ is called a **binomial coefficient**. Thus, we can relabel Table 5-1 as shown in Table 5-2.

Table 5-2

S	ways S can occur	$\binom{2}{S}$
2	WW	$\binom{2}{2} = 1$
1	WL, LW	$\binom{2}{1} = 2$
0	LL	$\binom{2}{0} = 1$

Understanding statistics

Now we have reformulated Marty's betting problem as a binomial problem. We want to find $P(S = 2)$, $P(S = 1)$, $P(S = 0)$. In order to do this we must talk about the probability of success on any one trial. We use the symbol p for this:

p = probability of success on any one trial of a binomial experiment

Since a "success" on 1 trial means "a red on 1 spin," and the probability of red on 1 spin is 1/3, we have $p = 1/3$. Similarly, probability of a "failure" on 1 trial is symbolized by q, and so we have $q = 2/3$.

To compute the probabilities $P(S = 2)$, $P(S = 1)$, and $P(S = 0)$, we may apply the following rule. Multiply $\binom{n}{S}$ by p for each success and by q for each failure. This is shown in Table 5-3.

Table 5-3

S	$\binom{2}{S}$ number of ways S can occur		$P(S)$	
2	1	$1 \cdot p \cdot p = 1p^2$	$= (1)(1/3)^2$	$= 1/9$
1	2	$2 \cdot p \cdot q = 2pq$	$= (2)(1/3)(2/3)$	$= 4/9$
0	1	$1 \cdot q \cdot q = 1q^2$	$= (1)(2/3)^2$	$= 4/9$

You see that we can now read the probabilities directly from this table.

(a) The probability that Marty wins twice is $P(S = 2) = 1/9$.

(b) The probability that Marty wins once is $P(S = 1) = 4/9$.

(c) The probability that Marty wins not at all is $P(S = 0) = 4/9$.

■ The histogram is the same as before (Figure 5-1).

The principle involved in binomial problems is that the probability of a *sequence of events* is found by *multiplying* the probabilities of the individual events. This is often called the **multiplication rule.** For example, the probability of Marty first winning and then losing, $P(WL)$, is $1/3 \times 2/3 = 2/9$. The probability of him first losing and then winning, $P(LW)$, is $2/3 \times 1/3 = 2/9$. Marty wins exactly once by either of these 2 sequences, so there are 2 ways he can win once, each with probability 2/9, for a total probability of 4/9. This is exactly what we computed in Table 5-3.

EXAMPLE 5-2 Suppose that Marty is going to play the spinner game 4 times. He can win 4, 3, 2, 1, or 0 times. Find the probability for each of these possible outcomes.

SOLUTION If we attempt to solve this problem by listing all the possible outcomes as we did in Method 1 in the previous example, we will have to list $3 \times 3 \times 3 \times 3 = 81$ different outcomes. Some of them would be (red, red, red, red), (red, red, blue, green), (red, red, green, blue). We can avoid this by solving the problem using Method 2 above. We treat it as a binomial problem, and concern ourselves just with wins and losses. Since Marty is going to play 4 times, he

The binomial distribution

will win either 4, 3, 2, 1, or 0 times. The possible ways he can win and lose are shown in Table 5-4. Recall that $p = 1/3$ and $q = 2/3$, because p and q are the probabilities of success and failure on any one trial.

Table 5-4

S	ways S can occur	number of ways S can occur $\binom{4}{S}$
4	WWWW	$\binom{4}{4} = 1$
3	WWWL, WWLW, WLWW, LWWW	$\binom{4}{3} = 4$
2	WWLL, WLWL, WLLW, LWWL, LWLW, LLWW	$\binom{4}{2} = 6$
1	WLLL, LWLL, LLWL, LLLW	$\binom{4}{1} = 4$
0	LLLL	$\binom{4}{0} = 1$

We compute the probabilities in Table 5-5.

Table 5-5

S	$\binom{4}{S}$			$P(S)$	
4	1	$1 \cdot pppp$	$= 1p^4$	$= 1(1/3)^4$	$= 1/81$
3	4	$4 \cdot ppp \cdot q$	$= 4p^3 q$	$= 4(1/3)^3(2/3)$	$= 8/81$
2	6	$6 \cdot pp \cdot qq$	$= 6p^2 q^2$	$= 6(1/3)^2(2/3)^2$	$= 24/81$
1	4	$4 \cdot p \cdot qqq$	$= 4pq^3$	$= 4(1/3)(2/3)^3$	$= 32/81$
0	1	$1 \cdot qqqq$	$= 1q^4$	$= 1(2/3)^4$	$= 16/81$

Now we can read the desired probabilities from the table.

(a) The probability that Marty wins all 4 times is $P(S = 4) = 1/81$.

(b) The probability that Marty wins 3 times is $P(S = 3) = 8/81$.

(c) $P(S = 2) = 24/81$.

(d) $P(S = 1) = 32/81$.

(e) $P(S = 0) = 16/81$.

The histogram for these probabilities is shown in Figure 5-2.

You should also notice how you can interpret the expressions in the $P(S)$ column of Table 5-5. Consider, for example, the $4p^3 q$ in the line corresponding to $S = 3$. The exponent of p is 3. This indicates that there were 3 successes. The exponent of q is 1. This means that there was 1 failure. The 4 indicates that there are 4 ways that this can occur, namely, WWWL, WWLW, WLWW, and LWWW. ∎

Figure 5-2

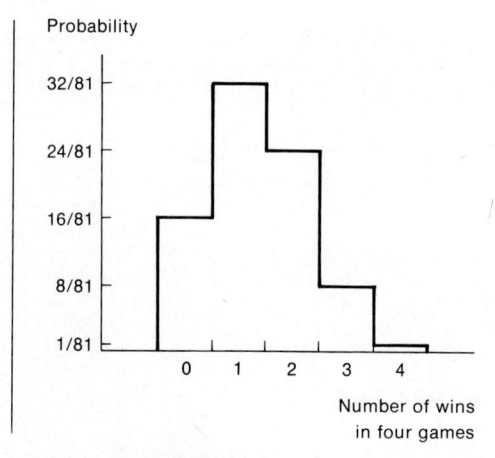

[Histogram: Probability vs Number of wins in four games; bars at 0: 16/81, 1: 32/81, 2: 24/81, 3: 8/81, 4: 1/81]

PASCAL'S TRIANGLE

Two men are gambling by rolling a die. If the die shows a 1 or a 2, Jerome wins. Otherwise Ramon wins. Jerome will roll 3 times. Find the probability that Jerome wins:

(a) All 3 times (b) Twice (c) Once (d) Not at all

Since he will play 3 times, we have $n = 3$. We let $p = P$(Jerome wins on any one roll) $= 2/6 = 1/3$. Therefore, $q = P$(Jerome loses on any one roll) $= 2/3$. We set up a table similar to Tables 5-4 and 5-5 in the previous example.

S	ways S can occur	$\binom{3}{S}$ number of ways S can occur	$P(S)$
3			
2			
1			
0			

To answer the four questions we need to fill in the last column in the table. We know that these answers will be found by multiplying the values from the third column by the appropriate number of p's and q's.

S	ways S can occur	$\binom{3}{S}$	$P(S)$
3			$\binom{3}{3} p^3$
2			$\binom{3}{2} p^2 q$
1			$\binom{3}{1} pq^2$
0			$\binom{3}{0} q^3$

The binomial distribution

Previously we have found the entries in the $\binom{n}{S}$ column by listing all the ways S can occur and then counting these ways. There is, however, a way to obtain these entries without counting. In this problem, for instance, the numbers are 1, 3, 3, 1. Thus, the table can be filled out as follows.

S	ways S can occur	$\binom{3}{S}$	$P(S)$
3		1	$1(1/3)^3 = 1/27$
2		3	$3(1/3)^2(2/3) = 6/27$
1		3	$3(1/3)(2/3)^2 = 12/27$
0		1	$1(2/3)^3 = 8/27$

Therefore, the answers to questions (a), (b), (c), and (d) are:

(a) $P(S = 3) = 1/27$ (b) $P(S = 2) = 6/27$
(c) $P(S = 1) = 12/27$ (d) $P(S = 0) = 8/27$

The numbers in column 3, $\binom{n}{S}$, are the number of ways that S can occur. We get these numbers from the triangle of numbers below, which is known as **Pascal's triangle**. Blaise Pascal (1623–1662) was a French mathematician, one of the founders of the science of probability. He constructed the following triangle of the numbers $\binom{n}{S}$, where n stands for the number of trials and S for the number of successes.

n trials	\multicolumn{6}{c	}{S successes}
	0	1	2	3	4	5
0	1					
1	1	1				
2	1	2	1			
3	1	3	3	1		
4	1	4	6	4	1	
5	1	5	10	10	5	1

In the above problem, since Jerome rolled the die 3 times, $n = 3$. Therefore, we look at the row corresponding to $n = 3$. There we find the entries 1, 3, 3, 1. Recall that in Example 5-2 Marty played a spinner game 4 times. Looking at the row of Pascal's triangle corresponding to $n = 4$, we see the entries 1, 4, 6, 4, 1. These are the values that we had calculated ourselves in that example.

It is easy to construct the triangle. In order to get each number, we add two numbers, the number directly above the one we want and the one above and to the left. As indicated, the 10 in the row corresponding to $n = 5$ comes from adding the 6 and the 4 which are in the row above it. We can continue this process to construct more rows. Note that each row starts and ends with a 1. Try constructing the row corresponding to $n = 6$. You can check your answer by turning to Table C-1 (see Appendix C).

EXAMPLE 5-3 Let us use Pascal's triangle in one further example. A genetic engineer is experimenting in an attempt to produce a certain protein. The genetics of the situation is such that the probability of a successful experiment is .7. The engineer has enough equipment and funds for 6 repetitions of the experiment. Let $p = P$ (success on any one experiment) $= .7$ and $q = P$(failure) $= .3$. Find the following probabilities.

(a) $P(S = 3)$ (b) $P(S > 3)$ (c) $P(S \geq 3)$ (d) $P(S \leq 3)$
(e) $P(1 \leq S \leq 4)$ (f) P(at most 3 successes)
(g) P(at least 3 successes)

SOLUTION We first list the number of successes and the corresponding line from Pascal's triangle for $n = 6$ (because there are 6 repetitions of the experiment).

$$\binom{6}{S}$$

S	number of ways S can occur	$P(S)$	
6	1	$1p^6 = 1(.7)^6$	$= .12$
5	6	$6p^5q^1 = 6(.7)^5(.3)$	$= .30$
4	15	$15p^4q^2 = 15(.7)^4(.3)^2$	$= .32$
3	20	$20p^3q^3 = 20(.7)^3(.3)^3$	$= .19$
2	15	$15p^2q^4 = 15(.7)^2(.3)^4$	$= .06$
1	6	$6p^1q^5 = 6(.7)(.3)^5$	$= .01$
0	1	$1q^6 = 1(.3)^6$	$= .00$

(a) $P(S = 3) = .19$

(b) In this problem $S > 3$ means $S = 4$, $S = 5$, or $S = 6$. To find $P(S > 3)$, we add $P(S = 4) + P(S = 5) + P(S = 6)$:

$$P(S > 3) = P(S = 4) + P(S = 5) + P(S = 6)$$
$$= .32 + .30 + .12$$
$$= .74$$

(c) $P(S \geq 3) = P(S = 3) + P(S > 3)$. We calculated $P(S > 3)$ in part (b), so we now have

$$P(S \geq 3) = .19 + .74 = .93$$

(d) $P(S \leq 3) = P(S = 3) + P(S = 2) + P(S = 1) + P(S = 0)$
$= 1 - P(S > 3)$

and so from part (b),

$$P(S \leq 3) = 1 - .74 = .26$$

(e) $P(1 \leq S \leq 4) = P(S = 1) + P(S = 2) + P(S = 3) + P(S = 4)$
$= .01 + .06 + .19 + .32 = .58$

(f) At most 3 successes means 0, 1, 2, or 3 successes. So we can write $P(S \leq 3)$. This is the same as part (d).

(g) At least 3 successes means 3, 4, 5, or 6 successes. We can write $P(S \geq 3)$. See part (c).

The binomial distribution

EXERCISES

5-1 In every course Stuart Dente takes, he will either pass or not pass. Still the question of whether or not Stu will pass exactly 3 of his 5 courses is not a binomial experiment. Why not?

5-2 Which of the following are binomial experiments?
(a) A box contains 500 red beads and 300 white beads. Ruby D. Redoux selects 20 beads at random and counts the red ones. She replaces the 20 beads and starts again. She repeats this over and over again.
(b) Same procedure as part (a), except that Ruby does not replace the 20 beads each time.
(c) Slim Jim is on a diet. He has decided to eat twice as much at 1 meal each day. Each day he will pick the meal by spinning a spinner divided into 3 equal parts labeled breakfast, lunch, and dinner.
(d) A card is drawn from an ordinary shuffled deck to see if it is a face card or not. The card is replaced, and the draw is repeated many times.
(e) We tag the ears of 60 rabbits in a colony of rabbits which at this time has 100 members. On the first day of each month, we select 1 rabbit at random and note if it has an ear tag.

5-3 For each part below, use Pascal's triangle to find the value of a.

(a) $\binom{18}{7} = a$ (b) $\binom{a}{4} = 70$ (c) $\binom{14}{a} = 3432$

(d) $\binom{11}{a} = 330$ (e) $\binom{n}{0} = a$ (f) $\binom{n}{n} = a$

(g) $\binom{n}{1} = a$ (h) $\binom{a}{S} = 10$

5-4 Use Pascal's triangle to solve the following.

(a) $\binom{2}{0} + \binom{2}{1} + \binom{2}{2} =$

(b) $\binom{3}{0} + \binom{3}{1} + \binom{3}{2} + \binom{3}{3} =$

(c) $\binom{4}{0} + \binom{4}{1} + \binom{4}{2} + \binom{4}{3} + \binom{4}{4} =$

(d) $\binom{n}{0} + \binom{n}{1} + \binom{n}{2} + \cdots + \binom{n}{n} =$

5-5 Write the first four numbers of the row corresponding to $n = 21$ in Pascal's triangle. Use Table C-1.

5-6 List the 32 possible outcomes of success S or failure F in a binomial experiment with 5 trials. Do your results agree with Pascal's triangle for $n = 5$?

5-7 In Example 5-2 we stated that there were 81 possible outcomes in terms of color. Have your kid brother list them.

5-8 In an experiment of rolling 1 die several times, we group the 6 possible outcomes into two categories of "3" and "not 3." Is the experiment binomial?

5-9 A fair coin is tossed 3 times.

(a) Complete the chart below.

S = number of heads	$\binom{3}{S}$ number of ways S can occur from Pascal's triangle	$P(S)$
3	1	
2	3	
1	3	
0	1	

(b) Draw the histogram depicting these results.
What is the probability of getting heads:
(c) All 3 times?   (d) Exactly twice?   (e) At least once?
Find:
(f) $P(S = 0)$ (g) $P(S < 2)$ (h) $P(S \leq 2)$

5-10(a) A true-false quiz is given with 5 questions. To pass you need at least 4 right. You guess every answer. (1) What is the probability that you pass? (2) If the class is large and if everyone else in the class is also guessing, about what percentage of the class will probably pass?
(b) If in each pregnancy the probability of having a girl is 1/2, what is the probability that a childless couple planning to have 5 children will have at least 4 girls? What percentage of couples with 5 children would you expect to have 4 or 5 girls?
(c) In one binomial experiment, $p = .5$, $q = .5$, and $n = 5$. Find $P(S \geq 4)$. In what percentage of the results would you expect to have 4 or more successes?

5-11(a) Eileen Dover needs 70 percent to pass an examination in abnormal physiology. The exam consists of 10 true-false questions. If she guesses blindly at all the questions, what is the probability that she will pass?
(b) Her brother Ben also needs 70 percent to pass an examination in nematode anatomy. However, Ben's test will consist of 10 multiple-choice questions where each question has 5 possible answers. If he guesses blindly at all the questions, what is the probability that he will pass?

5-12 A bent coin has a probability of falling heads equal to .4. This coin is tossed 5 times.
(a) What is the probability of getting at least 3 heads?
(b) What is the probability of getting at most 3 heads?

5-13 When a thumbtack is tossed in the air, it will land point up or not. The probability of landing point down is .21 (for a No. 35 tack). Three tacks are tossed. Find the probability of:
(a) All landing point up.   (b) All landing point down.
(c) At least one landing point up.

5-14 According to a recent magazine article, if a child is conceived through artificial insemination, the probability that the child will be a boy is .8. Ms. Clark plans on having two children by this method. Find the probability that:
(a) Both are girls.   (b) At least one is a girl.

5-15(a) In the game of Montezuma's Revenge you must toss a run of more heads than your opponent using an ordinary coin. Michelle has scored a run of 3 heads in a row. Mike must now score a run of 4 heads to beat her. What is the probability that Mike will beat Michelle? that he will not beat her?

The binomial distribution

(b) In an alternative form of the game, a die is used. Felix rolled the same number 3 times in a row. Felicia must now score a run of 4 rolls of any one number. What is the probability that she will lose? (Why should you use $n = 3$ and not 4?)

5-16 A space hero, Luke Warmwater, has 4 functioning rockets left on his ship. The ship has been damaged, and the probability that any individual rocket will fire is .20. To escape sudden death (and survive to the next episode), he needs at least one of the rockets to fire. He pushes all 4 firing buttons. What is the probability that he will be around for the next movie?

5-17 *Medical Screening Test* A screening test for a disease is a simple procedure which indicates if a patient is *likely* to have a disease. A person whose results on the screening test are "positive" is likely to have the disease, and would be asked to submit to further tests in order that a proper diagnosis be made. This means that some people who screen "positive" do not, in fact, have the disease. These people are "false positives." The false positive *rate* is the *proportion* of people who screen positive but who do not, in fact, have the disease.

Suppose the false positive rate for a certain screening test is .10.
(a) What does this mean? About how many false positives would you expect in 10 positive screenings?
(b) What is the probability that of 10 patients who screen positive, more than 1 is false positive?
(c) Which is more likely, that there are 0 or 1 false positives in 10 positive screenings or that there are more than 1 false positives?

5-18 Consider the following spinner game. A wheel has 4 sections marked as shown. Each afternoon, Algernon, a student, spins the spinner. It stops at random on 1 of the sections. (If it stops on a line, it does not count.) Let us suppose that if it stops on the section P, then Algernon goes out to play that evening; if it stops on a section S, then Algernon stays home to study that evening. Algernon spins the spinner twice.

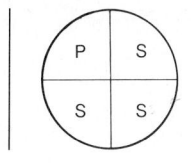

What is the probability that he:
(a) Plays both times?
(b) Does not play at all?
(c) Plays at least once?
(d) Plays less than 2 days?

5-19 Algernon's friend, Morgel, decides to use the spinner-game method to decide how to spend his evenings, too. He makes a spinner wheel as shown, where 4 of the sections are marked P for play and 1 section is marked S for study. Morgel decides to spin the wheel 7 times so that his entire week's schedule will be worked out.

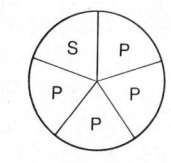

What is the probability that Morgel will end up:
(a) Going out to play all 7 evenings?
(b) Studying exactly 3 evenings?
(c) Not playing at all?
(d) Studying less than 4 evenings?

5-20 Morgel's friend, Chastain, wants to play the spinner game too. His spinner looks like the one shown below, where J is for Jay's Joint and K is for Ken's Kubicle, his two favorite night spots. If the spinner lands on J, he spends the evening at Jay's. If the spinner lands on K, he spends the evening at Ken's. There is a 4-day holiday coming up, and so Chastain spins the spinner 4 times to determine his plans.

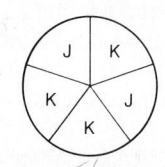

What is the probability that:
(a) He will spend 1 evening at Jay's and 3 evenings at Ken's?
(b) He will spend 2 evenings at Jay's and 2 at Ken's?
(c) He will spend all 4 evenings at the same night spot?

5-21 Assistant Professor Ratso, a leading experimental psychologist, is in the habit of sending rats through mazes. She predicts that a rat reaching the end of a T-shaped maze is more likely to turn left than right. She believes that the proportion of rats which turn left is .65. If this is true and she sends 6 rats down a maze, what is the probability that:
(a) 3 will turn left and 3 will turn right?
(b) Fewer than 5 will turn left?
(c) All will turn left or all will turn right?

5-22 An emergency life support system has 4 batteries. The probability of any battery failing is .01. What is the probability that:
(a) None fails?   (b) All fail?   (c) More than 2 fail?

5-23 A manufacturer claims that 4 of 5 dentists recommend sugarless gum for their patients who chew gum. Assuming that this claim is true, find the probability that in a randomly selected group of 20 dentists, 16 or more will recommend sugarless gum for their patients who chew gum.

5-24 Suppose that 2 baseball teams, A and B, are equally matched. The outcome of each game between them can be considered as a random variable, with the probability of team A winning equal to .5.
(a) If they play 6 games, what is the probability that each team will win 3 games?
(b) If they play a series of games where the first team to win 4 games wins the series, what is the probability that team A will win the series *on the seventh game*? *Hint:* Use your answer to part (a).

5-25 Chris and Evonne, 2 tennis players, are playing a game. The probability that Chris wins a point is .8. The game is currently stopped with "advantage Evonne" (she needs 1 point to win). What is the probability that Chris will win the next 3 points consecutively, and thus win the game?

The binomial distribution

5-26 Peter Pedaler must bike 2 miles to get the Sunday newspaper. If he gets to the store too early, the papers have not arrived. If he gets there too late, they are all sold. He has learned that if he arrives at 8:30 he has an 85 percent chance of getting the paper. What is the probability that he will get the paper at least 6 of the next 8 Sundays if he shows up at 8:30?

5-27 Chef Victoir knows how to prepare 2 yummy dishes: Quiche Fillisse and Eggs Ari-Bari. A patron of his restaurant, Mme. Sharonne, has noticed that 63 percent of the time Victoir prepares Eggs Ari-Bari. If Mme. Sharonne enters the restaurant 6 times, what is the probability that she will get 3 meals of each dish?

5-28 An element of Pascal's triangle is referred to as $\binom{n}{S}$. For example, $\binom{6}{2} = 15$. This can be calculated with the formula

$$\binom{n}{S} = \frac{n!}{S!(n-S)!}$$

where $n! = 1 \times 2 \times 3 \times \cdots \times n$. (Your calculator may have an $x!$ button.)

(a) Verify that $\binom{6}{2} = \frac{6!}{2!(6-2)} = 15$.

(b) Verify that $\binom{14}{9} = 2002$.

(c) Calculate the value of $\binom{35}{3}$.

(d) Calculate the value of $\binom{50}{47}$.

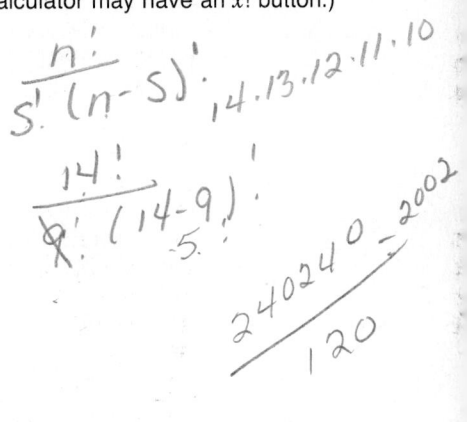

BINOMIAL PROBABILITY TABLES

(This section may be omitted without loss of continuity.)

Now that you understand the theory, the ideas behind the binomial distribution, let us look at another approach. In Exercise 5-10 we saw that different problems have the exact same computational solution. This leads us to the idea of listing the results of such computations in tabular form. Thus if you look at Table C-2 you will find the solution to Exercise 5-10 by locating the two numbers for $S = 4$ and $S = 5$ corresponding to $n = 5$, $p = .50$. These numbers are .1563 and .0313. Since their sum is .1876, the solution to the problem is about 19 percent.

Statisticians often use tables of binomial probabilities to solve problems. Note that, as with any table, round off errors do occur. For example, if we calculate the solution to Exercise 5-10 directly, we obtain $.15625 + .03125 = .1875$, and not .1876. Also, since the table cannot possibly contain every feasible value of the probability p, it is often necessary to be satisfied with approximate tabular values. As an example, $\binom{6}{4}(.59)^4(.41)^2$ is approximately equal to .3055. However, $p = .59$ is not in our table. The closest value is .60.

Understanding statistics

Looking in our table for the entry corresponding to $p = .60$, $n = 6$, we find for $S = 4$ the solution to be .3110. If the researcher does not need extremely accurate results, then the approximate value can be found very quickly from the table. (Of course the proliferation of hand-held calculators which calculate exponents quickly has reduced the dependence of statisticians on such tables.)

EXAMPLE 5-4 Let us redo parts (a), (b), and (c) of the genetics experiment in Example 5-3, where $p = .7$ and $n = 6$.

SOLUTION
(a) To find $P(S = 3)$, we simply look in our table at the entry corresponding to $p = .70$, $n = 6$, $S = 3$. The solution is quickly found to be .1852, or .19.

(b) To find $P(S > 3)$, we add the entries corresponding to $S = 4, 5$, and 6. The solution is $.3241 + .3025 + .1176 = .7442$, or about .74.

(c) To find $P(S \geq 3)$, we again add $P(S = 3)$ to the answer in part (b), obtaining $.1852 + .7442 = .9294$, or .93.

■

EXAMPLE 5-5 Sundat Motors had to replace the frimfram on the left widget in 39 percent of last year's models. Acme Trucking bought 12 Sundats at various times last year. What is the probability that Acme will need to replace the frimfram in 6 trucks?

SOLUTION
(a) The solution is $P(S = 6) = \binom{12}{6}(.39)^6(.61)^6 = .1675$, or about 17 percent.

(b) Using our tables we can approximate the solution. Since .39 is not in the table, we will use $p = .40$, $n = 12$, and $S = 6$. The solution is found to be .1766, or about 18 percent.

Depending on the accuracy that the statistician needs, this second, quicker solution may be close enough for her purposes.

■

STUDY AIDS

VOCABULARY

1. Binomial variable
2. Binomial coefficient
3. Successful outcome
4. Pascal's triangle
5. Trial

SYMBOLS

1. n
2. S
3. $\binom{n}{S}$
4. p
5. q

The binomial distribution

EXERCISES

Solve any of the Exercises 5-11 to 5-27 by using the binomial probability table.

CLASS SURVEY

1 Let $p = P$(a student in class is female).
(a) Evaluate p
(b) Imagine that each student's name is printed on a piece of paper and that all these papers are then placed in a large box and thoroughly mixed up. Withdraw one name at random, record it, and replace it in the box. Do this 2 more times. What is the probability that all 3 will be female? male? the same sex?

2 (a) Which of the questions we asked in the survey could be used for binomial experiments?
(b) Using one of your answers to part (a), construct an exercise similar to the previous class exercise. That is, define p, evaluate p, and calculate the probability of some specific event of your choice.

$P = 0.6$
$n = 16$
$S =$ win the game.
$P(S \geq 10) = \binom{16}{10}(0.6)^{10}(0.4)^6$
$= 0.1983.$
$P(S = 16) = \binom{16}{16}(0.6)^{16}(0.4)^0$
$= 0.00028.$

The Normal Distribution

6

Consider a problem based on the drawing of a card at random from a shuffled deck of cards. Suppose we have defined success as "drawing a club." Our experiment consists of drawing 1 card, seeing if it is a club, replacing it, re-shuffling, drawing the next card, and so on 5 times all told. This is a binomial experiment, with $n = 5$ and $p = .25$. Consequently, if we did a binomial analysis we would get the histogram of Figure 6-1.

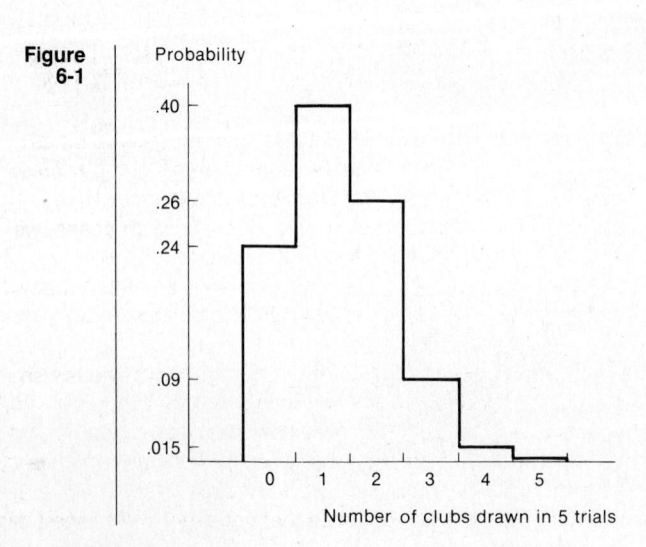

Figure 6-1

Number of clubs drawn in 5 trials

The normal distribution

If we were interested in a similar experiment, but with $n = 10$ or $n = 20$, we would get the histograms shown in Figures 6-2 and 6-3.

Figure 6-2

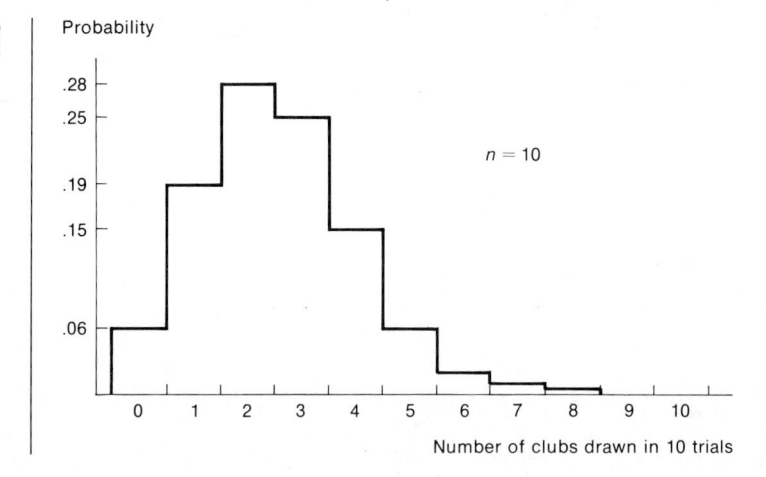

[Histogram: Probability vs Number of clubs drawn in 10 trials, $n = 10$. Values approximately: .06, .19, .28, .25, .15, .06, and small values for higher numbers.]

Figure 6-3

[Histogram: Probability vs Number of clubs drawn in 20 trials, $n = 20$. Values approximately: .01, .03, .07, .13, .17, .20, .17, .11, .07, .03, .01.]

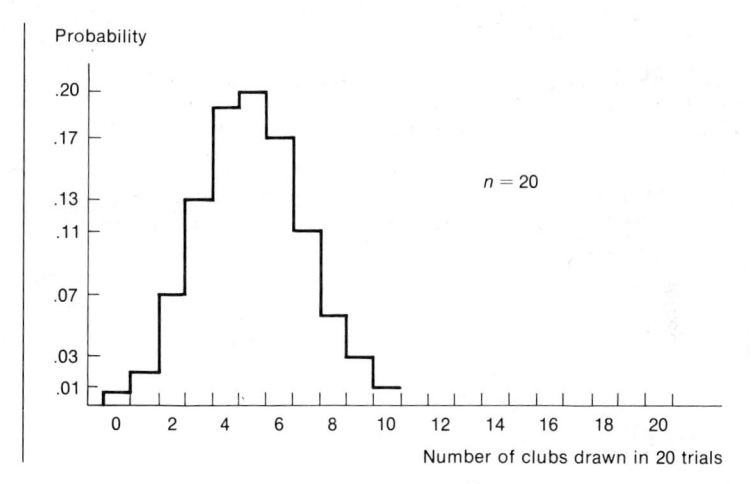

If we were to repeat this card-drawing binomial experiment 100 times ($n = 100$), there would be 101 possible outcomes ranging from 0 to 100 clubs. On ordinary-size paper, the histogram for the distribution of successes would have intervals so narrow that they would be difficult to draw. In such cases, we will often draw a smooth curve to approximate the shape of the histogram. The curve would be drawn by placing a dot over the midpoint of each interval at the proper height (the same height we would make the bar if we *were* drawing the histogram).

For example, if we are approximating the three histograms previously shown, we would get the curves that are shown in Figures 6-4, 6-5, and 6-6. Fortunately, according to mathematical theory, we know that under certain common conditions, the curve used to approximate a binomial histogram will be *bell-shaped,* and more specifically, that it will be very close to a particular bell-shaped curve called the **normal curve.** (Note that not every bell-shaped curve

Figure 6-4

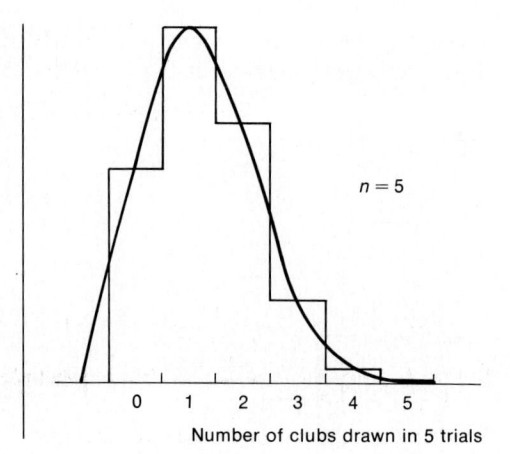

n = 5

Number of clubs drawn in 5 trials

Figure 6-5

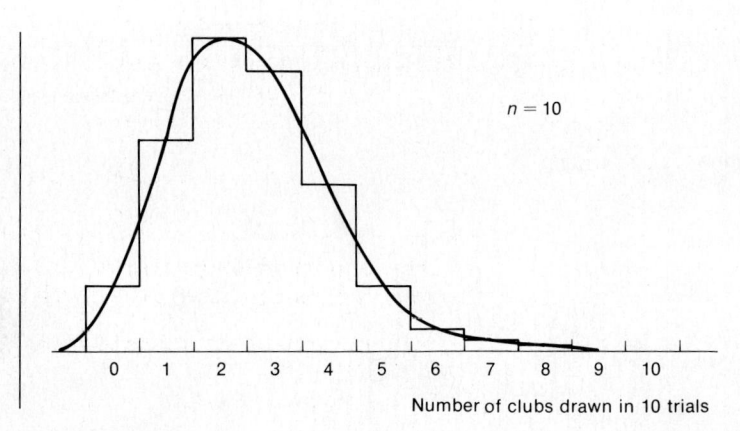

n = 10

Number of clubs drawn in 10 trials

Figure 6-6

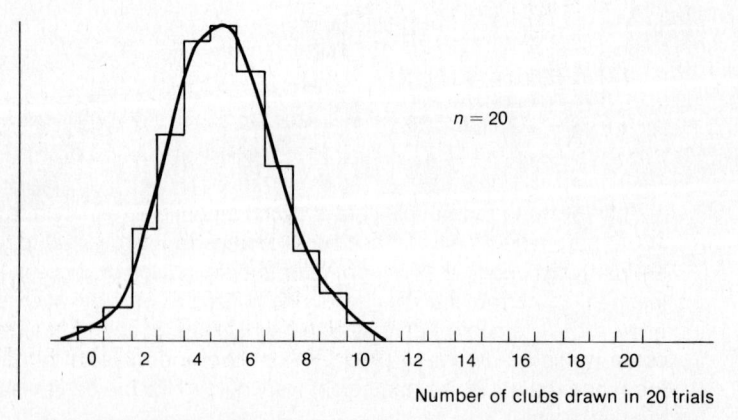

n = 20

Number of clubs drawn in 20 trials

is normal.) The normal curve pictures a distribution of numbers called the **normal distribution.** This means that if we are faced with a binomial problem, we might be able to use a normal distribution instead. Because all the necessary data for the normal distribution have already been tabulated, using a normal distribution saves a great deal of time at the expense of only a little accuracy.

Figure 6-7

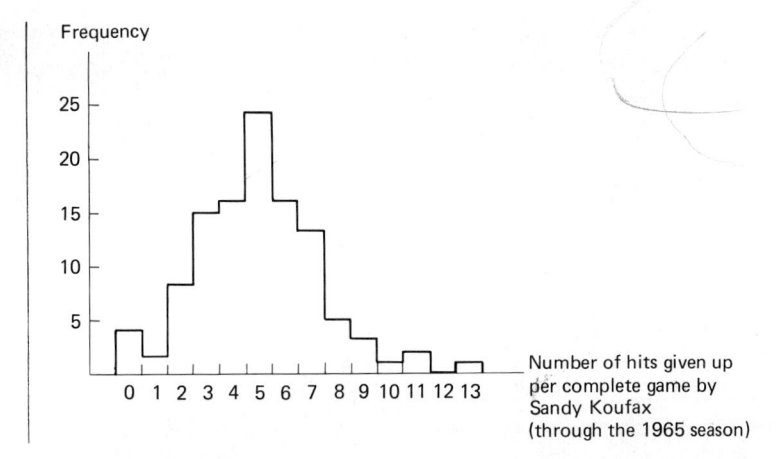

Frequency

Number of hits given up per complete game by Sandy Koufax (through the 1965 season)

The normal distribution is the most important distribution in statistics. One of its earliest uses was to approximate the binomial distribution, and this use is still very important. However, it has other applications, some of which we will study in this text. In fact, many collections of raw data in real experiments have been found to be approximately normally distributed.

Figure 6-7 illustrates an example from major league baseball records that looks approximately normal.

THE THEORETICAL NORMAL CURVE

The normal distribution and its graph (the normal curve, Figure 6-8) have the following important properties.

1. There is symmetry about the mean (i.e., the left and right halves of the curve are mirror images of each other). Therefore, the mean equals the median.

2. The scores that make up a normal distribution cluster about the middle.

3. The range of scores is unlimited, but only a very small percentage of the scores, fewer than 3 in 1000, are more than 3 standard deviations away from the mean.

For our purposes the normal distribution will be defined by a table of z scores and percentile ranks. We saw in Chapter 3 that a table of z scores and per-

Figure 6-8

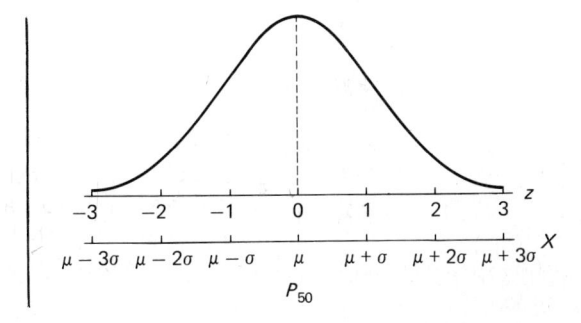

Understanding statistics

centile ranks *does* determine the shape of a distribution; the more z scores and percentile ranks we have, the more accurate the representation of the distribution. In your own work it is not necessary to draw a *precise* graph for the normal curve. You can represent it by sketching any reasonably bell-shaped curve, because all the numbers used in calculations based on the normal curve are recorded accurately in Tables C-3 and C-4.

USE OF THE NORMAL CURVE TABLE

Table C-4 gives a list of z scores from -4 to $+4$. For each z score the table also gives the area under the curve to the left of the z score. We shall illustrate the use of this table in the following examples.

EXAMPLE 6-1 In a normal curve, find the area to the left of $z = 1$.

SOLUTION For $z = 1$, we look in the table under $z = 1$ and we read Area $= .8413$. Sometimes we write Area$(z = 1) = .8413$. This means that 84.13 percent of the area under the normal curve is to the left of $z = 1$, as seen in Figure 6-9. Some interpretations of this are the following.

Figure 6-9

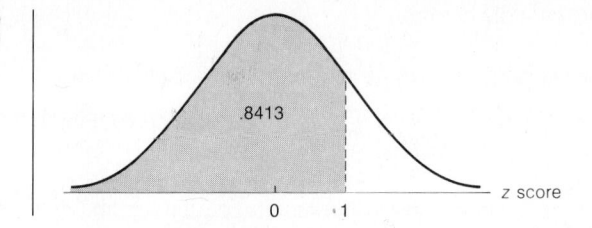

1. 84.13 percent of the members of the normal population described here have z scores less than 1.

2. A member of the normal population who has a z score of 1 also has a percentile rank of 84.

3. The probability of randomly selecting a member from this normal population with a z score less than 1 is .8413.

EXAMPLE 6-2 In a normal curve, find the area to the *right* of $z = -1.65$.

SOLUTION Looking up $z = -1.65$, we read Area$(z = -1.65) = .0495$. Since this is the area to the *left* of z, subtract .0495 from 1.0000, the total area, to find what remains on the right. Thus $1.0000 - .0495 = .9505$. This is shown in Figure 6-10.

Some interpretations of this are the following.

1. 95.05 percent of the members of the normal population described here have z scores greater than -1.65.

The normal distribution

Figure 6-10

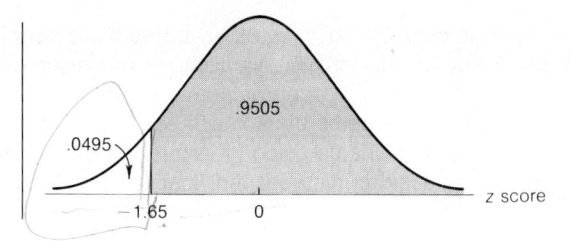

2. The probability of randomly selecting a member from this normal population with a z score greater than −1.65 is .9505.

EXAMPLE 6-3 Find the area under a normal curve between $z = -1.65$ and $z = 1$.

SOLUTION From Figure 6-11, we have

$$\text{Area}(z = 1) = .8413$$
$$\text{Area}(z = -1.65) = .0495$$
$$.7918$$

Figure 6-11

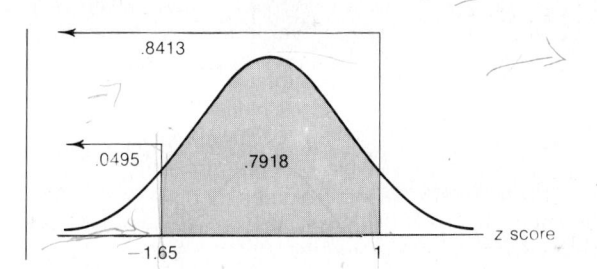

We find the area by subtracting. Thus, the area between is .7918, as shown in Figure 6-11.

In the preceding examples we have interpreted area as a percentage of raw scores, as a percentile rank, and as a probability. If we start with any one of these three quantities, we can reverse the preceding procedures and find the corresponding z score.

EXAMPLE 6-4 In a normal distribution 10.2 percent of the members of the population have z scores less than a certain z score. Find that z score.

SOLUTION We use the symbol $z_?$ for this unknown z score. The problem is to find $z_?$, as illustrated in Figure 6-12.

Figure 6-12

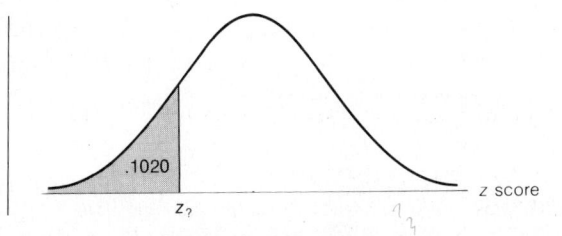

Understanding statistics

We write 10.2 percent as .1020, because the table records areas to four decimal places. Thus, in the table we find the corresponding z score, $z_? = -1.27$.

EXAMPLE 6-5 In a normal distribution a certain z score $z_?$ has the property that if we pick another z score at random, the probability that the second z score is greater than $z_?$ is .95. Find $z_?$.

SOLUTION This score is illustrated in Figure 6-13.

Figure 6-13

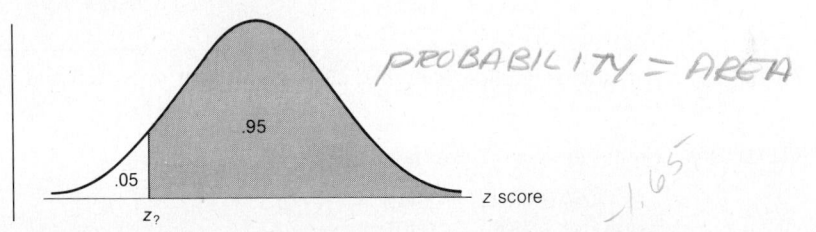

PROBABILITY = AREA

1.65

Since 95 percent of the z scores must be greater than $z_?$, then 5 percent must be less than $z_?$. We look in the table for an area equal to .0500. This exact value is not there. However, we do find an area of .0495, which corresponds to a z score of -1.65. Hence $z_?$ is approximately -1.65.

THE NORMAL CURVE TABLE AND RAW SCORES

In most problems we will deal primarily with raw scores. Sometimes we will be given raw scores and will be asked to make a statement about percentage, percentile rank, or probability. In other problems we will be given the percentage, percentile rank, or probability, and will be asked to find a raw score. Table C-4 does not contain raw scores, but only z scores and percentages. In order to do such problems we must convert raw scores to z scores and vice versa.

EXAMPLE 6-6 It was found that the weights of a certain population of laboratory rats are normally distributed,† with $\mu = 14$ ounces and $\sigma = 2$ ounces. We denote such a population by $ND(\mu = 14, \sigma = 2)$.

The expression $ND(\mu = 14, \sigma = 2)$ tells us three things:

1. The distribution is normal.
2. The mean is 14.
3. The standard deviation is 2.

(a) One of the rats weighs 12 ounces. What is the percentile rank for this weight?

(b) In such a population, what percentage of the rats would we expect to weigh between 10 and 15 ounces?

†Obviously this must be only *approximately* normal. It is customary in much statistical literature to refer to such distributions as "normal" when the context makes it clear that the distributions must necessarily be only approximations to the normal distribution.

The normal distribution

SOLUTION (a) Recall that the formula for converting raw scores to z scores is

$$z = \frac{X - \mu}{\sigma}$$

Thus, the z score for 12 ounces is

$$z_{12} = \frac{12 - 14}{2} = \frac{-2}{2} = -1$$

In Table C-4 we find Area($z = -1$) = .1587. This is shown in Figure 6-14, where we have drawn two horizontal axes, one for the z scores and one for the corresponding raw scores.

Figure 6-14

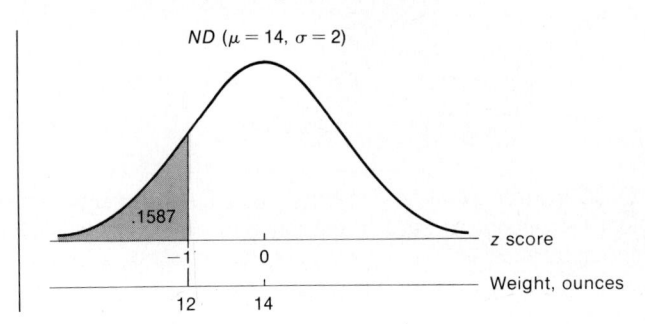

ND ($\mu = 14$, $\sigma = 2$)

.1587

z score: −1, 0
Weight, ounces: 12, 14

This means that 15.87 percent of the rats weigh less than 12 ounces. Rounding off, we say that 12 ounces has a percentile rank of 16, and we write P_{16} = 12 ounces, or PR_{12} = 16.

(b) We draw and label a rough sketch of this population (Figure 6-15).

Figure 6-15

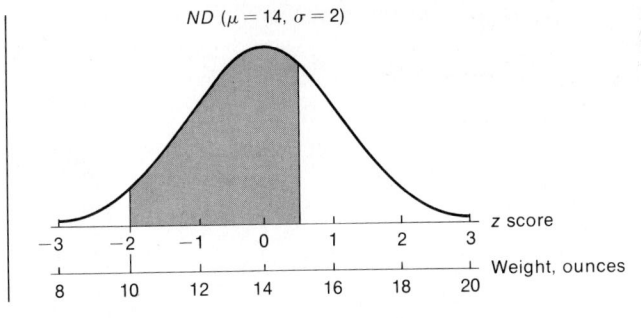

ND ($\mu = 14$, $\sigma = 2$)

z score: −3, −2, −1, 0, 1, 2, 3
Weight, ounces: 8, 10, 12, 14, 16, 18, 20

Since $\mu = 14$, 14 corresponds to $z = 0$. The standard deviation is 2 ounces. Therefore the weight increases by 2 ounces, each time the z score increases by 1.

We want to compute the area between 10 and 15 ounces. Converting the raw scores to z scores, we get:

$$z_{15} = \frac{15 - 14}{2} = \frac{1}{2} = 0.5 \quad \text{and} \quad z_{10} = \frac{10 - 14}{2} = \frac{-4}{2} = -2$$

In Table C-4 we find

Area($z = .5$) = .6915

Area($z = -2$) = .0228.

Understanding statistics

Subtracting .6915 − .0228, we get .6687. Therefore, we expect about 67 percent of the population of rats to weigh between 10 and 15 ounces, as shown in Figure 6-16.

Figure 6-16

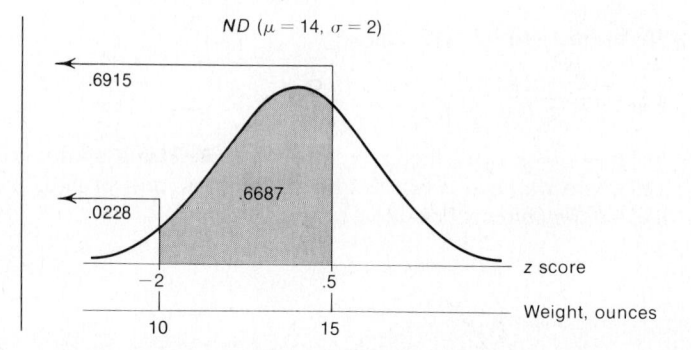

ND ($\mu = 14$, $\sigma = 2$)

.6915
.0228
.6687

z score: −2, .5
Weight, ounces: 10, 15

EXAMPLE 6-7 A quality control technician is testing the accuracy of a certain type of resistor which is supposed to have a resistance of 14 ohms. He finds that the distribution of resistances is approximately normal, with μ about 14.06 ohms and σ about 1.73 ohms.

(a) What is the probability that one of these resistors picked at random will have a resistance of 16 ohms or more?

(b) What is the probability that a resistor picked at random will have a resistance within 1 ohm of the mean? That is, find $P(13.06 \leq X \leq 15.06)$.

SOLUTION (a) This situation is illustrated by the graph in Figure 6-17.

Figure 6-17

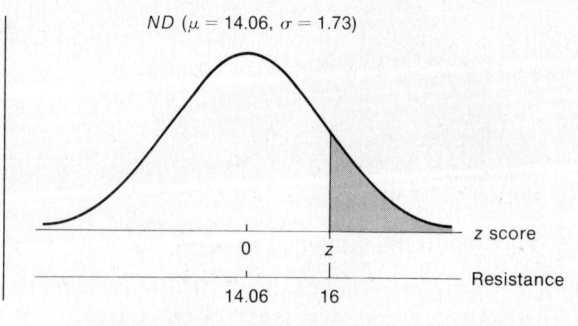

ND ($\mu = 14.06$, $\sigma = 1.73$)

z score: 0, z
Resistance: 14.06, 16

We first find the area to the *left* of 16 by converting 16 to a z score:

$$z_{16} = \frac{16 - 14.06}{1.73} = \frac{1.94}{1.73} = 1.12$$

In Table C-4 we find Area($z = 1.12$) = .8686. Our problem can now be represented as in Figure 6-18.

Hence, the probability that one of these resistors picked at random will have resistance 16 ohms or more is 1.0000 − .8686 = .1314, or about .13. If we use X to stand for the resistance of any resistor that we pick, then we can write $P(X \geq 16) = .13$.

The normal distribution

Figure 6-18

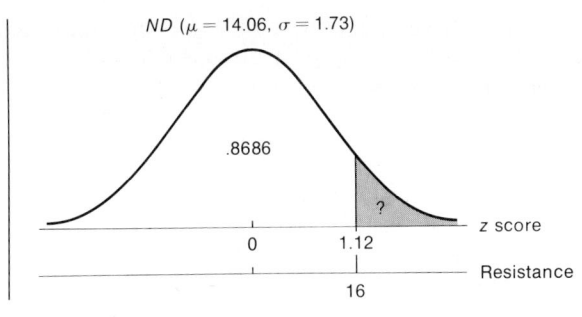

ND ($\mu = 14.06$, $\sigma = 1.73$)

.8686

?

0   1.12   z score

16   Resistance

(b) This situation is illustrated in Figure 6-19.

Figure 6-19

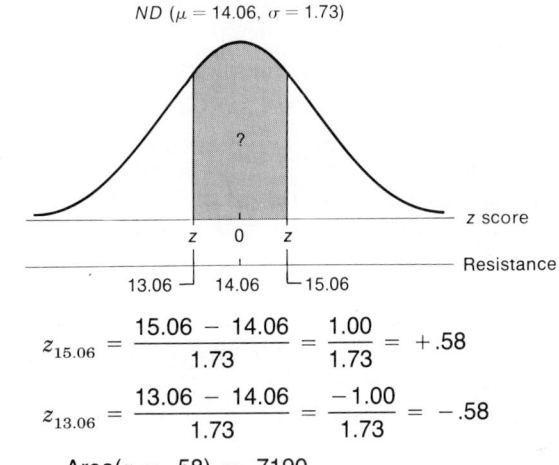

ND ($\mu = 14.06$, $\sigma = 1.73$)

?

z 0 z   z score

13.06   14.06   15.06   Resistance

$$z_{15.06} = \frac{15.06 - 14.06}{1.73} = \frac{1.00}{1.73} = +.58$$

$$z_{13.06} = \frac{13.06 - 14.06}{1.73} = \frac{-1.00}{1.73} = -.58$$

Area($z = .58$) = .7190
Area($z = -.58$) = .2810

Subtracting, we get .4380 or about .44.
Therefore, $P(13.06 \leq X \leq 15.06) = .44$.

EXAMPLE 6-8 Professor Frankenstein raises laboratory vampire bats. The lengths of their left fangs are normally disturbed with $\mu = 28$ mm and $\sigma = 4$ mm.

(a) The Professor's favorite vampire, Sheldon, has a left fang the length of which has a percentile rank of 84 (Figure 6-20). How long is Sheldon's left fang?

Figure 6-20

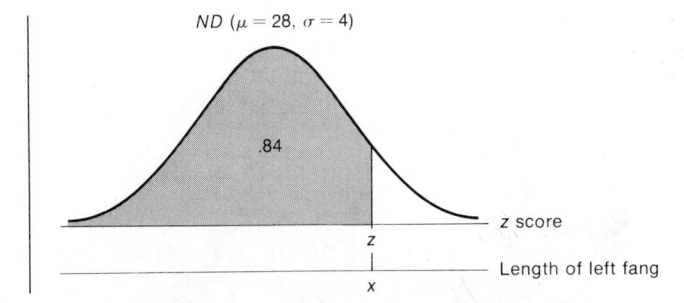

ND ($\mu = 28$, $\sigma = 4$)

.84

z   z score
x   Length of left fang

(b) Professor Frankenstein knows that a bite from a vampire bat whose left fang length is in the top 5 percent will result in instant death. Find the length of the left fang which cuts off the top 5 percent.

SOLUTION (a) The problem is where to draw the vertical line so that 84 percent of the area will be to the left of that line. We search Table C-4 for an Area entry of .8400 (or the closest entry to .8400). The closest entry is .8389, which corresponds to $z = .99$, and so the vertical line is drawn at $z = .99$. We then convert this z score to a raw score. Recall that the formula for converting z scores to raw scores is

$$X = \mu + z\sigma$$

Thus,

$$X = 28 + .99(4) = 28 + 3.96 = 31.96$$

Therefore, Sheldon's left fang is about 32 mm long.

(b) This situation is illustrated in Figure 6-21.

Figure 6-21

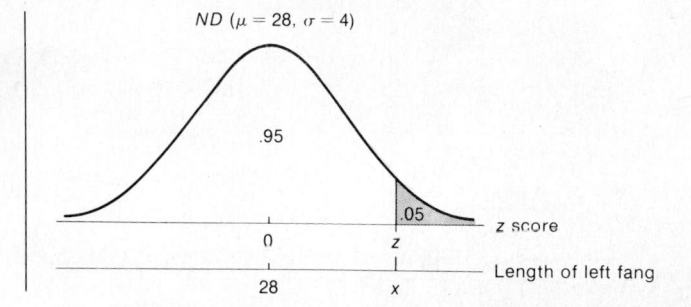

Since the area given in Table C-4 is to the left, we look up $1 - .0500 = .9500$. Since .9500 is not exactly in the table, we choose .9505, which corresponds to $z = 1.65$. Converting this z score to a raw score, we have

$$X = \mu + z\sigma$$
$$= 28 + 1.65(4) = 34.6 \text{ mm, or about } 35 \text{ mm}$$

Thus, a left fang of length 35 mm cuts off the top 5 percent, that is, about 5 percent of the vampire bats have left fangs more than 35 mm long.

STUDY AIDS

VOCABULARY

1. Normal curve   2. Normal distribution

SYMBOLS

1. Area($z =$)   2. $ND(\mu =$, $\sigma =$)

EXERCISES

6-1 A normal distribution is in theory an infinite distribution. Most distributions that we encounter in statistics are finite. Suppose you toss a coin one million times and count the number of heads that occur. There are only one million

The normal distribution

and one possible outcomes. What do we mean, then, when we say that the distribution of these outcomes is normal?

6-2 A distribution of the diameters of pizza pies is symbolized by $ND(\mu = 18$ inches, $\sigma = 2$ inches). What can you say about this distribution?

6-3 In a normal distribution of weights of third-grade boys find the percentage of weights whose z scores are:
(a) Less than $z = 2.33$.
(b) More than $z = 1.65$.
(c) Between the mean and $z = 2.5$.
(d) Between $z = 0$ and $z = -1.6$.
(e) Less than $z = -1.6$.
(f) More than $z = -1.6$.
(g) Between $z = 2$ and $z = 2.5$.
(h) More than 1 standard deviation from the mean.

6-4 In a normal distribution of lengths of Venusian middle legs find the percentage of lengths whose z scores are:
(a) Above $z = -1.96$.
(b) Below $z = 0.23$.
(c) Between $z = -2.13$ and $z = -1.45$.
(d) Above $z = 3$.
(e) Within 2 standard deviations of the mean.
(f) Above $z = 6$.
For the same distribution, find:
(g) $P(z < .5)$ (h) $P(z > 1.5)$
(i) $P(.3 < z < .7)$ (j) $P(z > 7)$

6-5 In a normal distribution of diameters of bolts find the percentage of diameters whose z scores are:
(a) Above $z = 1.96$.
(b) Below $z = -2$.
(c) Not between $z = -1$ and $z = +1$.
For the same distribution, find:
(d) $P(z < -3)$ (e) $P(z > 0.15)$
(f) $P(z > 1.5)$ (g) $P(2 < z < 3)$
(h) $P(-3 < z < -2)$

6-6 (a) For a normal distribution of bushels of corn per acre, find the percentile rank of $z = 2$ plus the percentile rank of $z = -2$.
(b) What happens if you replace the 2s in part (a) by 3s?

6-7 In a normal distribution of automobile gasoline mileage,
(a) Which z score has a percentile rank of 2?
(b) Which z score has a percentile rank of 95?
(c) Which z score has a percentile rank of 50?
(d) Which z scores cut off the middle 70 percent of the population?
(e) If parts (a) through (d) of this exercise had been about a normal distribution of weights of bags of sugar, what would the four answers have been? Why?

6-8 In a normal distribution,
(a) Which z score cuts off the top 10 percent?
(b) Which z score cuts off the bottom 20 percent?
(c) Which z scores cut off the middle 30 percent?
(d) If $P(z < z_?) = .45$, find $z_?$.
(e) If $P(z > z_?) = .35$, find $z_?$.
(f) If $P(-z_? < z < z_?) = .40$, find $z_?$ and $-z_?$.

Understanding statistics

6-9 In a normal distribution, what z score has a percentile rank equal to:
(a) 1 (b) 5 (c) 10
(d) 50 (e) 95 (f) 99

6-10 For a normal distribution, find the z score that cuts off the top
(a) 5 percent. (b) 2.5 percent.
(c) 1 percent. (d) .5 percent.

6-11 Label the raw scores corresponding to the z scores given on the axis on this graph.

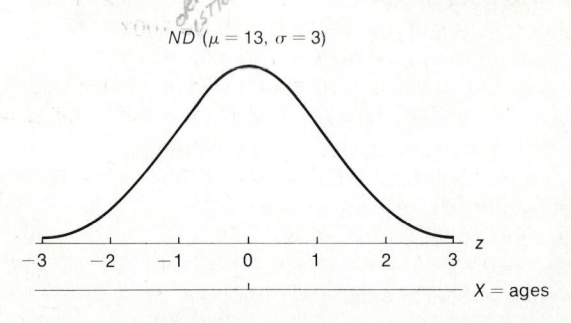

ND ($\mu = 13$, $\sigma = 3$)

−3 −2 −1 0 1 2 3 z

X = ages

6-12 Label the z scores corresponding to the raw scores given on this graph.

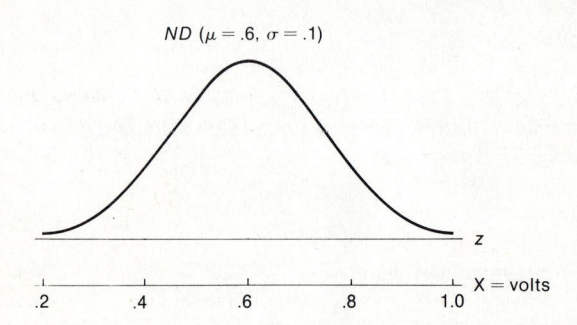

ND ($\mu = .6$, $\sigma = .1$)

.2 .4 .6 .8 1.0 z

X = volts

6-13 Boris Gudenuv entered a Russian literature contest. Last year's competition results are symbolized by $ND(\mu = 230, \sigma = 11)$. If the results are similar this year and Boris wants to score in the top 10 percent, what score will be good enough for Gudenuv?

6-14 The distribution of raw scores on a test of moral fitness is normal, with mean equal to 0 and $\sigma = 10$. What *raw* score cuts off the
(a) Top 5 percent? (b) Bottom 5 percent?
(c) Bottom 2.5 percent? (d) Top 2.5 percent?

6-15 This year's distribution of the earnings of the Aunty Pasto franchises was $ND(\mu = \$24,500, \sigma = \$320)$. The earnings of the Aunt Chilada franchises were $ND(\mu = \$23,900, \sigma = \$600)$.
(a) What is the probability that the owner of an Aunty Pasto franchise earned more than $25,200?
(b) What is the probability that the owner of an Aunt Chilada franchise earned over $25,200?

6-16 An electronics technician repeats an experiment many times, each time recording a voltage reading. The technician finds that the collection of readings is approximately normally distributed, with μ about 74 volts and σ about 6 volts.

The normal distribution

(a) What percentage of the readings were between 70 and 80 volts?
(b) What is the probability that a reading taken at random will be over 86 volts?
(c) What is the probability that a random reading will be *outside* the range of 69 to 79?

6-17 The results on a certain blood test performed in a medical laboratory are known to be normally distributed, with $\mu = 60$ and $\sigma = 18$.
(a) What percentage of the results are between 40 and 80?
(b) What percentage of the results are between 76 and 78?
(c) What percentage of the results are above 100?
(d) What percentage of the results are below 60?
(e) What percentage of the results are between 78 and 80?
(f) What percentage of the results are outside the "healthy range" of 30 to 90?
(g) What is the probability that a blood sample picked at random will have results in the "healthy range" of 30 to 90?
(h) Which test result has a percentile rank of 5?

6-18 A study was done to see how many hours during the school day high school seniors spend thinking about sex. The results were normally distributed with $\mu = 2.7$ hours and $\sigma = .6$ hour. What percent of these students think about sex:
(a) More than 1 hour per school day?
(b) More than 4.5 hours?
(c) Between 2 and 3 hours?

6-19 At an urban hospital the weights of newborn infants are normally distributed, with $\mu = 7$ pounds, 2 ounces, and $\sigma = 15$ ounces. Let X be the weight of a newborn infant picked at random. Find the following probabilities.
(a) $P(X \geq 8 \text{ pounds})$
(b) $P(X \leq 5 \text{ pounds, 5 ounces})$
(c) $P(6 \text{ pounds} \leq X \leq 8 \text{ pounds})$
(d) What infant weight is at the 70th percentile? (That is, find P_{70}.)
(e) Let W be a fixed weight. The probability is .70 that a baby picked at random weighs less than W. Find W. [That is, find W such that $P(X < W) = .70$.]
(f) Find W such that $P(X < W) = .10$.

6-20 The lifetimes of a certain brand of movie floodlights are normally distributed, with $\mu = 210$ hours and $\sigma = 56$ hours. Let X be the lifetime of a light picked at random. Find the following probabilities.
(a) $P(X \geq 300)$
(b) $P(X \leq 100)$
(c) $P(100 \leq X \leq 300)$
(d) The company guarantees that its light will last at least 120 hours. What percentage of the bulbs do they expect to have to replace under this guarantee?

6-21 It happens that income for junior executives in a large retailing corporation is normally distributed, with $\mu = \$32,800$ and $\sigma = \$3000$.
(a) There is an unspoken agreement that you have "arrived" if you are in the top 15 percent. What salary must a junior executive earn in order to arrive?
(b) The highest 25 percent get keys to the executive washroom. Victoria earns $35,080. Does she have a key?
(c) Due to a recession the bottom 5 percent may be let go. What salary cuts off the bottom 5 percent?

(d) The top 20 percent go out to lunch. The bottom 30 percent bring their lunches in interoffice envelopes. The remaining 50 percent bring their lunches in attaché cases. Find the two salaries that separate these three categories.

6-22 True or false? For a normal curve,

(a) The area between $z = 0$ and $z = 1$ equals the area between $z = 0$ and $z = -1$.

(b) The area between $z = 0$ and $z = 1$ equals the area between $z = 1$ and $z = 2$.

(c) The percentile rank of $z = 1$ equals the percentile rank of $z = -1$.

6-23 The graphs below represent the distribution of incomes in two populations. Discuss any differences between the two distributions.

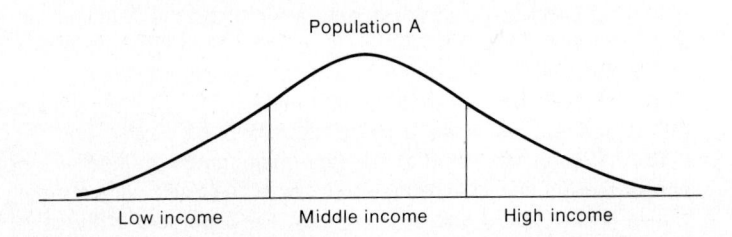

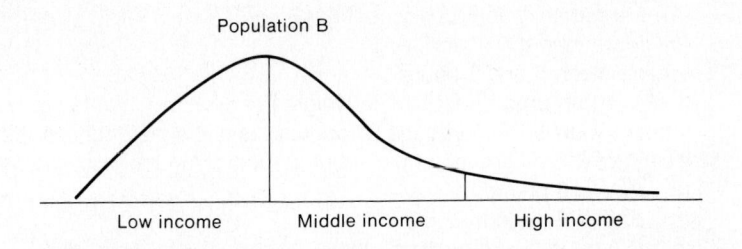

Population A

Low income   Middle income   High income

Population B

Low income   Middle income   High income

6-24 The graphs below represent the distribution of grades at a large university. One graph is for freshman-level courses. The other is for advanced courses. Discuss any differences between the two distributions of grades.

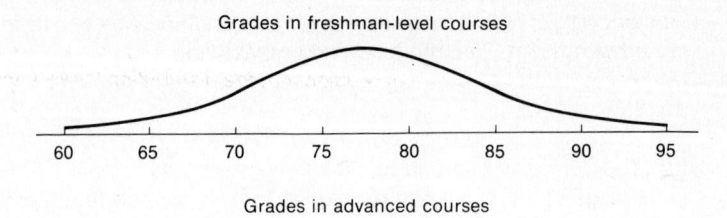

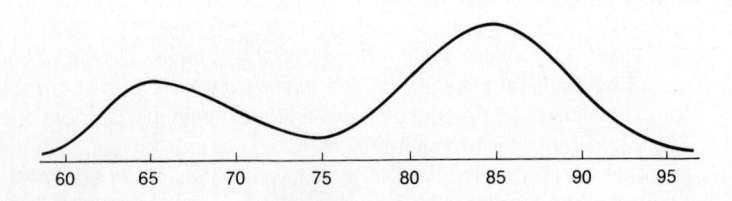

Grades in freshman-level courses

60   65   70   75   80   85   90   95

Grades in advanced courses

60   65   70   75   80   85   90   95

The normal distribution

6-25 In a distribution of grades, $\mu = 80$ and $\sigma = 10$. Yet it is *not* true that 34 percent of the distribution is between 80 and 90. In fact, we cannot find what percentage of the grades lie between 80 and 90 from the above information. Why? DON'T KNOW IF ITS NORMAL/CAN'T DO

6-26 A distribution of professors' salaries at a large college has $\mu = \$17{,}000$ and $\sigma = \$2400$. What percentage of the professors earn between $17,000 and $19,000? DON'T KNOW IF ITS NORMAL/CAN'T DO

6-27 Cat Moran, the sailing instructor at Marny's Marina, has noticed that on a beginner's first "solo" the boat invariably tips over. She has observed that the time it takes them to capsize is normally distributed, with the mean equal to 21 minutes and the variance equal to 16 minutes.
(a) What is the percentile rank of someone who manages to stay afloat ½ hour?
(b) Is it more likely that a beginner will tip over during the first 15 minutes or during the second 15 minutes?

6-28 A technical writer wants to write a brochure describing a new steam iron. He finds that 15 irons, when filled with 7 ounces of distilled water, will operate for an average of 17 minutes before all the water is gone. He also finds that the distribution of times is approximately normal, with $s = 2$ minutes. He decides to write that the irons will operate for "about 20 minutes." What percentage of the people who use this model iron will find that their iron does not live up to the description in the brochure? (That is, their iron operates *less* than 20 minutes.)

6-29 Here are the data for the Sandy Koufax example in the text (Figure 6-7):

number of hits per game	0	1	2	3	4	5	6	7	8	9	10	11	12	13
frequency	4	2	8	15	16	24	16	13	5	3	1	2	0	1

We have found the mean and the standard deviation of these 110 numbers. Sandy gave up an average of 5.0 hits per game with a standard deviation of 2.3.
(a) From the data calculate both the z score for 6 hits and the percentile rank for 6 hits.
(b) Find the percentile rank in Table C-4 for the z score you just calculated in part (a). Compare the two percentile ranks you have found.
(c) Repeat parts (a) and (b) for several other numbers of hits. How does this confirm or deny the visual impression that the distribution looked approximately normal?

CLASS SURVEY

1 Do the heights of the females in your sample appear to be normally distributed?

2 Do you think the heights of all the females in your school would be approximately normal?

Approximation of the Binomial Distribution by Use of the Normal Distribution

7

Dr. I. M. Normal, the noted research psychologist, is doing an experiment with rats. Each rat is sent through a T-shaped maze. She assumes that due to chance factors alone it is equally likely that the rat will go left or right. Note that this is a binomial experiment because there are two outcomes, and if we consider "left" as success, then P(rat goes left) $= p = 1/2$ and P(rat goes right) $= q = 1/2$.

Dr. Normal sends 20 rats through the maze and notes that 9 of them turn left. She then has one of her lab assistants repeat this experiment with 20 rats 200 times. The assistant records the 200 numbers, indicating how many rats turned left in each experiment. A partial list of the results is given in the table, with S representing the number of successes.

experiment number	number of successes S
1	9
2	10
3	11
4	10
5	9
6	16
.	.
.	.
.	.
200	8

Approximation of the binomial distribution by use of the normal distribution

The assistant computes the mean and the standard deviation for these 200 numbers. He finds that the mean is about 10 and the standard deviation is about 2.3.

If you consider sending 20 rats through a T-shaped maze a binomial experiment, with $p = 1/2$ and $n = 20$, and you repeat this experiment many times, you might expect the mean number of left turns to be about 10. Statisticians have shown mathematically that if a binomial experiment is repeated over and over a large number of times, then theoretically the mean number of successes equals np, and their standard deviation equals $\sqrt{npq}$.

We write

$$\mu_S = np \quad \text{and} \quad \sigma_S = \sqrt{npq}$$

The assistant repeated this experiment a large number of times (200). Therefore, these formulas should give us approximately the same results that he got experimentally.

$$p = 1/2 \quad \text{and} \quad q = 1/2$$
$$\mu_S = np = 20(1/2) = 10$$

and $\sigma_S = \sqrt{npq} = \sqrt{20(1/2)(1/2)} = \sqrt{5} = 2.24$

EXAMPLE 7-1 For the above experiment:

(a) What is the probability that more than 16 rats turn left?

(b) What is the probability that 16 or fewer rats turn left?

SOLUTION 1 A solution using Pascal's triangle

S	number of ways S can occur	$P(S)$	
20	1	$1(1/2)^{20}$	= .000001
19	20	$20(1/2)^{19}(1/2)$	= .000019
18	190	$190(1/2)^{18}(1/2)^2$	= .000181
17	1140	$1140(1/2)^{17}(1/2)^3$	= .001087

(a) Adding these probabilities, we get $P(S > 16) = .0013$, which rounds to .001, or about one chance in a thousand.

(b) We could continue the above table for values of $S = 16, 15, 14, \ldots, 0$ and compute the corresponding probabilities. If we added these probabilities together, the result would be .9987. Of course, we would most likely have solved the problem by simply subtracting $1 - .0013 = .9987$, or about .999.

SOLUTION 2 An approximate solution using the normal curve

Under certain conditions (to be discussed later) the binomial distribution is very close to the normal distribution. Since $n = 20$ and $p = 1/2$, we have $\mu_S = np = 10$ and $\sigma_S = \sqrt{npq} = \sqrt{5} = 2.24$. The question is, where do we draw the line that cuts off the area corresponding to more than 16 rats (see Figure 7-

Figure 7-1

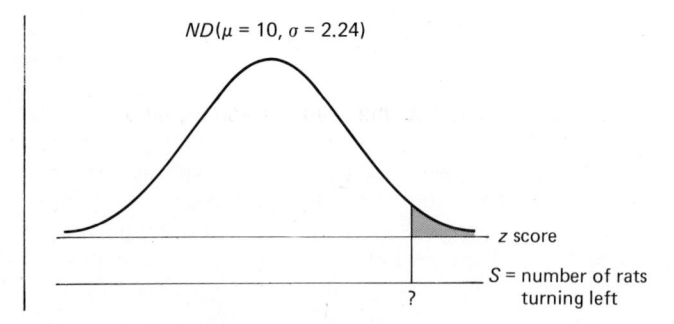

ND(μ = 10, σ = 2.24)

z score
S = number of rats turning left

1)? Since the number of successes is a whole number and the normal distribution takes into account all numbers including fractions, we have to adapt one to the other. The first problem is to find $P(S > 16)$, and the second problem is to find $P(S \leq 16)$. These two probabilities should add up to 1.

If we used $S = 17$ and up to solve the first problem, and then we used $S = 16$ and down to solve the second problem, as we did in the solution by Pascal's triangle, the two answers would not add up to 1. This is because the gap from 16 to 17 would not be included (Figure 7-2).

Figure 7-2

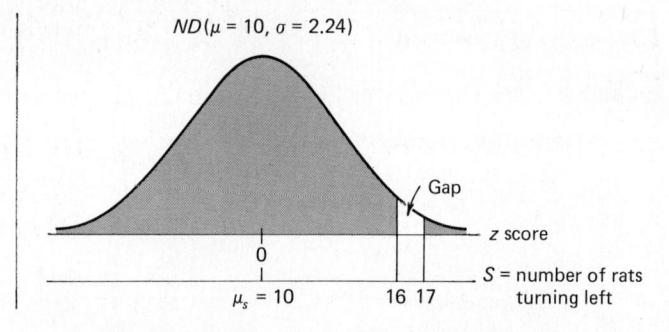

ND(μ = 10, σ = 2.24)

Gap
z score
$\mu_s = 10$ 16 17
S = number of rats turning left

To avoid this error, we draw the line halfway between 16 and 17, at 16.5 (Figure 7-3), which separates the favorable outcomes, 17, 18, 19, and 20, from the unfavorable ones, 0, 1, 2, ..., 16. Recall that 16.5 is the boundary on the binomial histogram which we are approximating by this normal curve. Now to estimate $P(S > 16)$ where S is a variable in a *binomial* distribution, we compute $P(S \geq 16.5)$ where S is a variable in a *normal* distribution. Since $S = 16.5$ is a raw score,

Figure 7-3

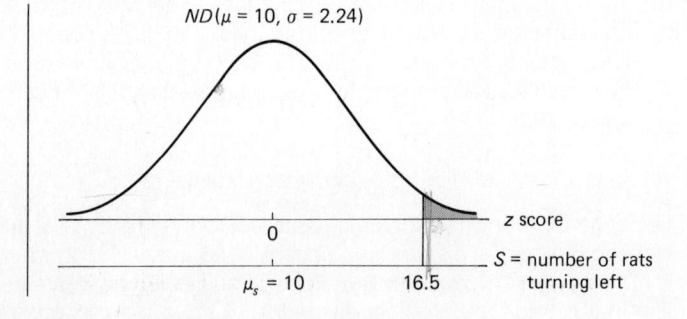

ND(μ = 10, σ = 2.24)

z score
$\mu_s = 10$ 16.5
S = number of rats turning left

Approximation of the binomial distribution by use of the normal distribution

$$z_{16.5} = \frac{S - \mu_S}{\sigma_S} = \frac{16.5 - 10}{2.24} = 2.90$$

In Table C-4 we find that the area corresponding to $z = 2.90$ is .9981 (see Figure 7-4). Therefore, $P(S \geq 16)$ is approximately $P(S \geq 16.5) = 1 - .9981 = .0019$, or about .002. Notice that the answers in Solution 1 and Solution 2 differ by only .001.

Figure 7-4

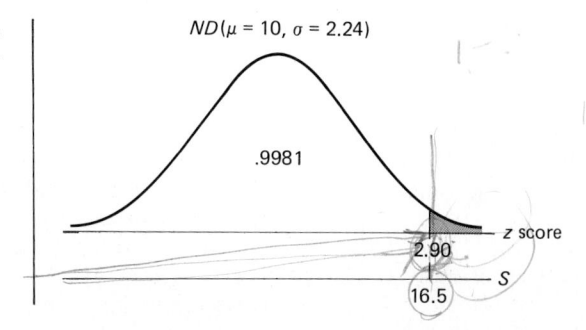

$ND(\mu = 10, \sigma = 2.24)$

.9981

z score
2.90
S
16.5

The second solution gives a result which is accurate enough for many practical problems and which is much easier to calculate. What little we lose in accuracy is more than compensated for by the ease in computation, especially when n is large. ∎

CONDITIONS FOR APPROXIMATING A BINOMIAL DISTRIBUTION BY USE OF A NORMAL DISTRIBUTION

When can we say that this approximation is fairly accurate? Let us consider the histograms for two binomial distributions.

EXAMPLE 7-2 For $n = 11, p = 1/2$, we get

S	$\binom{11}{S}$ number of ways S can occur	$P(S)$	
11	1	$1(1/2)^{11}$	= 1/2048
10	11	$11(1/2)^{10}(1/2)$	= 11/2048
9	55	$55(1/2)^9(1/2)^2$	= 55/2048
8	165	$165(1/2)^8(1/2)^3$	= 165/2048
7	330	$330(1/2)^7(1/2)^4$	= 330/2048
6	462	$462(1/2)^6(1/2)^5$	= 462/2048
5	462	$462(1/2)^5(1/2)^6$	= 462/2048
4	330	$330(1/2)^4(1/2)^7$	= 330/2048
3	165	$165(1/2)^3(1/2)^8$	= 165/2048
2	55	$55(1/2)^2(1/2)^9$	= 55/2048
1	11	$11(1/2)(1/2)^{10}$	= 11/2048
0	1	$1(1/2)^{11}$	= 1/2048

The histogram for these probabilities appears in Figure 7-5.

Figure 7-5

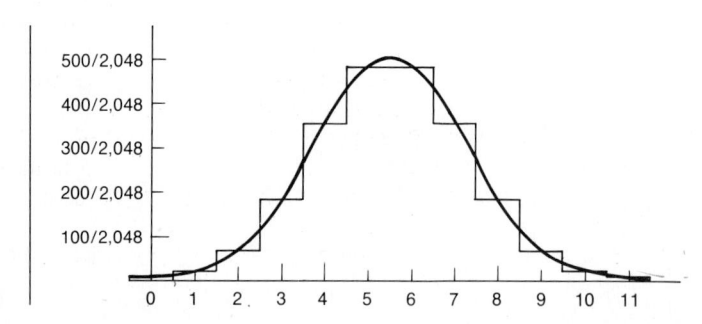

[Figure 7-5: histogram with normal curve overlay, y-axis from 100/2,048 to 500/2,048, x-axis from 0 to 11]

EXAMPLE 7-3

For $n = 3$, $p = 1/4$, we have

S	$\binom{3}{S}$ number of ways S can occur	$P(S)$
3	1	$1(1/4)^3 = 1/64$
2	3	$3(1/4)^2(3/4) = 9/64$
1	3	$3(1/4)(3/4)^2 = 27/64$
0	1	$1(3/4)^3 = 27/64$

The graph for these data is shown in Figure 7-6.

Figure 7-6

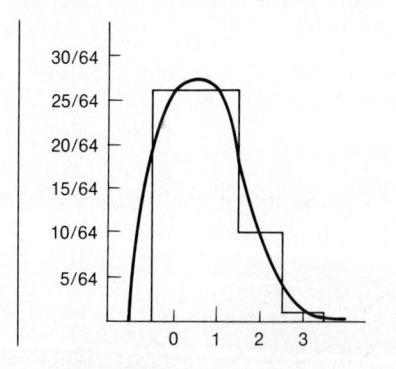

[Figure 7-6: histogram with curve, y-axis from 5/64 to 30/64, x-axis from 0 to 3]

The smooth curve sketched through the first histogram looks approximately normal, while the second curve does not. It does not look symmetrical. This second curve appears chopped off on the left because the mean = np = $3(1/4) = 3/4$ is too close to 0. Whenever np is too small, this occurs. On the other hand, if np is too large, the mean will be too close to n and the curve will appear chopped off on the right. In this case nq will be too small.

A good rule of thumb that will assure us that the binomial histogram is approximately normal is when *both* np and nq are greater than 5. If np or nq is smaller than or equal to 5, this approximation may not be close enough, and we would solve the problem by use of Pascal's triangle. In Example 7-2, np = 5.5 and nq = 5.5. Both of these are greater than 5, while in Example 7-3 np = 3/4, which is too small.

Approximation of the binomial distribution by use of the normal distribution

EXAMPLE 7-4 Let us consider the following example which further illustrates how we can handle the problem of approximating a binomial distribution by use of a normal distribution. A commuter drives to work during the morning rush hour. She must drive through a heavily traveled railroad crossing. From the railroad timetable she has figured that the gate is closed 30 percent of the time. Due to traffic conditions her time of arrival at the crossing is random.

(a) Find the probability that on any given day she arrives when the gate is open.

Next month she will drive to work 19 times. Let S equal the number of times that she will be successful in arriving at the crossing when the gate is open.

(b) Find $P(S < 12)$. $S \leq 11.5$

(c) Find the probability that S is at least 15; that is, find $P(S \geq 15)$.

(d) Find $P(14 \leq S \leq 18)$.

(e) Find $P(S = 16)$.

(f) Certain numbers of successes are so high, and therefore so unlikely, that there is only about a .05 probability that they will occur. Which numbers of successes are these? In symbols this would be:
Find $S_?$ such that $P(S > S_?)$ is approximately .05.

SOLUTION

(a) $p = P(\text{gate is open}) = 1 - .30 = .70 = .7$

(b) $q = .3$
$n = 19$
$np = 19(.7) = 13.3$
$nq = 19(.3) = 5.7$

As both 13.3 and 5.7 are greater than 5, we can use the normal distribution as an approximation to the binomial.

$\mu_S = np = 13.3$
$\sigma_S = \sqrt{npq} = \sqrt{(13.3)(.3)} = \sqrt{3.99} = 2.00$

Because these are whole numbers, we find $P(S < 12)$ by using the boundary of 11.5, which separates the favorable outcomes 0, 1, 2, ..., 11 from the unfavorable outcomes 12, 13, ..., 19. This is shown in Figure 7-7.

$$z_{11.5} = \frac{11.5 - 13.3}{2} = \frac{-1.8}{2} = -.90$$

The area to the left of $-.90$ is Area($z = -.90$) = .1841, or about .18. Therefore, $P(S < 12)$ is approximately $P(S \leq 11.5) = .18$.

(c) To find $P(S \geq 15)$ we use the boundary 14.5, which separates the favorable outcomes from the unfavorable ones (Figure 7-8).

$$z_{14.5} = \frac{14.5 - 13.3}{2} = \frac{1.2}{2} = .60$$

Understanding statistics

Figure 7-7

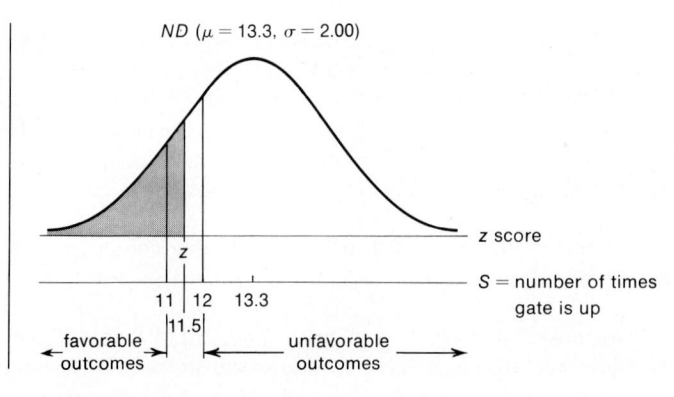

ND ($\mu = 13.3$, $\sigma = 2.00$)

11 | 12   13.3
 |11.5|

favorable outcomes ← | → unfavorable outcomes

z score
S = number of times gate is up

Area($z = .60$) = .7257

Our answer is $P(S \geq 15)$, which is approximately $P(S \geq 14.5)$, $= 1 - .7257 = .2743$, or about .27.

Figure 7-8

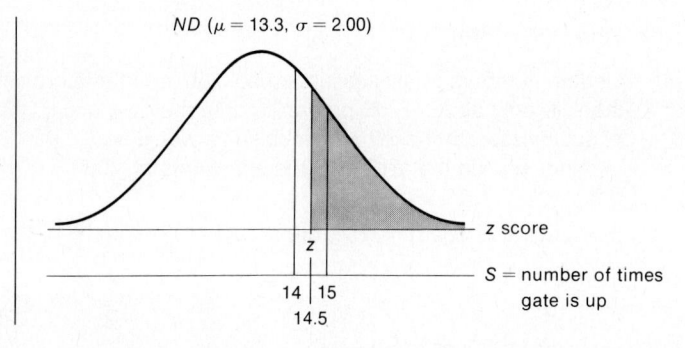

ND ($\mu = 13.3$, $\sigma = 2.00$)

14 | 15
 14.5

z score
S = number of times gate is up

(d) To find $P(14 \leq S \leq 18)$ we will compute $P(13.5 \leq S \leq 18.5)$. This area is shown in Figure 7-9.

$$z_{18.5} = \frac{18.5 - 13.3}{2} = 2.6$$

$$z_{13.5} = \frac{13.5 - 13.3}{2} = .1$$

Area($z = 2.6$) = .9953
Area($z = .1$) = .5398

Figure 7-9

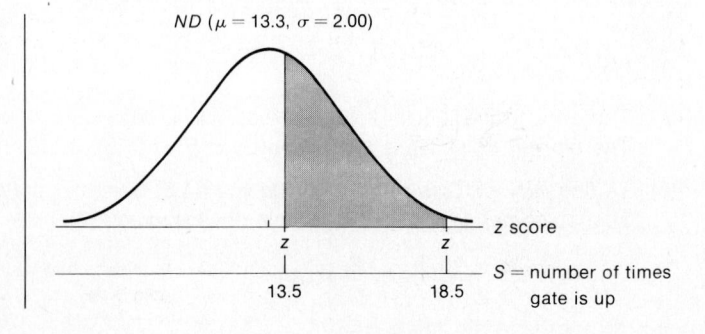

ND ($\mu = 13.3$, $\sigma = 2.00$)

13.5   18.5

z score
S = number of times gate is up

Approximation of the binomial distribution by use of the normal distribution

The area between 13.5 and 18.5 is found by subtracting these two areas, and the answer is approximately $P(13.5 \leq S \leq 18.5) = .4555$, or about .46 (Figure 7-10).

Figure 7-10

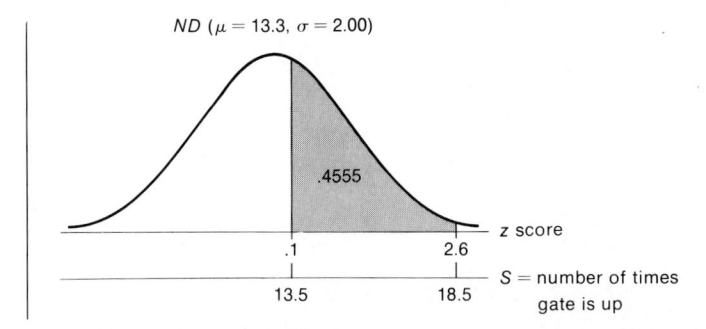

ND ($\mu = 13.3$, $\sigma = 2.00$)

.4555

.1 2.6   z score

13.5 18.5   S = number of times gate is up

(e) To find $P(S = 16)$ we use the boundaries which cut off 16, namely, 15.5 to 16.5 (Figure 7-11).

Figure 7-11

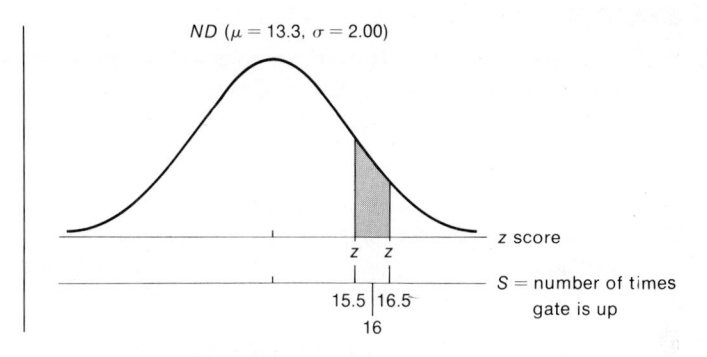

ND ($\mu = 13.3$, $\sigma = 2.00$)

z z   z score

15.5 | 16.5
16   S = number of times gate is up

$z_{16.5} = 1.6$ and $z_{15.5} = 1.1$

Area($z = 1.6$) = .9452

Area($z = 1.1$) = .8643

$P(S = 16)$ is approximately $P(15.5 \leq S \leq 16.5) = .0809$, or about .08.

(f) To find the number of successes $S_?$ such that $P(S > S_?) = .05$, we look for Area equal to .9500 in Table C-4. This area is depicted in Figure 7-12.

Figure 7-12

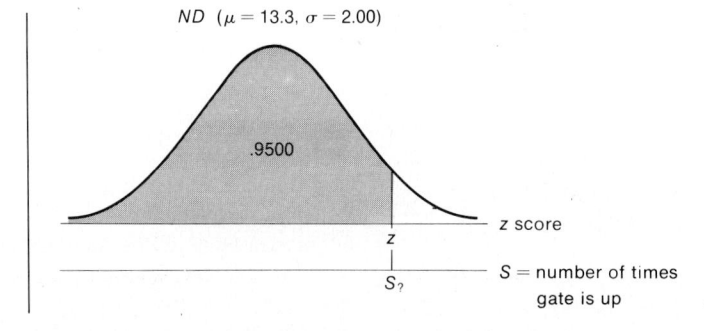

ND ($\mu = 13.3$, $\sigma = 2.00$)

.9500

z   z score

$S_?$ S = number of times gate is up

The area .9500 leads to a z score of 1.65. To convert this z score to a raw score of a number of successes use the formula $S_? = \mu_S + z\sigma_S$. This yields

$$S_? = 13.3 + 1.65(2) = 13.3 + 3.30 = 16.6$$

Thus outcomes 17, 18, and 19 occur only about 5 percent of the time.

STUDY AIDS

SYMBOLS

1. n

FORMULAS

1. $\mu_S = np$ 2. $\sigma_S = \sqrt{npq}$ 3. $S_? = \mu_S + z\sigma_S$

EXERCISES

7-1 Consider the following binomial problem. Over the years the percentage of freshmen who pass Calc I with Prof Tomchick is 0.430. If 12 freshmen are in her class, what is the probability that 10 or more will pass? If we solve by Pascal's triangle, the answer is .005. Using the binomial probability table (with $p = .40$) we get .003. The normal approximation yields .006. Which answer is most accurate? least accurate? Comment.

7-2 Do this problem two ways, first, by the binomial techniques using Pascal's triangle or the binomial table, and second, by normal approximation. Letitia tosses a coin to decide whether she will eat in a Chinese restaurant (heads) or a Greek restaurant (tails). She wishes to make plans now for her next 12 meals out. What is the probability that she will go to a Chinese restaurant more than 9 times? Comment on your answers. *Hint:* "More than 9" means 10, 11, or 12.

7-3 Kent, a clerk in a supermarket, can work with either of two part-time accountants, Lois or Lane. Clerk Kent decides whom he will work with each day by using the spinner illustrated.

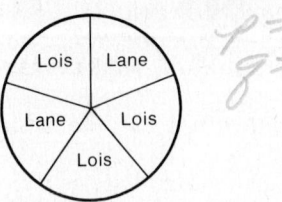

(a) Find $p = P$(he works with Lois).
(b) Find q.
If the supermarket man uses the spinner 14 times to determine which accountant he will work with each day for the next 2 weeks,
(c) Justify the use of the normal approximation to the binomial distribution.
(d) Find μ_S.
(e) Find σ_S.
(f) Find the probability that Lois works with the superman 7 or fewer times.

Approximation of the binomial distribution by use of the normal distribution

7-4 A pair of loaded dice has $p = P(\text{rolling a 7}) = .3$. If the dice are rolled 20 times.
(a) Justify the use of the normal approximation to the binomial distribution.
(b) Find μ_S.
(c) Find σ_S.
(d) Find the probability of rolling more than three 7s.
(e) Find $P(S > 9)$, where S is the number of 7s rolled.
(f) If $P(S > S_?)$ is approximately equal to .99, find the whole number $S_?$.

7-5 A game is played by two persons who each show 2, 1, or 0 fingers at the same time. If the total number of fingers is odd, then the person predesignated as "odd" wins. If the total number of fingers is even, then the other person, designated "even," wins.
(a) List the 9 possible outcomes.
(b) Show $P(\text{odd}) = 4/9$.
If Mario and Gina play the game 18 times, find the probability that odds occur:
(c) More than 14 times.
(d) Exactly 9 times.
(e) Either 7, 8, or 9 times.

7-6 It is known that 30 percent of the autos produced by the Necromate Auto Company are defective when they come off the assembly line. (So we say that the probability that a car off the line is defective is .30.)
(a) What is the probability that if 900 of these cars are picked at random, then more than 40 percent of the 900 cars will be defective? (This means that more than 360 cars will be defective.)
(b) What is the probability that more than 33 percent of the 900 cars will be defective?
(c) What is the probability that less than 28 percent of the 900 cars will be defective?

7-7 The probability that a 40-year-old man will have died before age 60 is .17. An insurance company insures 1200 men of age 40.
(a) What is the probability that fewer than 15 percent of these men will have died before age 60?
(b) What is the probability that more than 20 percent of the men will have died before age 60?

7-8 Under certain conditions, the probability that a tadpole survives to mature into a frog is .10. If we have 100 tadpoles,
(a) What is the probability that more than 14 survive?
(b) What is the probability that none survive? *Hint:* You may interpret "none" to mean "less than 1."
(c) If the probability of at least $S_?$ tadpoles maturing is about .95, find the whole number $S_?$.

7-9 The morbid *Journal of Morbidity Tables* indicates that the probability of a teenager contracting terminal acne is .12. If 100 teenagers are checked at random, what is the probability that more than 20 will have contracted this disease?

7-10 For a certain disease half the victims recover in one week with no treatment. Dr. Quack invents a treatment for this disease. (The probability of recovering with his treatment in one week is still .5. It is no better than doing nothing.) He treats 20 patients. Of them, 12 recover in a week. He claims that this proves his treatment is good. What is the probability that at least 12 people out of 20 would have recovered with no treatment?

Understanding statistics

7-11 Rosencrantz tossed a sixpence 10 times. It fell heads up each time. He claims that the coin is a fair one. Is that possible? Do you believe him? (Justify your answer mathematically.)

7-12 Refer to the rat problem in the beginning of this chapter. What would your reaction be if you performed this experiment 3 times, and each time more than 16 rats turned left?

7-13 Bob and Gene play a game by drawing 1 card from an ordinary deck and then replacing it. If the card is a spade, Bob wins. If not, then Gene wins. On the last 24 draws, Bob won 18 times.
(a) Using $p = P(\text{Bob wins}) = 1/4$, find the probability that Bob would win 18 or more times in 24 plays.
(b) Using $p = P(\text{Gene wins}) = 3/4$, find the probability that Gene would win 6 or fewer times in 24 plays.

7-14 Do Exercise 5-10(b) again by using the normal approximation. Explain any differences in your answers to the problem by the binomial method and by the normal approximation. We restate the problem.

If in each pregnancy the probability of having a girl is 1/2, what is the probability that a childless couple planning to have 5 children will have at least 4 girls? What percentage of couples with 5 children would you expect to have 4 or 5 girls? (The answer by the binomial method was .1875.)

7-15 Do Exercise 5-23 again by using the normal approximation. Explain any differences in your answers to the problem by the binomial method and by the normal approximation. We restate the problem.

A manufacturer claims that 4 of 5 dentists recommend sugarless gum for their patients who chew gum. Assuming that this claim is true, find the probability that in a randomly selected group of 20 dentists, 16 or more will recommend sugarless gum for their patients who chew gum. (The answer by the binomial method was .63.)

7-16 A large distribution of the number of peanuts in a package of Sower's peanuts is approximately normal, with $\mu = 120$ and $\sigma = 15$. Suppose that one package is selected at random, and X represents the number of peanuts in the package.
(a) Find $P(X < 100)$.
(b) Find the number A for which $P(X > A)$ is approximately equal to .35. *Hint:* The number of peanuts is a whole number.
(c) Repeat parts (a) and (b) above if the distribution is the weight in grams of a package of Sower's peanuts. *Hint:* Weight is continuous.

7-17 Suppose that you have data from a continuous distribution. The distribution is approximately normal, with a mean of 14 and a standard deviation of 3.
(a) If a number X is selected at random from the distribution, find $P(X > 18)$.
(b) The value of A in this distribution is such that $P(X < A) = .4$. Find A.
(c) Repeat parts (a) and (b) above if the distribution consists of whole numbers only.

7-18 It is known that 20 percent of EZ brand zippers are defective. If many random samples of 100 zippers are examined,
(a) Find the mean number of defectives in these samples.
(b) Find the standard deviation of the number of defectives.
(c) Find the probability that more than 25 zippers in a random sample will be defective.

Approximation of the binomial distribution by use of the normal distribution

(d) The probability that more than $S_?$ zippers in a random sample will be defective is approximately equal to .30. Find the whole number $S_?$.

7-19 A carnival packages prizes in identical unmarked boxes in such a way that 9 of 10 boxes contain a prize worth 10 cents and 1 of 10 contains a prize worth 50 cents. There are hundreds of such boxes all mixed up in bins at the prize booth. For 30 cents, Sylvia may pick out any box and keep the prize.

(a) If she takes one chance, what is the probability that her prize will be worth less than what she paid to play?

(b) If 500 people each play once on a Saturday evening, what is the probability that more than 12 percent will win "good" prizes?

(c) How many times does Sylvia have to play to have better than a 50 percent chance of getting *at least one* 50-cent prize? *Hint:* This means that the probability of *no* "good" prizes must be less than 50 percent.

7-20 G. Whilakers runs an amusement parlor. One popular game there is called Skee-Boll, in which a player can win by getting over 200 points. Mr. Whilakers has noticed that, overall, in 80 percent of the games the player wins. On the average night, 600 games are played. What is the probability that there will be over 500 winners?

7-21 (a) The probability that a person in a certain community is a carrier for the fatal Tay-Sachs disease is 1 in 30. If 300 people are examined in this community, what is the probability that more than 10 will be found to have the disease?

(b) If 2 people from this community marry, what is the probability that they will both be carriers?

(c) If both parents are carriers, there is a 1/4 probability that their baby will have the fatal disease. What is the probability if 2 people in this community marry that their first child will have the disease? *Hint:* This is a sequence of events. First the 2 people would have to both be carriers, then they would have to produce a diseased baby. Use the multiplication rule mentioned in Chapter 4 for the probability of a sequence of events.

7-22 A recent election in the city of Gotham was contested. The facts were these: 10,000 citizens voted; 4900 voted for Elizabeth Goodsoul, while 5100 voted for James Badguy. The election was challenged, and 1000 voters were found to have voted illegally. Ms. Goodsoul asked for a reelection, but Judge Blank ruled against her request. He reasoned that it was not likely that a reelection among the legal voters would change the final outcome.

Using $p = P(\text{a voter votes for Goodsoul}) = \dfrac{4900}{10{,}000} = .49$, and assuming that $n = 1000$, perform the following calculations. If we randomly throw out 1000 votes, Goodsoul would need at least 4501 of the remaining 9000 to win. This means that of the 1000 votes being randomly thrown out less than 400 of hers would have to be thrown out. Find the probability that less than 400 of the 1000 thrown out would be her votes.

SAMPLE TEST FOR CHAPTERS 5, 6, AND 7

1 (a) The weights of the costumes worn at Disneyland are approximately normally distributed. What percentage weigh more than 1½ standard deviations above the average?

(b) If the mean weight is 6.7 pounds and $\sigma = 2.3$ pounds, how heavy are the top 10% of the costumes?
(c) What weight cuts off the lightest 15% of the costumes?
(d) What is the probability that a costume selected at random weighs between 7 and 8 pounds?

2 If the average employee at Friendly's Emporium is paid $3.10 per smile with a standard deviation of $1.29, is it possible that the top 6% of the employees are paid less than $4.00 per smile?

3 One line of Pascal's triangle shows 1, 4, 6, 4, 1. What do the 4s signify?

4 Evaluate:

(a) $\binom{10}{3} =$

(b) $\binom{5}{x} = 10$

(c) $\binom{125}{0} =$

(d) $\binom{119}{119} =$

(e) $\binom{117}{1} =$

(f) $\binom{22}{2} =$

Solve both Problem 5 and Problem 6 three ways:
(a) *By using Pascal's triangle.*
(b) *By use of the binomial probability tables.*
(c) *By the normal approximation.*
Comment on your solutions.

5 A fair coin is tossed 18 times. Find the probability of getting either 8, 9, or 10 heads.

6 The probability that one landing light burns out on the XL-12 rocket to the moon is 0.42. Find the probability that all 12 landing lights are operating.

7 The publisher of *Honesty* magazine claims that 70 percent of its subscribers are under age 30. A random sample of 500 subscribers shows that only 63 percent of the sample are under 30. The publisher says that you only sampled some of the subscribers, and that a second sample might yield 77 percent.

(a) Presuming that his original claim is true, what is the probability of getting 63 percent or less in a sample of 500 subscribers?
(b) Is it possible that the publisher's claim is correct?
(c) Would you believe the publisher?
(d) It is possible that the sample was not random, but had some hidden bias. How could this be determined?

Hypothesis Testing: Binomial One-Sample

8

This chapter brings together many of the ideas we have discussed so far. The material is of the utmost importance. Hypothesis testing is one of the most widely used procedures in statistics. At the conclusion of this chapter you will be in a position to understand hypothesis testing and to perform your own experiments.

STATISTICAL HYPOTHESES

Consider the following five questions.

1. What percentage of coupons printed in the newspapers are redeemed?
2. How much more effective is prescription A than prescription B?
3. Is it true that 30 percent of shoppers buy their favorite brand of toothpaste, regardless of price?
4. Is this die biased in favor of 3?
5. Do boys and girls score differently on the verbal portion of the SAT tests?

These questions fall into two types. Questions 1 and 2 ask for a *numerical* response. The last three questions require a *yes-no* response. Statisticians often treat these yes-no questions by first formulating a pair of opposite statements, called *hypotheses*. A statistical hypothesis is a statement about a population. An experimenter attempts to prove or disprove the statement "beyond a reasonable doubt" by analyzing a sample from that population. In questions 3, 4, and 5 they would get the following pairs of hypotheses.

3. Let $p = P$(shopper buys favorite brand of toothpaste regardless of price). Then the two hypotheses would be:

$H1$: 30 percent of shoppers buy their favorite brand regardless of price, $p = .30$

$H2$: the percentage who stick with their brand is different from 30 percent, $p \neq .30$

4. Let $p = P$(the die shows a 3 in one toss). Our two hypotheses would be:

$H1$: the die is fair, $p = 1/6$

$H2$: the die is biased for 3s, $p > 1/6$

Note that we are not considering the possibility that p could be less than 1/6. Strictly speaking, $H1$ and $H2$ in this example are not exact opposites. The opposite of $H2$: $p > 1/6$ is H: $p \leq 1/6$, that is, p is *equal* to 1/6 or *less than* 1/6. But it often happens in a real experiment that certain conceivable alternatives are just not considered. In this case, if someone has become suspicious that a die is showing too many 3s, the behavior of the die itself indicates that there is no reason to try to establish that it is getting *too few* 3s. We just want to decide whether or not it is getting *more* 3s than are expected of a fair die.

5. Let μ_B equal the average score for boys, and μ_G equal the average score for girls. Our hypotheses would then be:

$H1$: boys and girls score the same on the verbal portion of the SAT tests, that is, $\mu_B = \mu_G$

$H2$: boys and girls score differently on the verbal portion of the SAT tests, that is, $\mu_B \neq \mu_G$

TESTING THE NULL HYPOTHESIS

Suppose you were interested in the question, is this a fair coin? A fair coin is one for which the probability of heads on a single toss is .50. Therefore, if we let $p = P$(a head appears on one toss), then our two hypotheses are $H1$: $p = .50$ and $H2$: $p \neq .50$. We could attempt to establish the truth of either one of these two hypotheses, for if one is true, the other is false, and vice versa.

It turns out that it is much easier to test the first one: $p = .50$. Notice that if we assume that $H1$ is true, and if we toss the coin, say, 80 times, then we know ahead of time *what to expect*. We expect to get about 40 heads. We wouldn't be suspicious if the outcome were 38 heads, but an outcome of 72 heads would certainly arouse our suspicions. On the other hand, if we tried to test $H2$ by tossing the coin 80 times, we would not know what to expect, because $H2$ does not give us a specific value of p with which to work. Would we expect 50 heads? 15 heads?

Statisticians usually test the hypothesis which tells them *what to expect* by giving a specific value to work with. They refer to this hypothesis as the **null hypothesis** and symbolize it as H_0. The null hypothesis is often the one that assumes fairness and honesty. It seems to look at the world through rose-colored glasses. This die is *fair*. This newspaper claim is *true*. This theory is *correct*. The opposite hypothesis is called the **alternative hypothesis** and is symbolized by H_a. This hypothesis, however, is often the one that is of interest.

Hypothesis testing: binomial one-sample

We *suspect* that the die is biased, that the newspaper erred, that the theory is wrong. It is often this suspicion that moved us to investigate the question in the first place. Some statisticians refer to H_a as the **motivated** hypothesis.

Go back to the two hypotheses for questions 3, 4, and 5 above. Can you pick out which hypothesis is the null hypothesis? That is, which one, by giving you a specific value to work with, would tell you most clearly what to *expect* if it were true?

For question 3, if you interviewed 200 shoppers, how many would you expect to respond, "Yes, I buy my favorite brand of toothpaste regardless of price"? If $H1$ is true, you would expect about 60 shoppers to say yes, since 30 percent of 200 is 60. If $H2$ is true, you do not know with much precision what to expect, except that it is not 60.

For question 4, if you rolled the die 60 times, how many 3s would you expect to occur? For question 5, suppose that you tested a group of 50 boys and a group of 50 girls and found the difference between the average score of each group. What should you expect the difference to be?

In each of these examples the null hypothesis would be $H1$, because if $H1$ were true, we would know what to expect. In example 3 we would expect about 30 percent of 200, or 60, shoppers to say yes. In example 4 we would expect about 1/6 of 60, or ten 3s to occur in 60 tosses of a fair die. In example 5 we would expect the difference between the two means to be near 0.

To summarize, the null hypothesis gives us a specific value of a population parameter on which to base our expectations. This is reflected by the appearance of an equals sign ($=$) when we use symbols to write it. The alternative hypothesis in symbols is often characterized by the appearance of a not equal sign ($\neq$) or an inequality sign ($>$, $<$).

ONE-TAIL AND TWO-TAIL TESTS

If you suspect that a certain null hypothesis is false, you can formulate three different alternatives. Suppose that you read in *Pets* magazine that 34 percent of the people in Goatemola own more than 2 pets, and you wonder if that same percentage would hold true in your locality, North Southtown. Your null hypothesis would be that the 34-percent figure *is* true.

Let $p = P$(a person in North Southtown has more than 2 pets). Then H_0 is $p = .34$.

Your alternative hypothesis could be any one of the following.

1. If you think that p is greater than .34, you would write

   $H_a: p > .34$

2. If you suspect that p is less than .34, you would write

   $H_a: p < .34$

3. If you don't have any idea whether p is larger or smaller than .34, then you could write $p \neq .34$.

In the first instance, you would only be interested in values of p greater than 34 percent, and in the second instance you would only be interested in values of p less than 34 percent. These are referred to as **one-tail tests,** since the values of interest are in only *one* direction away from 34 percent. The third

Understanding statistics

instance, however, is referred to as a **two-tail test**, since values far from 34 percent in *either* direction are of interest to you in your experiment.

Note that we have formulated our hypotheses so that the equals sign (=) always appears in the null hypothesis, while either the less than (<) or the greater than (>) sign appears in the alternative hypothesis for a one-tail test. The alternative hypothesis for a two-tail test always contains the not equal sign (≠). The choice of either a one-tail or a two-tail test is determined by what the statistician is interested in finding out.

EXAMPLE 8-1 Formulate the null hypothesis and the alternative hypothesis for each of the following.

(a) Is the average life span of a dog more than 13 years?

(b) Is the proportion of 18-year-old drivers who have accidents the same as the proportion of 26-year-old drivers who have accidents?

(c) What percentage of people born with Down's syndrome can be taught to read?

(d) Do teenage girls receive a smaller weekly allowance than teenage boys?

SOLUTION (a) Let μ = the average life span of dogs,

$H_0: \mu = 13$
$H_a: \mu > 13$ (one-tail test)

(b) Let $p_1 = P(\text{18-year-old driver has an accident})$
$p_2 = P(\text{26-year-old driver has an accident})$

$H_0: p_1 = p_2$ or $p_1 - p_2 = 0$
$H_a: p_1 \neq p_2$ or $p_1 - p_2 \neq 0$ (two-tail test)

(c) This calls for a numerical response, not a hypothesis test.

(d) Let μ_1 = average weekly allowance for teenage girls and μ_2 = average weekly allowance for teenage boys,

$H_0: \mu_1 = \mu_2$ or $\mu_1 - \mu_2 = 0$
$H_a: \mu_1 < \mu_2$ or $\mu_1 - \mu_2 < 0$ (one-tail test)

■

EXERCISES

8-1 A claim is made that 40 percent of all viewers watch the TV show "Mathematics and Humor." The null hypothesis to test this claim would be $p = .40$. What would the motivated hypothesis be for:

(a) A statistics student who was considering testing the truth of the claim?
(b) A prospective sponsor who was considering advertising on the show?
(c) Al Jebra, the show's star, who is considering asking for a raise?

In Exercises 8-2 to 8-8 formulate the two hypotheses where applicable, and decide whether the situation calls for a one-tail or a two-tail test.

8-2 Is the average work week in Centerville less than 40 hours?

8-3 Do more than 10 percent of pet owners own goldfish?

8-4 What is the average jail sentence for bank robbery?
8-5 Do 12 percent of the students in your school major in mathematics?
8-6 Is the average height of 6-year-old male horses the same as the average height of 6-year-old female horses?
8-7 What is the average weight of adult baboons?
8-8 Do a greater percentage of 20-year-old females diet than do 20-year-old males?

DECISION RULES

At the beginning of an experiment you should formulate the two opposing hypotheses. Then you should state what evidence will cause you to say that you think the alternative hypothesis is the true one. This statement is called your **decision rule**. When the evidence supports the alternative hypothesis, we say that we "reject the null hypothesis." When the evidence does not support the alternative, we say that we "fail to reject the null hypothesis."

EXAMPLE 8-2 Guildenstern suspects that a certain coin is biased for heads. She decides to test it by tossing it 40 times. Her null hypothesis is

H_0: coin is fair, $p = P(\text{heads}) = .5$

Her alternative hypothesis is

H_a: coin is biased for heads, $p > .5$

She reasons that if it is fair, then she should get about 20 heads. She makes the following decision rule: if the 40 tosses produce 25 or more heads, then conclude that the coin is biased in favor of heads. If she lets S stand for the number of heads she will get, then her decision rule is: if S is more than 25, ■ reject H_0.

EXAMPLE 8-3 Amir Treifel is testing the null hypothesis that the mean income of sheiks is 1.5 million dollars per year, H_0: $\mu = 1.5$ million dollars; his alternative hypothesis is H_a: $\mu \neq 1.5$ million dollars. He decides that he will reject his null hypothesis if the mean income for a random sample of sheiks turns out to be either less than 1 million or more than 2 million dollars. If he lets m stand for the mean of his sample, then, briefly, his decision rule is: if m is less than 1 ■ million or more than 2 million dollars, reject H_0.

STATISTICAL ERRORS

One basic idea which is inseparable from hypothesis testing is that you can almost never have *absolute* proof as to which of the two hypotheses is the true one. For example, in the case of testing a coin to see if it is fair, you must realize that the very definition of "fair coin" makes it impossible to completely test a coin. Recall that a fair coin is one for which the probability of heads on a single toss equals .5. But, "a probability equal to .5" means that the coin comes up heads one-half the time *in the long run,* and "in the long run" means you have to keep tossing the coin *forever.* So any time *you* test a coin, no matter how many times *you* toss it, it's only a small portion of how many times it *could* be tossed.

Suppose our friend Guildenstern tosses her coin 40 times and gets 30 heads. She will say the coin is not fair (remember her decision rule). Who knows what *might* happen if she *kept on* tossing the coin? It's possible that Guildenstern does in fact have a fair coin, but her run of extra heads fools her. If so, the data have led her to make a mistake through no fault of her own. This is an unavoidable possibility in all hypothesis testing, because we look at only a sample and not the whole population. We can *reduce* the probability of this happening by collecting more data, but we have to stop sometime, and so there will always remain some possibility of a sampling error.

When you are testing a null hypothesis, you are trying to decide if it is true or false. However, since statistical hypothesis testing is based on sample information and you cannot be absolutely sure that your decision is correct, you really are faced with four possible situations.

1. H_0 is true and the sample data lead you correctly to decide that it is true.
2. H_0 is true but by bad luck the sample data lead you mistakenly to think that it is false.
3. H_0 is false and the sample data lead you correctly to decide that it is false.
4. H_0 is false but by bad luck the sample data lead you mistakenly to think that it could be true.

In the first and third situations above, you have been led to make a correct decision. In the second situation you rejected a *true* null hypothesis. We refer to this as a **Type I error**. In the last situation, you failed to reject a *false* null hypothesis. Statisticians call this a **Type II error**. This can be summarized in Table 8-1.

Table 8-1

	you fail to reject H_0	you reject H_0
H_0 is true	correct	Type I error
H_0 is false	Type II error	correct

We use the first letter of the Greek alphabet, alpha (α), to represent the probability of making an error of Type I. Similarly, the second letter, beta (β), represents the probability of an error of Type II.

EXAMPLE 8-4 The sheriff in a gambling town prefers to believe that the roulette wheel is honest rather than take the chance of arresting the croupier and being charged with false arrest. If we think of the statement, "the wheel is honest" as a null hypothesis, does the sheriff prefer to make a Type I or a Type II error?

SOLUTION H_0: wheel is honest
H_a: wheel is crooked
Type I error: Sheriff claims the wheel is crooked, but it is honest
Type II error: Sheriff accepts the wheel as honest, but it is really crooked

■ The sheriff prefers to make a Type II error.

EXERCISES

8-9 Mary did a statistical analysis for her avionics instructor and ended up rejecting her null hypothesis. Mart says: "You could not possibly have made a

Type II error." Marv interjects: "Therefore you must have made a Type I error."
Are the boys correct? Explain.

In Exercises 8-10 to 8-17 think of statements such as "this person is innocent," "this drug is safe," or "this product is priced fairly" as null hypotheses.

8-10 If a judge prefers to let the guilty go free rather than run the risk of punishing the innocent, does he prefer to make Type I or Type II errors?

8-11 If a school principal prefers to believe that a student is guilty unless the accused student can prove his or her innocence, does the principal prefer to make a Type I or a Type II error?

8-12 If a pharmaceutical firm prefers to not sell a drug rather than risk selling one with bad side effects, does the firm prefer to make Type I or Type II errors?

8-13 If a person reacts positive to a medical screening test for a disease, he is given further tests to decide if he has the disease. Typically these tests are more difficult, time consuming, and expensive to administer than the screening test.

In such a screening test two "errors" similar to Type I and Type II can occur. The test may indicate that a well person is sick. It may indicate that a sick person is well.

(a) If you were designing the screening test, which error would you be more concerned about? Does it depend on the disease?

(b) A screening test for glaucoma measures the pressure of some fluid in the eye. If the pressure is over a certain value, say 1000, that counts as "positive," and the person is given further tests. Discuss what would happen if the test were changed so that the new cutoff was 900. The figure shown will be helpful in solving the problem.

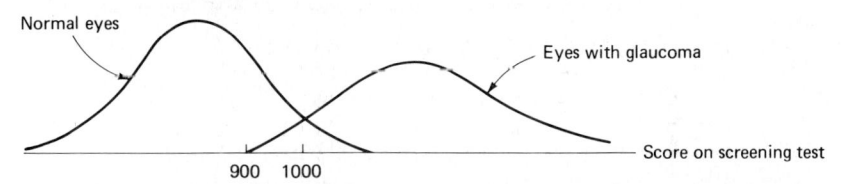

8-14 A basketball coach screens new candidates by having them shoot 30 shots from a distance of 20 feet. She rejects the player if less than 20 shots are good. If the coach decides to have the candidates shoot from a distance of 25 feet, is she increasing or decreasing α?

8-15 A biology teacher is performing an experiment which is to show that a new curriculum is superior to what had been done in the past. The hypotheses are:

H_0: new curriculum is equal in value to former method

H_a: new method is superior in value to former method

In the eyes of the school board, how would the value of α depend on whether the new curriculum utilizes the present laboratory equipment or demands a large expense for new equipment?

8-16 A penny and a centavo, which are fair coins, and a sen and a pfennig, which are biased for tails, are all tossed 40 times. The results are listed below.

	number of heads	conclusion
penny	21	fair
centavo	33	biased for heads
sen	18	fair
pfennig	6	biased for tails

Understanding statistics

(a) Which situation is a Type I error?
(b) Which situation is a Type II error?

8-17 Detective Sgt. Wednesday of the rackets squad confiscated 4 spinners, each divided into 7 equal areas. Of them, 2 are honest and 2 are biased. Bored while waiting for the grand jury hearings to begin, she begins spinning each one 70 times. For each spinner she records the number of times area 6 wins. In table form, list the four situations that might occur and indicate the Type I and Type II errors.

MORE ABOUT TYPE I ERRORS

EXAMPLE 8-5

In Example 8-2 Guildenstern used the following decision rule: I will conclude that the coin is biased if I get 25 or more heads in 40 tosses. Of course, it is possible for even a fair coin to fall heads 25 or more times just by chance, but it is not highly probable. If the coin were fair and fell heads more than 25 times, she would mistakenly call the coin biased and thereby make a Type I error. Find the probability that using her decision rule Guildenstern will mistakenly decide that a fair coin is biased. That is, find $P(S \geq 25)$.

SOLUTION

This is a binomial experiment with $n = 40$.

H_a: coin is biased for heads, $p = P(\text{heads}) > .5$ (one-tail test)
H_0: coin is fair, $p = P(\text{heads}) = .5$

We take the value of p from the null hypothesis H_0:

$p = P(\text{heads}) = .5$ and $q = .5$

Since $np = 20 > 5$ and $nq = 20 > 5$, the normal approximation may be used with

$\mu = np = 20$ and $\sigma = \sqrt{npq} = \sqrt{10} = 3.16$

If S represents the number of heads, then we want to find $P(S \geq 25)$. Since we are using the normal approximation, we will find $P(S > 24.5)$. This is shown in Figure 8-1.

Figure 8-1

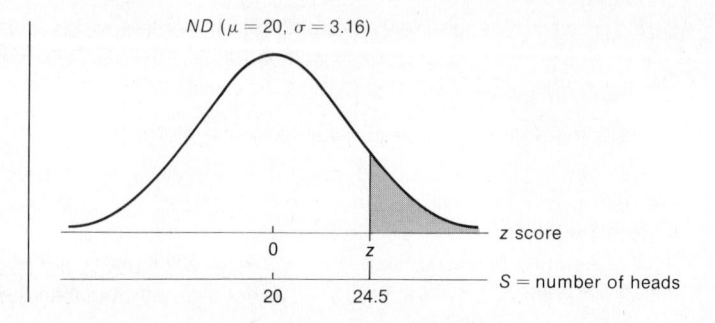

$$z_{24.5} = \frac{24.5 - 20}{3.16} = \frac{4.5}{3.16} = 1.42$$

Area($z = 1.42$) = .9222

Hypothesis testing: binomial one-sample

Therefore, $P(S \geq 25) = 1 - .9222 = .0778$, or only about 8/100. In other words, the probability of a Type I error is about .08, and we would write $\alpha = .08$. Some researchers refer to this as the **p value** for the experiment. If Guildenstern tosses this coin 40 times and gets 25 or more heads, she will reject the null hypothesis. In doing so she can be pretty confident that the outcome is due to the fact that the coin is biased, and that the result is not just the chance outcome of a fair coin.

Remember that any time we reject a null hypothesis, no matter what decision rule we use, there is always some probability, no matter how small, that the rejection is in error due to bad luck.

FINDING THE DECISION RULE THAT WILL CORRESPOND TO A GIVEN VALUE OF α

In Example 8-5, α was .08. Now perhaps Guildenstern would not be satisfied with a probability of error as large as .08. She might wish to do the problem in reverse. That is, she would *first* state how large a risk of a Type I error she is *willing* to accept (say, .01). This is called setting the **significance level** of the test (at .01). *Then* she would compute a decision rule to satisfy that requirement. The decision rule will tell her how many heads must appear in order to reject the null hypothesis so that the probability of a Type I error will be no larger than .01. We illustrate that procedure now (Figure 8-2).

Figure 8-2

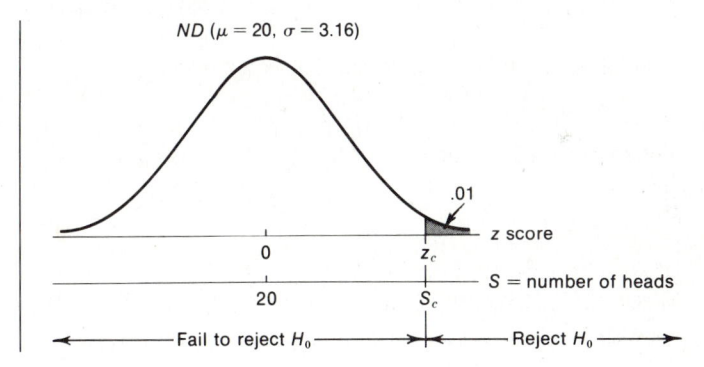

$ND (\mu = 20, \sigma = 3.16)$

.01

z score — z_c

S = number of heads — S_c

20

Fail to reject H_0 ← → Reject H_0

Since we wish α to be .01, we look up the area of $1 - .01$, or .99. (Recall this is a one-tail test on the right.) This area corresponds to $z_c = 2.33$. We use the subscript c to indicate that this is a **critical value**. A critical value is the value that marks the start of the rejection region.

We transform this z score into a number of heads as follows:

$$S_c = \mu + z_c \sigma$$
$$= 20 + 2.33(3.16) = 20 + 7.4 = 27.4$$

Therefore, if Guildenstern sets $\alpha = .01$, she will have to get more than 27.4 heads, that is, at least 28 heads, to reject the null hypothesis and conclude that the coin is biased for heads.

EXAMPLE 8-6 Tenacious Tom, while recovering from an unfortunate accident to his hands, is once more trying to make a pair of loaded dice. He has altered one die in various secret ways. He thinks that the probability of getting a 6 on this die is now changed, but, because of his inexperience, he is not sure if it will now get too many 6s or not enough 6s. He decides to roll it 60 times and count the number of 6s that appear. What can he use for the decision rule if he decides to test the die at the .05 significance level?

SOLUTION In order to determine the decision rule for this hypothesis test, he first states the hypotheses in terms of $p = P(6)$:

H_a: die is biased, $p \neq 1/6$ (two-tail test)
H_0: die is fair, $p = 1/6$

Assuming that the null hypothesis is true,

$p = P(6) = 1/6$
$q = 5/6$
$n = 60$

Since $np = 10 > 5$ and $nq = 50 > 5$, the normal approximation may be used:

$\mu = 10$ and $\sigma = \sqrt{8.33} = 2.89$

Because he suspects that the die could be biased either for *or* against 6, he might end up with either too few or too many 6s. He divides the significance level of 5 percent into the two tails of the normal distribution as shown in Figure 8-3.

Figure 8-3

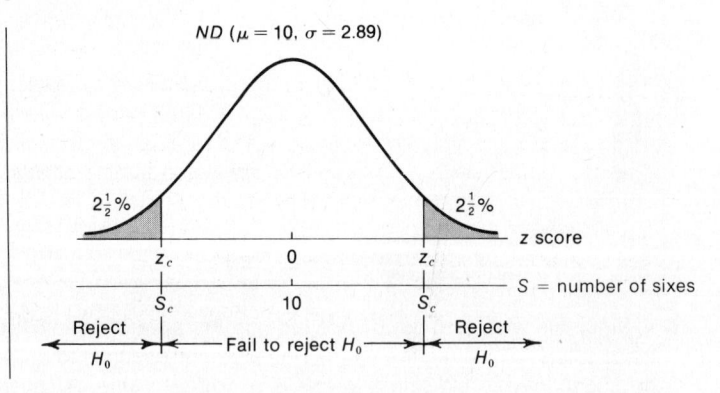

Looking up areas of .025 and .975, he gets the critical z scores as ± 1.96. To convert these to numbers of 6s he uses

$S_c = \mu + \sigma z_c$
$= 10 + 2.89(\pm 1.96) = 10 \pm 5.7 = 15.7$ and 4.3

Therefore his decision rule will be: reject the null hypothesis if he gets less than 4.3 sixes or more than 15.7 sixes. If this happens, he can conclude that the die is biased. The probability of making a Type I error is .05.

■

An experimenter can set the significance level at any value, but in most published reports it is more or less standard to use .01 and .05 as significance

Hypothesis testing: binomial one-sample

levels. These probabilities of error are small enough to be considered reasonable in most circumstances.

EXERCISES

8-18 You decide. Which option would you choose?
(a) A coin is tossed 100 times, resulting in 97 heads.

option 1	option 2
It is a fair coin, and I have just witnessed a very extraordinary event.	It is not a fair coin, and I have just witnessed an ordinary event for the tosses of this biased coin.

(b) A sample of 500 teenagers shows that 25 percent more girls smoke than boys.

option 1	option 2
There is no difference in the smoking habits of boys and girls and this sample is highly unlikely to have been picked randomly.	This sample is representative of all teenagers. I conclude that it is highly unlikely that the same percentage of boys and girls smoke.

(c) A sample is taken from a population, and the sample statistic is nowhere near some hypothetical population parameter.

option 1	option 2
This was a highly unlikely event, and so I have just witnessed a near miracle, but I see no reason to reject the proposed parameter.	This was a representative sample, an ordinary event, from a population with a different parameter. The proposed parameter is most likely not correct.

8-19 Danny Dropout and Sally Student disbelieve a newspaper claim that says 30 percent at Happy High cut class at least once a week. They think 30 percent is too high. They decide to test this using a random sample of 200 students. Danny says, "I'll reject the claim if our sample contains less than 50 cutters." Sally, who studied statistics, says, "I'll use $\alpha = .05$." Who has the larger risk of making a Type I error?

8-20 A coin is suspected of being biased.
(a) State the two hypotheses.
(b) The coin is tossed 40 times. How many or how few heads are needed to establish statistically that the coin is biased if $\alpha = .05$?
(c) Repeat part (b) with $n = 100$.

8-21 A researcher is beginning to suspect that less than 3 percent of all genitz are pibled.
(a) State the two hypotheses.
(b) If 1000 genitz are sampled, how few must be pibled to convince us that the suspicion is correct at the .01 significance level?

8-22 A statistician is testing the claim that a certain population is one-fourth male, three-fourths female. She is going to select 80 persons at random from the population. Her decision rule is: if she gets fewer than 15 or more than 25 males, she will reject the claim that $p = P(\text{male}) = .25$. Suppose it is true that p does equal .25. What is the probability that her results will lead her to mistakenly reject that fact? That is, what is α? (Be sure to draw a clearly labeled normal curve.)

8-23 The local chapter of Women Against the Exploitation of Females (WAEF) is picketing an "art" theater. They claim that at least 75 percent of the people who view such films are men. A newspaper reporter believes that this figure is too high. He gathers a random sample of 100 people from the audiences at the theater. He will use as his decision rule: if the sample contains fewer than 60 men, reject the claim made by WAEF.
(a) State the motivated hypothesis.
(b) State the null hypothesis.
(c) Draw the normal curve including the z score line, the line for the number of successes, and a shaded rejection region.
(d) Find the probability of a Type I error.

8-24 Governor U. R. Careless claims that the ILLR railroad is seldom late. A company official claims that only 4 percent of its trains arrive more than 5 minutes late. A statistician hired by an irate commuter group gathers a random sample of 500 train arrivals. According to the company figures he would expect about 20 trains to be late. He is willing to give the company the benefit of the doubt, and so his decision rule is: if more than 40 trains are late, reject the company's claim.
(a) State the motivated hypothesis.
(b) State the null hypothesis.
(c) Draw the normal curve including the z score line, the line for the number of successes, and a shaded rejection region.
(d) Find the probability of a Type I error.
(e) If the decision were: reject H_0 if more than 30 trains are late, find α.
(f) If in part (e) you change the number 30 to 50, would α increase or decrease?
(g) What decision rule gives $\alpha = .05$?

8-25 The population of Smalltown is 42 percent female. A statistician is trying to establish the claim that more than 42 percent of the Republicans in Smalltown are women. Assuming that sex has nothing to do with political affiliation, then $p = P(\text{a Republican is female}) = .42$. We select 100 Republicans at random. Using the .01 significance level, find the decision rule for the number of women that must occur in the sample before we reject the assumption.

8-26 A newspaper article states that 60 percent of children in the age bracket 1 to 4 who die do so as a result of motor accidents. Doubting that this is true, a public health official gathered information on the cause of death of 30 randomly selected children.
(a) State the motivated hypothesis.
(b) State the null hypothesis.
(c) How many or how few of the 30 deaths should be due to motor vehicles in order to reject the claim of the article at the .01 significance level?

8-27 A manufacturer claims that the mixed nuts that he sells have only 30 percent peanuts. We open a large bag and select 100 nuts at random and find that 36 of them are peanuts. If p does equal .30, what is the probability that 100 nuts picked at random contain 36 or more peanuts? Would you be willing to accuse the manufacturer of a false claim?

8-28 A gambler tosses a *fair* coin 10 times and gets 8 heads. He makes a Type I error in stating that the coin is biased. Find the probability that a fair coin tossed 10 times will produce 8 or more heads.

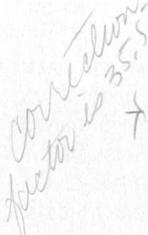

Hypothesis testing: binomial one-sample

MORE ABOUT TYPE II ERRORS

(*This section may be omitted without loss of continuity.*)

We have just discussed the Type I error, which occurs when statistical evidence leads us to reject a null hypothesis when in reality the null hypothesis is true. Recall that the Type II error occurs when the null hypothesis is false but the statistical evidence is not strong enough to indicate it. That is, we have mistakenly failed to reject the null hypothesis. The probability of a Type II error is denoted by β. If the value of p given in the null hypothesis is wrong, then some other particular value of p is correct. Recall that the alternative hypothesis did not give us one specific value of p. Therefore, for each possible value of p there is a corresponding value of β.

EXAMPLE 8-7 Suppose you are asked to test a coin. You decide to toss it 60 times, reasoning that if it is fair, you will get about 30 heads. You choose for your decision rule: the coin is not fair if the number of heads is less than 26 or more than 34.

If, unknown to you, the coin *is* biased and $p = P(\text{heads}) = .6$, then what is the probability that your experiment will yield between 26 and 34 heads anyway? That is, what is the probability that you commit a Type II error?

SOLUTION

$p = .6$

$q = .4$

$n = 60$

$np = 36 > 5$

$nq = 24 > 5$

Therefore, the normal approximation may be used:

$\mu = 36$

$\sigma = \sqrt{14.4} = 3.79$

$P(26 \leq S \leq 34)$ is approximately equal to $P(25.5 < S < 34.5)$; see Figure 8-4.

Figure 8-4

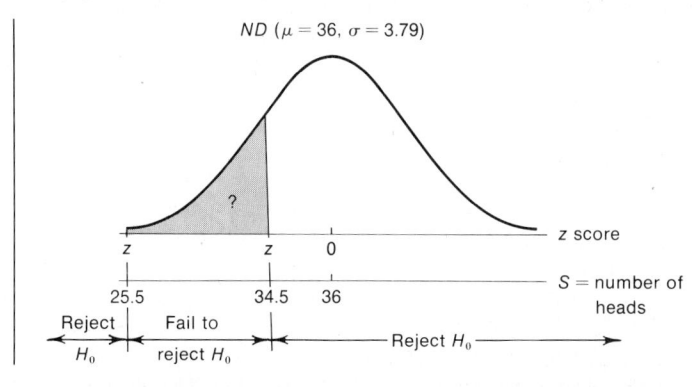

$$z_{34.5} = \frac{34.5 - 36}{3.79} = -.40$$

$$z_{25.5} = \frac{25.5 - 36}{3.79} = -2.77$$

Area($z = -.40$) = .3446

Area($z = -2.77$) = .0028

■ Therefore, $P(26 \le S \le 34) = .3446 - .0028 = .3418$, or about .34.

The probability is .34, or about 1/3, that we would accept the statement "this coin is fair," when in fact the coin is biased with $p = .6$. We write $\beta = .34$. You see that our decision rule is not very powerful for distinguishing between fair coins and coins biased with $p = .6$. You will notice that we needed the probability $p = .6$ in order to compute β. We cannot compute a β until we choose a specific value of p from the alternative hypothesis. We could repeat the above calculation for other values of p, such as .7, .8, etc., in order to get an idea of the range of possible values for the Type II error. This contrasts with the case of the Type I error, where, as soon as the decision rule is made, α is known.

EXAMPLE 8-8 An experimenter believes that a coin is biased in favor of heads. He wants to test the claim "the coin is fair" at the .05 significance level by tossing the coin 40 times.

(a) Find α.

(b) Find the decision rule.

(c) Find β.

SOLUTION (a) $\alpha = .05$ because the experimenter *chose* it that way.

(b) $H_a: p = P(\text{heads}) > .5$
$H_0: p = P(\text{heads}) = .5$

$p = .5$ (from the null hypothesis "the coin is fair")
$q = .5$
$n = 40$
$np = 20 > 5$
$nq = 20 > 5$
$\mu = 20$
$\sigma = \sqrt{10} = 3.16$
$\alpha = .05$

The normal approximation may now be used, as shown in Figure 8-5. Looking up the area .9500, we find $z_c = 1.65$. Converting this z score to a raw score, we have

$$S_c = 20 + 1.65(3.16) = 25.2$$

Therefore, his decision rule is: reject the null hypothesis that the coin is fair if the coin turns up heads more than 25.2 times. In this problem, since we are using the continuous curve to approximate a distribution of whole numbers, we would have to interpret "more than 25.2 heads" as any number of heads from 26 up. However, for the purposes of evaluating β

Hypothesis testing: binomial one-sample

Figure 8-5

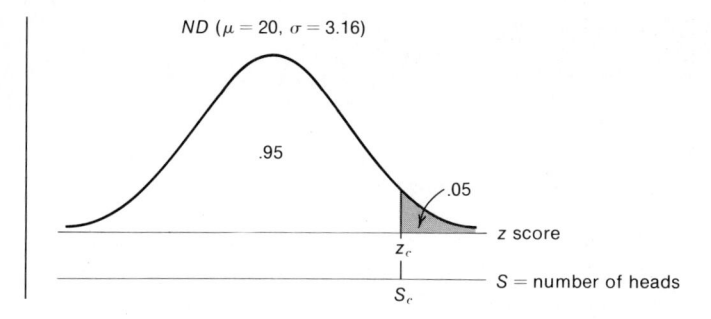

ND ($\mu = 20$, $\sigma = 3.16$)

.95

.05

z_c — z score

S_c — S = number of heads

it is all right to continue to think of S_c as 25.2. This simplifies the computations and makes very little difference in our final results. Later on, in problems involving continuous variables from the beginning, we will be able to proceed in exactly the same manner.

(c) A specific value of β cannot be computed until we have decided upon a specific value of p from the alternative hypothesis. ∎

EXAMPLE 8-9 In Example 8-8, suppose that in reality p were approximately .7. Now compute β, the probability that the evidence will indicate that this biased coin is *fair*.

SOLUTION The decision rule comes from our choice of α and was computed in part (b) of Example 8-8 under the assumption that $p = .5$ (see Figure 8-6). The decision rule was: reject H_0, the claim that the coin is fair, if the coin turns up heads more than 25 times. But if we now suppose that $p = .7$, we get a different distribution:

Figure 8-6

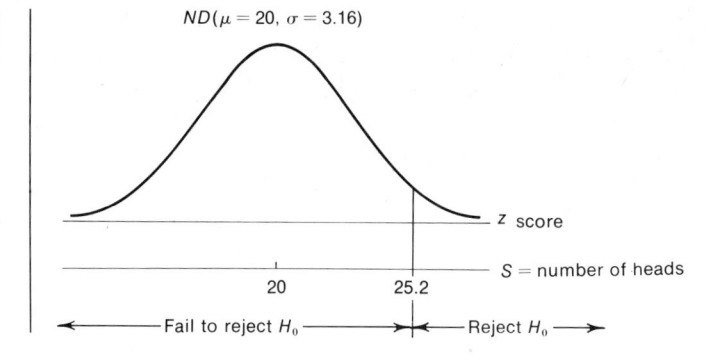

ND($\mu = 20$, $\sigma = 3.16$)

20   25.2

←—— Fail to reject H_0 ——→ ←— Reject H_0 —→

z score

S = number of heads

$p = .7$
$q = .3$
$n = 40$
$np = 28 > 5$
$nq = 12 > 5$
$\mu = 28$
$\sigma = \sqrt{8.4} = 2.9$

Understanding statistics

Figure 8-7

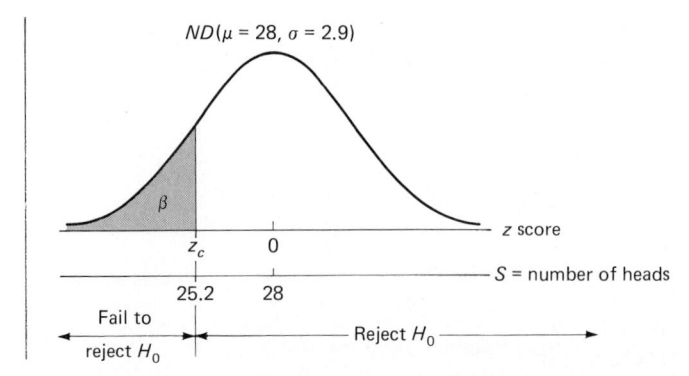

$ND(\mu = 28, \sigma = 2.9)$

z_c : 25.2
0 : 28

Fail to reject H_0 | Reject H_0

z score
S = number of heads

Recall that in this problem β measures the probability that a *biased* coin will produce results that are *in support* of the *null* hypothesis. Therefore, we want to measure the area of the graph that corresponds to those outcomes that support the null hypothesis that the coin is fair, namely, the area to the *left* of 25.2 (see Figure 8-7):

$S_c = 25.2$

Converting 25.2 to a z score in order to measure the area, we have

$$z_{25.2} = \frac{25.2 - 28}{2.9} = \frac{-2.8}{2.9} = -.97$$

Area($z = -.97$) = .1660 = .17

Therefore $\beta = .17$.

Using this decision rule, there is a probability of .17 that the experimenter will mistakenly decide that the coin is fair if in fact it is biased with $p = .7$.

In the example just discussed, if we compute β for $p = .6$ and $p = .8$, we get values of β equal to .65 and .004, respectively. Further, if we start over again with $\alpha = .01$, then we find that $S_c = 20 + 2.33(3.16) = 27.4$. For *this* decision rule the values of β for $p = .6, .7,$ and .8 turn out to be .86, .42, and .03. We can summarize these results in Table 8-2.

Table 8-2

Values of β for $n = 40$

true value of p	$\alpha = .05$	$\alpha = .01$
.6	.65	.86
.7	.17	.42
.8	.004	.03

Thus you can see by reading across each row of this table that *for a fixed value of n*, if we decrease α, we will increase β, and vice versa.

Now let us see what happens to β when we change sample size. If we keep $\alpha = .05$ and let $n = 60$, the values of β can be found as follows. First, we establish the decision rule assuming that the null hypothesis ($p = .5$) is true. We get

$S_c = 30 + 1.65(3.87) = 36.4$

Hypothesis testing: binomial one-sample

We will reject H_0 if we see more than 36.4 heads. Now that we have a decision rule, we can compute β. The calculation for $p = .7$ is

$$np = 42$$
$$nq = 18$$
$$z_{36.4} = \frac{36.4 - 42}{3.55} = -1.58$$
$$ND_S(\mu = 42, \sigma = 3.55)$$
$$\beta = .06$$

Similarly, values of β for $p = .6$ and $.8$ can be found. We summarize these results in Table 8-3.

Table 8-3 Values of β for $\alpha = .05$

true value of p	$n = 40$	$n = 60$
.6	.65	.54
.7	.17	.06
.8	.004	.0001

You can see that increasing the sample size decreases the value of β without increasing the value of α. In performing a statistical experiment, if you have the time, you can often make both α and β as small as you want by using a sufficiently large value of n. For example, suppose you have $H_0: p = .5$ and you want to be reasonably sure to reject H_0 if the true value of p is anything higher than .6 (a one-tail test). Then by making the sample large enough, you can have, for instance, *both* α and the *maximum* β equal to .05. (You might try to find how big a sample is needed to do this. From Table 8-3 you can see that n must certainly be bigger than 60.)

EXERCISES

8-29 In Exercise 8-19 Danny Dropout said, "I'll reject the claim if the sample contains less than 50 cutters." The problem essentially asked you to find α. Why can we not simply ask you to find β?

8-30 Verify in Table 8-2 the value of β for $n = 40$, $p = .6$, $\alpha = .05$.

8-31 Verify in Table 8-2 the value of β for $n = 40$, $p = .8$, $\alpha = .05$.

8-32 Verify in Table 8-3 the value of β for $n = 60$, $p = .6$, $\alpha = .05$.

8-33 Verify in Table 8-3 the value of β for $n = 60$, $p = .8$, $\alpha = .05$.

8-34 Connie Consumer claims that 30 percent of Never Fail Brand spark plugs are defective. Sparky the mechanic claims that the rate of defectives is lower. They decide to examine a random sample of 100 NFB plugs.
(a) Using $\alpha = .05$, find their decision rule.
(b) If in fact 20 percent of the NFB plugs are defective, what is the probability that the evidence will mistakenly lead them to believe that Connie is correct?

8-35 Repeat Exercise 8-34 if $H_a: p > .30$.

8-36 Repeat Exercise 8-34 if $H_a: p \neq .30$.

8-37 Refer to the story of Tenacious Tom in Example 8-6. If p is really .35, what is the probability that Tom will be misled into believing the dice are fair?

THE POWER OF A TEST

Statisticians refer to the value of $1 - \beta$ as the **power** of a test. The power of a test is a measure of how good the test is at rejecting a false null hypothesis. The more "powerful" a test is (the closer the value of $1 - \beta$ is to 1), the more likely the test is to reject a false null hypothesis.

An important part of statistical theory deals with the problem of finding a decision rule that will make a hypothesis test as powerful as possible for any given value of α. The original theoretical work in this area was done by J. Neyman and E. S. Pearson in the 1930s.

EXAMPLE 8-10

Suppose we have a binomial test where

H_0: $p = .6$
H_a: $p < .6$
$\alpha = .05$
$n = 50$

Then, for example, we can compute the decision rule and values of β corresponding to $p = .5$ and .4. The values of the power of the test would then be given in Table 8-4.

Table 8-4

p	β	power
.5	.56	.44
.4	.10	.90

EXERCISES

8-38 Verify the power of the above test for $p = .4$.

8-39 AGRI magazine recently reported the results of two studies on fertilizer effectiveness on corn. Dr. Bulschmidt reported that he "failed to reject" his null hypothesis at the .05 level. In the second article Dr. Senserd reported that using a test with significance level .05 and a power better than 96 percent, he "accepted the null hypothesis."
(a) Why have they both reported their results correctly?
(b) What is the difference in what they have learned?

8-40 In Example 8-10, change H_a to $p > .6$ and find the power of this new test for the indicated values of p.

p	β	power
.5		
.6		
.7		
.8		
.9		

8-41 Change H_a to $p \neq .6$ in Example 8-10 and find the power of this new test for the indicated values of p.

Hypothesis testing: binomial one-sample

p	β	power
.4		
.5		
.7		
.8		

8-42 In Middlesex, Massachusetts, 100 extraterrestrial humanoids were sampled to test whether more than 20 percent were neither male nor female.
(a) Find the critical number of successes for $\alpha = .01$.
(b) If the true fraction is 25 percent, find β.
(c) If the true fraction is 18 percent, find β.
(d) If the true fraction is 15 percent, find β.

8-43 A coin is tossed 50 times to test the null hypothesis that the coin is fair.
(a) If α is set at .05, find the two critical values for the decision rule.
(b) If it happens to be true that $p = P(\text{heads}) = .8$, find β.
(c) If the coin is tossed 100 times instead of 50 times, do (a) and (b) again.
(d) How is β affected by the increase in the sample size?

8-44 Let us reconsider Exercise 8-20(b). We decided that if we got more than 26 heads or fewer than 14 heads in 40 tosses, we would call the coin biased. If the true proportion of heads was really $p = .25$, find the probability of a Type II error.

8-45 A cancer research group wished to test the hypothesis that 40 percent of all college students smoke cigarettes. Using a .05 significance level it was found that the critical points of the decision rule were 13.5 and 26.5 smokers for a random sample of 50 students. If the true proportion of smokers was 45 percent, find the probability of a Type II error.

8-46 Repeat Exercise 8-45, but this time find the probability of a Type II error if the true proportion of smokers was 70 percent.

8-47 Max, the bartender at Bernie's Paragon Café, claims that only one-fourth of the customers can differentiate between Northern Comfort and Rot Gut. Don, the barfly, makes a bet that the figure is higher. They gather a random sample of 25 customers. Each customer is given a glass of each liquor and asked to identify both of them.
(a) Using $\alpha = .05$, how many customers would have to be correct for Don to win the bet?
(b) After the experiment is over, Don feels that p is probably between .40 and .50. Assuming that this is true, what was the largest value of β in the hypothesis test of (a)?

8-48 A manufacturer of magic kits makes two types of "coins." One has $p = P(\text{heads}) = .8$, and the other is a fair coin with $p = P(\text{heads}) = .5$. An employee inadvertently mixed 100 of the fair coins with 200 of the biased coins. His employer instructs him to sort them. Since the coins look alike, the employee decides to toss each coin 30 times. The fair coin he expects to come up heads about 15 times. On the other coin he expects about 24 heads. He decides that if he tosses 19 or fewer heads, he will call the coin fair, but if he tosses 20 or more heads, he will call the coin biased. When he has finished sorting the coins into two piles,
(a) About how many coins are in each pile?

Understanding statistics

(b) About how many fair coins are in the biased pile? (How many Type I errors?)
(c) About how many biased coins are in the fair pile? (How many Type II errors?)
(d) How could he improve on his results?

8-49 A counterfeiter has a pile of mixed counterfeit and real coins that look alike. A detective, Michelle Ignatius Gallagher, is on his trail. Needing to buy some food, the counterfeiter wishes to use only real coins so that he will not get caught. He decides to toss each coin 20 times. On the real coins he expects about 10 heads, but the counterfeit coins are not well balanced and do not have $p = P(\text{heads}) = .5$. He decides to consider a coin real if he gets 9, 10, or 11 heads and counterfeit otherwise.

(a) If he sorts through 100 coins this way, about what percentage of the real coins will he reject?
(b) What percentage of the counterfeit coins will he accept?
(c) He decides to move his hideout, and he wants to transport as many real coins as possible. He decides he will separate them by tossing each one 20 times. Which of the following three decisions should he make to ensure the largest percentage of good coins in the "real" pile: (1) Leave the decision rule at 9 to 11 heads. (2) Reduce it to 10 heads only. (3) Increase it to 5 to 15 heads.

HYPOTHESIS-TESTING PROCEDURES

The procedure by which statisticians analyze data in order to decide when the evidence is strong enough to support the motivated hypothesis is called a **hypothesis test.** We will now summarize and formalize the material we have discussed so far. To test the validity of a statistical claim, the statistician formulates two opposite hypotheses: the alternative (or motivated) hypothesis and the null hypothesis. Often the statistician considers it likely that the alternative hypothesis is true, although it is important not to let any preconceived ideas interfere with carrying out the experiment in an objective and unbiased manner. The experiment is set up as follows to see if there is sufficient evidence to prove beyond a reasonable doubt that the motivated (or alternative) hypothesis is true by showing that the null hypothesis is very probably false.

1. Based on some prior experience or idea, you are motivated to make a claim which you desire to prove by an experiment. That is, you *state the motivated hypothesis* H_a about some population in which you are interested.

2. For statistical purposes you test the opposite hypothesis. Therefore, *you state the null hypothesis* H_0.

3. You *select the significance level* α (probability of a Type I error). That is, you are stating how large a risk you are willing to take in making the error of claiming that your motivated hypothesis is true. You may select any significance level you wish. In most published statistics the authors have used $\alpha = .05$ or $\alpha = .01$. These are generally accepted standards.

4. You *choose the size of the random sample.* This choice is often determined by the amount of time and/or money that you have to do the experiment,

Hypothesis testing: binomial one-sample

and the availability of subjects. Ease of computation might also be a factor in the selection of n.

5. Based on your motivated hypothesis and your choice of significance level, you *calculate the decision rule.* Your decision rule will have one or two critical points, depending on whether the motivated hypothesis is one-tailed or two-tailed.

6. *Select a random sample* from the appropriate population and *obtain your data.*

7. Based on the experimental outcome and the previously calculated decison rule, you will *make one of two decisions.*
   (a) Reject the null hypothesis and claim that your motivated hypothesis was correct.
   (b) Fail to reject the null hypothesis; you have been unable to prove that the motivated hypothesis is correct. Since we have not determined the power of the test, we do not wish to state that the null hypothesis is true. If the power is low, there would be a correspondingly large probability of a Type II error. So we use the phrase "fail to reject H_0" rather than "accept H_0." Some authors say that they "reserve judgment" in cases where they do not reject the null hypothesis. For example, we can never get (in a hypothesis test) sufficient proof to show absolutely that a null hypothesis, say, p equals .5, is true. "Equals" is a very strong word. Perhaps p really equals .5001, or .5000001. We could hardly expect to distinguish these from exactly .5.

Describe your findings clearly, being sure to specify the population involved and the meaning of your decision.

MORE HYPOTHESIS-TESTING EXAMPLES

EXAMPLE 8-11 A manufacturer claims that less than 2 percent of the women who use his birth control pill suffer from side effects. We have a feeling that this estimate is too low. We decide to test his claim at the .01 significance level using a sample of 900 randomly selected women. Find the decision rule.

SOLUTION Let p equal the probability that a randomly selected user of the pill suffers from a side effect.

Step 1 State the motivated hypothesis. This hypothesis states that more than 2 percent of the users suffer side effects (one-tail test). H_a: $p > .02$.

Step 2 State null hypothesis. This hypothesis states that the percentage suffering side effects is not greater than 2 percent, or that the probability of selecting a woman at random who suffers side effects is only .02. H_0: $p = .02$.

Step 3 $\alpha = .01$.

Step 4 $n = 900$. We can use the normal approximation to the binomial because $np = 900(.02) = 18 > 5$ and $nq = 900(.98) = 882 > 5$.

Step 5 Calculate the decision rule. Assuming that the null hypothesis is true, we have $\mu = 18$ and $\sigma = \sqrt{17.64} = 4.20$ (see Figure 8-8). Looking up the z

Understanding statistics

Figure 8-8

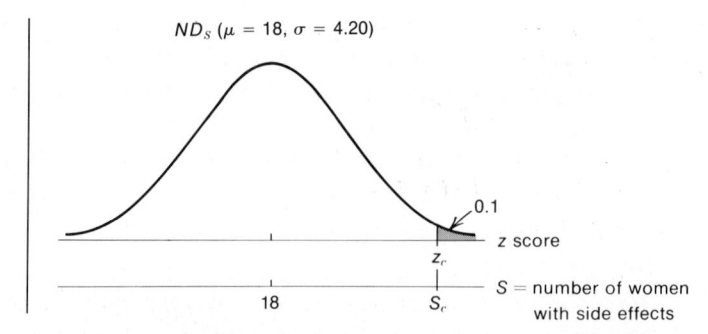

ND_S ($\mu = 18$, $\sigma = 4.20$)

S = number of women with side effects

score corresponding to area = .99, we find $z_c = 2.33$. Using our formula for converting z scores to raw scores, we have

$$S_c = \mu_S + z_c \sigma_S$$
$$= 18 + 2.33(4.20) = 18 + 9.8 = 27.8$$

We now have the normal distribution shown in Figure 8-9. With 27.8 as our critical value, our decision rule is as follows. If in our sample we find more than

Figure 8-9

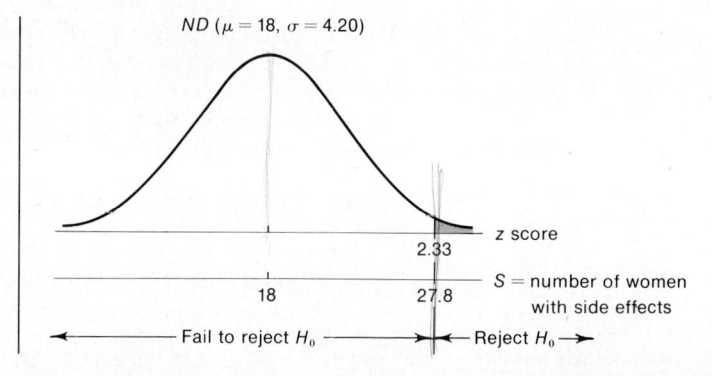

ND ($\mu = 18$, $\sigma = 4.20$)

S = number of women with side effects

Fail to reject H_0 — Reject H_0

27.8 women (that is, 28 or more) who have side effects, we reject the null hypothesis and say that the manufacturer's claim is too low. If we find fewer than 27.8 women with side effects, we will not reject the null hypothesis and we will have failed to prove the manufacturer wrong.

Step 6 Suppose we selected the sample and found 23 women with side effects.

Step 7 Since 23 is less than 27.8, according to our decision rule, we fail to reject the null hypothesis and thus have been unable to prove the manufacturer wrong at the .01 significance level. ∎

EXAMPLE 8-12 Dr. Bunny Hassenpfeffer, a noted biologist, is trying to change the coloring of rabbit offspring by diet. It is known that in a particular breed of rabbits 30 percent are pure white and the rest are spotted. A large group of this breed of rabbits is fed a special diet. Subsequently, a random sample of 100 offspring are selected and their coloring is noted. Dr. Hassenpfeffer decides to use the .05 significance level. Find her decision rule.

Hypothesis testing: binomial one-sample

SOLUTION **Step 1** She hopes that diet will affect the coloring of the offspring, but she is not sure whether it will produce more white or fewer white than usual in this particular breed of rabbits. In either case $p = P$(a white offspring) will be different from .30. That is, $H_a: p \neq .30$, a two-tail test.

Step 2 $H_0: p = .30$. That is, the proportion of white rabbits among the offspring will still be 30 percent.

Step 3 $\alpha = .05$.

Step 4 $n = 100$.

Step 5 Calculate the decision rule. Assuming that the null hypothesis is true, we have $p = .30$, $q = .70$. Since $np = 30 > 5$ and $nq = 70 > 5$, we may use the normal approximation, with $\mu = 30$ and $\sigma = \sqrt{21} = 4.58$, as shown in Figure 8-10.

Figure 8-10

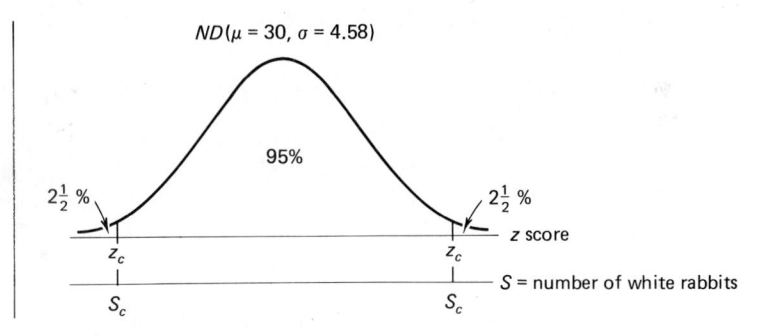

Looking up areas .025 and .975, we find that the corresponding z scores are -1.96 and $+1.96$. Converting these z scores to raw scores, we get

$$S_c = \mu + z_c \sigma$$
$$= 30 + (\pm 1.96)(4.58) = 30 \pm 8.98 = 21.02 \quad \text{and} \quad 38.98$$

Figure 8-11

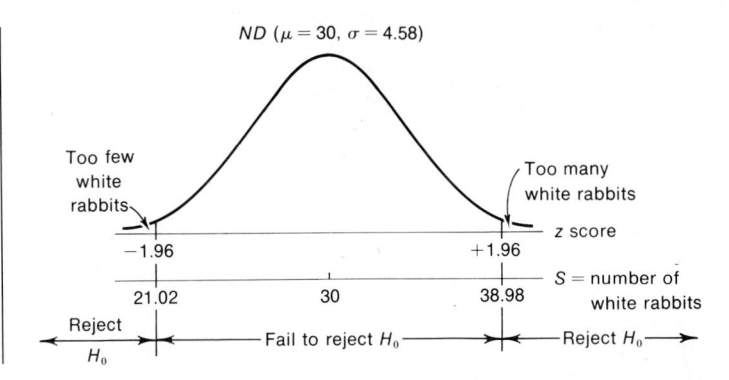

We now have the situation shown in Figure 8-11. Thus if she gets fewer than 21.02 or more than 38.98 white rabbit offspring, she will reject the null hypothesis and claim that her idea of diet affecting the coloring is true for this particular breed. Otherwise she will say that she has been unable to prove such a relationship at the .05 significance level.

Step 6 Suppose that she found only 15 of the 100 offspring to be pure white.

Step 7 Since 15 is less than 21.02, by the decision rule she will reject the null hypothesis. She will say that she has proved beyond a reasonable doubt that diet does affect the coloring of this breed of rabbit. It apparently reduces the number of white rabbits.

ANOTHER APPROACH

In Example 8-12, in order to come to a conclusion we had to compare our sample outcome (15 rabbits) with the critical z scores (± 1.96). We chose to convert the z scores to raw scores. Some statisticians prefer to convert raw scores to z scores. In this example they would get

$$z_{15} = \frac{X - \mu}{\sigma} = \frac{15 - 30}{4.58} = \frac{-15}{4.58} = -3.2$$

Since this is to the left of $z = -1.96$, we reject the null hypothesis.

COMMON SENSE AND STATISTICS*

When the American statesman Henry Clay (1777–1850) uttered the words "Statistics are no substitute for common sense," he may not have been referring to the conclusion of a hypothesis test, but his words certainly do apply. A statistical test of a hypothesis is merely a *model*—and often a very simplified model—of a real world situation. Just as the real problem must be translated into some mathematical model, so in turn the model conclusion must be interpreted in light of the real world, and perhaps in conjunction with factors that were not included in the model. This process is often ignored by students.

Suppose, for example, a test of some statistical hypothesis leads to the decision rule, "I will reject the null hypothesis at the .01 significance level if my outcome is greater than 61.1." Suppose further that your sample yields an outcome of 61.0. The textbooks say that you cannot reject the null hypothesis on the basis of this evidence; but some students react differently.
Alice: "Well I think that it's close enough."
Berry: "Can't you change the significance level to .05?"
Stosh: "Suppose we do a one-tail test instead of a two-tail test?"
Debbie: "Can we repeat the experiment with another sample?"
Annette: "I think we should gather a larger sample."

These students are looking for an exact numerical procedure which will make decisions for them, but it's not that simple. Here is an analogy which may clarify what's involved.

Suppose you want to paint your bedroom. You compute the area: 721 square feet. You read the label on the can of paint: "The contents of this can covers 350 square feet." Being a mathematical genius, you calculate that you need 2.06 cans. Now, how many cans of paint do you buy—2 or 3? The mathematical *model* of your real world problem provides an exact answer, namely, 2.06 cans. Clearly 2 cans will not be enough; you must get 3 cans.

But the real world intrudes with other issues: Is the paint on sale? Is the

*Reprinted with permission from *MATYC JOURNAL*, vol. 13, no. 2.

paint store around the corner or 12 miles down the road? Are you a sloppy painter or a neat one? Are you a pessimist or an optimist? Do you believe that the number—350 square feet—on the label is an average or a minimum? Will you be satisfied if the last 20 feet are covered only by a thin coat? Are you putting a dark color over a light color, or vice versa?

The mathematical model provides *guidelines to the ultimate decision, but the risks involved in a wrong decision must be considered.* If you buy only 2 cans of paint, you may have to go back to the store; they could be out of your color; the price may have gone up. If you buy 3 cans of paint, you still may have to go back to the store; and they might not make refunds.

In the same way a statistical test of a hypothesis is only a model of a real-world situation. It can, at best, only provide *guidelines* for decisions about actual problems. Thus our response to Alice, Berry, Stosh, Debbie, and Annette is, "You know, you have something there; you may be right. Let's consider the consequences of the possible errors involved in a mistaken decision."

STUDY AIDS

VOCABULARY

1. Null hypothesis
2. Alternative hypothesis (motivated hypothesis)
3. Hypothesis test
4. One-tail test
5. Two-tail test
6. Decision rule
7. Type I error
8. Type II error
9. Significance level
10. Critical value
11. Power

SYMBOLS

1. H_a
2. H_0
3. Alpha (α)
4. Beta (β)
5. S_c, z_c

FORMULAS

1. $\mu_S = np$
2. $\sigma_S = \sqrt{npq}$
3. $S_c = \mu_S + z_c \sigma_S$
4. Power $= 1 - \beta$

EXERCISES

8-50 Complete the following table.

	symbol	pronunciation	meaning
(a)	H_a		
(b)		H sub zero	
(c)		alpha	
(d)			probability of a Type II error

8-51 In the magazine article, "Common Sense and Statistics," which was reprinted above, one paragraph asks questions: Is the paint on sale? etc. What are some analogous questions that a statistician would consider in making a

practical decision as a result of the hypothesis tests found in the following exercises?
(a) 8-59   (b) 8-63   (c) 8-64

8-52 Lucky Larry notices that one of his coins comes up tails almost all the time. He decides to do a hypothesis test. He uses $H_0: p = .80$, $H_a: p > .80$. He rejects H_0 at the .01 significance level, and is therefore convinced that his coin comes up tails more than 80 percent of the time. Using this information, he makes several wagers and loses $500. Could it be that the coin was fair after all?

8-53 A quality control engineer at the Acame Bindery in South Jersey inspects every 25th book coming off the assembly line. His decision rule is: if more than 1 percent of the sample inspected is defective, the run is rejected. A recent run passed inspection and was shipped to Ketchum Book Distributors in Philadelphia. There it was discovered that most of the books were defective. How could this have happened?

8-54 For the following hypotheses decide if you would use a one-tail or a two-tail test. For each one state a null and an alternative hypothesis. In the case of a one-tail test, state in which tail the rejection region would be.
(a) The coin is biased.
(b) The coin is biased for heads.
(c) The new procedure will reduce the number of defective parts produced on the assembly line.
(d) The manufacturer's claim about the percentage of people with iron deficiency is false.
(e) The manufacturer's claim about the percentage of people with iron deficiency is too high.
(f) The vaccine will reduce the number of cases of measles.

8-55 Follow the instructions given in Exercise 8-54.
(a) Less than 3 percent of the rattlesnakes found on Park Avenue have broken fangs.
(b) More than 18 percent of the teachers at the On-Time Railroad Dispatchers School are tardy.
(c) Joining the sandworm pickers union will change the average takehome pay of a sandworm picker.
(d) Dr. Meany's practice of surprise tests in his course "Do it yourself open-heart surgery" increases his students' grades.
(e) Coach Aquanut's policy of having his basketball players wear flippers during practice sessions improves their average number of points per game.

8-56 Lorenzo Jones has invented a new process for manufacturing computer chips. He claims it is a better process than the current one. A manufacturer decides to compare the two processes in a pilot study and to analyze the results by a statistical hypothesis test. The manufacturer wants to use a two-tail test, but Lorenzo wants to use a one-tail test. Why would Lorenzo prefer a one-tail test?

For Exercises 8-57 to 8-70 perform hypothesis tests, that is:
(a) State the population and the motivated hypothesis.
(b) State the null hypothesis.
(c) Decide if you will use a one-tail or a two-tail test.
(d) Compute the decision rule.
(e) Interpret your conclusion in terms of the problem.

Hypothesis testing: binomial one-sample

8-57 At a large university, the dean of desks and chairs has always assumed that 10 percent of the students are left-handed. Martinique, a left-handed student, has been having trouble finding left-handed desks. She suspects that more than 10 percent of the students are left-handed. She takes a survey of 100 students picked randomly and finds 16 left-handers. Is this good evidence at $\alpha = .05$ to indicate that the dean is wrong?

8-58 Fargo North, decoder and cryptographer, stated that 40 percent of the letters in most messages are vowels. Using the first sentence of this exercise as a random sample, test the hypothesis that Fargo has overestimated the percentage of vowels. Use the .01 significance level.

8-59 A manufacturer of automobiles finds that 20 percent of the cars are unfit for delivery when they come off the assembly line. A worker proposes a new assembly technique which he claims will reduce the percentage of unfit cars. His technique is tried on 80 randomly chosen cars. It turns out that 3 are defective. Use $\alpha = .05$.

8-60 For the disease *Dandruffia terminata*, usually 68 percent of the victims recover without any treatment. The rest die within a short time. Dr. Ubaldo has a new drug which he hopes is a cure for the disease. He administers his drug to 64 patients picked at random from among victims of the disease, and 50 recover. Is this enough evidence to establish that the drug is effective? (Or is this result likely just by chance?) Test at the .05 significance level.

8-61 The usual dropout rate in the freshman class at Wealth College is 50 percent. A new dean of admissions claims that recent policies have lowered the dropout rate because in this year's class of 600 freshmen only 260 dropped out. Test at the .05 level. Do the statistics support the claim?

8-62 Using a particular type of bombing mechanism usually results in 70 percent of the bombs being on target. An engineer claims to have invented a more accurate mechanism. The new mechanism is used to drop 100 test bombs on a target; 75 hit the target. Is this enough evidence to establish that the new mechanism is better than the old one? Use $\alpha = .05$.

8-63 According to a nationwide poll, only 40 percent of the voters favor a health care bill that would benefit the poor. A certain politician believes that the percentage of voters who favor the bill is higher in his district. A random sample of 30 voters in his district is taken, and it is found that 14 support the health care bill. Is this sufficient evidence to support the politician's belief at the .01 significance level?

8-64 A certain medicine is said to be at least 90 percent effective in giving relief to people with allergic reactions to cats and dogs. Dr. Kay Nyne believes that this claim is incorrect. A random sample of 60 people with such allergies is selected from patients at an allergy clinic. What would you say about the claim at the .05 significance level if 58 people got relief?

8-65 Ralph complains that on the Saturday morning programs on WW-TV about 25 percent of the time is devoted to commercials. A student tests this one Saturday by switching on the TV at 50 randomly chosen times between 7 A.M. and noon. She finds that at 9 of these times she sees a commercial. Does this support Ralph's claim? (Use $\alpha = .05$.)

8-66 It is known that in July on a certain stretch of the northeast coast of the United States 60 percent of the sea gulls are Franklin's gulls. A birdwatcher goes out one morning and spots 80 gulls. He finds that 75 of them are Franklin's gulls. Give two possible explanations for this. Use $\alpha = .01$.

Understanding statistics

8-67 Two professors read in the paper that 60 percent of all college freshmen are more interested in being popular than in doing well at school. They think this is too high, so they go around and interview 100 freshmen. They find that 10 say they are more interested in being popular, while 90 say they are more interested in doing well. Test their results at the .05 significance level. Give several interpretations of your findings.

8-68 An economics student reads that in his county 35 percent of the employed earn more than $15,000 per year. He wants to see if the claim is accurate, so he mails out 500 questionnaires to people chosen at random from the phone book. He gets back 100 replies; 80 of them report incomes of more than $15,000. Give several interpretations of these results. He plans to test at the .05 significance level.

8-69 Under certain conditions the probability is .10 that a tadpole survives to mature into a frog. Now a scientist believes that he has found a way to place vitamins in the frog pond so that more tadpoles will survive. Using this new approach we take a random sample of 98 tadpoles, and test using the .05 significance level. For part (e), if 12 tadpoles survive, state whether you reject or fail to reject the null hypothesis, and explain what that means. If 27 tadpoles survive, state whether you reject or fail to reject the null hypothesis, and explain what that means.

8-70 Stan Sly claims that he can control the tosses of a fair coin. To see if he is correct, you take two ordinary coins and give him one. You toss one and ask him to try to toss the same thing, that is, if your toss results in a tail, then he is to try to toss a tail also. You repeat this experiment 18 times and Stan succeeds in tossing the same as you 15 times. Is this unusual at the .05 significance level?

8-71 With a fair pair of dice, doubles (two numbers the same) should come up about 1/6 of the time. Marsha, who is losing at Monopoly, has rolled 15 doubles in the past 60 rolls. She claims that the dice are biased.
(a) Find the probability that a pair of honest dice could produce 15 or more doubles in 60 rolls.
(b) Is this probability that you found in part (a) α, β, or neither?

8-72 How large a sample size would be needed to use a normal curve approximation to test the hypothesis that 3 percent of left-handed Carpathians have at least one green eye?

CLASS SURVEY

1. It seems reasonable to assume that the last digit of social security numbers is odd half the time and even half the time. (*Note:* Zero is even.) Test this assumption with $\alpha = .05$, using the data for this class.

2. Perform a similar hypothesis test on the fifth digit of your social security numbers. What happened?

FIELD PROJECTS

From this point on we will include suggestions for field projects. We hope that you will be able to do some during the semester. We suggest that you conduct field projects in two stages.

Hypothesis testing: binomial one-sample

1. Outline clearly what you *intend* to do. State the population or populations that you wish to sample. Describe your intended sampling procedure. Comment on its strengths and weaknesses. State your null and alternative hypotheses, the significance level, and the sample size. If you are going to ask questions of people in your sample, state these questions now *exactly* as you will ask them. If you are going to count something in your sample, state *exactly* what you are looking for. In any case, state how you will handle responses that do not fit into your predetermined categories. Give all this information to your instructor. After the instructor has approved it, then proceed with stage 2.

2. Perform the experiment as approved. Do the calculations. Submit your results with comments as to the strengths and weaknesses of the project as you actually carried it out.

EXAMPLE OF A FIELD PROJECT

A claim appears in a newspaper that 60 percent of Americans feel that the President is "doing a good job." A student doubts that this percentage is correct in his neighborhood, and so he sets up the following field project.

Stage 1

(a) *Population* All people 16 years old or older who live within 3 blocks of my house.

(b) *Sampling Procedure* There are 30 blocks in this neighborhood. I will pick 3 houses at random on each block. I will question only 1 person at each house. If no one is home, I will select another house on that block.

(c) *Questions to Be Asked*
   (1) I am doing a survey for my college class in statistics. Would you please answer the following questions?
   (2) Are you 16 years old or older?
   (3) Do you think that the President is doing a good job?
   I will continue until I get 90 yes responses to both questions (1) and (2), and either a yes or a no response to question (3). Since $np = 90(.6) = 54 > 5$, and $nq = 90(.4) = 36 > 5$, we can use the normal distribution.

(d) H_0: the fraction in my neighborhood is the same as the fraction of all Americans, $p = .60$

   H_a: the fraction in my neighborhood is not the same as the fraction of all Americans, $p \neq .60$ (two-tail test)

   I will use $\alpha = .05$.

(e) *Comments on Strengths and Weaknesses* Depending on when I am able to cover the houses, I may get more or less working people in my sample.

Stage 2   (Conducted after stage 1 was approved)

I went to 123 homes. In 12 homes no one answered the door. In 3 homes there was no one over 16 present. In 2 homes the answer to question (1) was no. In 4 homes people first answered yes to question (1) but changed their

Understanding statistics

minds after they heard question (3). In 12 homes the person was undecided on question (3). The remaining 90 responses were

yes	no
48	42

SOLUTION FOR THIS PROJECT

H_a: percentage in neighborhood is not the same as percentage of all Americans, $p \neq .60$ (two-tail test)

H_0: $p = .60$

$n = 90$

$\alpha = .05$ (two-tail test)

Since $np = 54 > 5$, $nq = 36 > 5$, we can use the normal approximation with $\mu = 54$ and $\sigma = \sqrt{21.6} = 4.6$.

$z_c = \pm 1.96$

The curve for this distribution is shown in Figure 8-12.

Figure 8-12

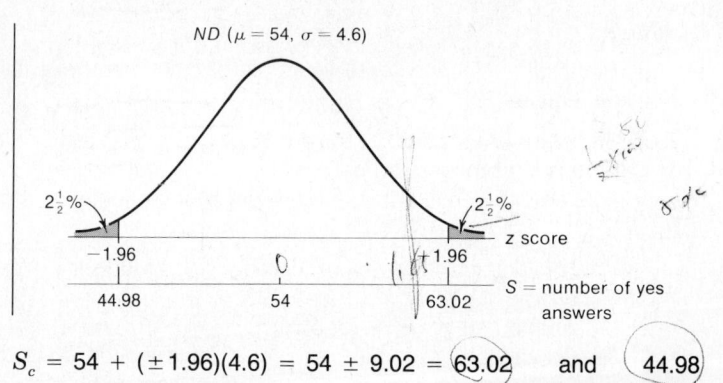

ND ($\mu = 54$, $\sigma = 4.6$)

$2\frac{1}{2}\%$ $2\frac{1}{2}\%$

−1.96   0   +1.96   z score

44.98   54   63.02   S = number of yes answers

$S_c = 54 + (\pm 1.96)(4.6) = 54 \pm 9.02 = 63.02$ and 44.98

Therefore, my decision rule is to reject the null hypothesis if I get more than 63 or fewer than 45 yes answers. Since the experimental outcome was 48 yes answers, I have failed to reject the null hypothesis. I was unable to show that the null hypothesis was false. Of the 33 nonresponses I have no evidence to indicate that they would differ markedly from the 90 who did respond. If there were a marked difference, it might change my conclusion.

SUGGESTED PROJECTS

Outline a one-sample binomial hypothesis test to be performed on some population of your choice. After your instructor has approved your outline, gather your data and perform the hypothesis test.

1. Select a reported fact from a newspaper or some other source as in the preceding example and test it on some population of your own choice.

2. Test a theoretical hypothesis concerning coins, dice, cards, etc. For example, does the American penny have a probability of coming up heads equal to .5? Get 100 pennies and toss them from a large container 10 times to test this hypothesis.

3. Perform any one-sample binomial test of your own choosing.

Hypothesis Testing: Binomial Two-Sample

9

SYMBOLS FOR ESTIMATES

In the previous chapter we have asked whether or not a sample could reasonably be expected to have been chosen at random from a given population. For example, if a population is 62 percent female, what is the probability that we could draw a random sample of 40 persons and find only 48 percent of the sample to be female?

Very often, however, we do not know the proportions that make up a given population. Do you know what proportion of college history majors are female? Do you know what percentage of people in your town are diabetics? Do you know what proportion of the people in North Carolina are Baptists?

If we wish to perform a statistical test, but we do not know what proportions to expect, we can estimate the proportions by drawing a random sample. For example, if you desire to know what percentage of the patrons of the school cafeteria are female, you could take a random sample of the patrons. If your sample has 80 persons and 60 are female, then your best **estimate** of the percentage of female patrons would be 60/80 = 75 percent. We do not claim that the probability of choosing a female is exactly .75, but only that .75 is a reasonable estimate of the true value based on what data we have.

When we use the symbol p for $P(x$ is female), then we are claiming that the true value of p is .75. In contrast to this we shall use the caret symbol ˆ (read: hat) when we may not have the true value, but only an estimate. In this case we would write $\hat{p} = .75$.

If we wish to find the mean age on campus, we could estimate it by taking a random sample of 50 students' ages. If we add these ages and divide by 50 we have the mean of the sample. This may not be the true value of the mean

Understanding statistics

of the population, but it would probably be a good estimate for it. If μ is the symbol for the true value of the unknown mean, then $\hat{\mu}$ (read: mu hat) is the symbol for our best estimate of μ.

population parameter	estimate from a sample
p	$\hat{p}$
μ	$\hat{\mu}$

A DISTRIBUTION OF DIFFERENCES

Imagine that for a project a class of 23 students wishes to compare two nearby campuses with regard to the percentage of students who support Senator Foghorn's policy on sales tax. Suppose that unknown to anyone, the true figure on campus 1 is 63 percent, that is, $p_1 = .63$. On campus 1 each of the 23 students gathers a random sample of 100 persons and computes the best estimate of p_1, which we call $\hat{p}_1$ (read: p sub one hat). Their results are as follows.

student	$\hat{p}_1$
1	.62
2	.61
3	.64
4	.63
5	.63
6	.60
7	.55
8	.70
9	.63
.	.
.	.
.	.
23	.62

If we added these 23 values of $\hat{p}_1$ and divided by 23, we would probably get a value very close to the actual value of .63.

Now suppose that each student gathers a second random sample of 100 persons from campus 2. (Unknown to anybody, the actual figure on campus 2 is 60 percent, that is, $p_2 = .60$.)

student	$\hat{p}_1$	$\hat{p}_2$
1	.62	.57
2	.61	.62
3	.64	.60
4	.63	.59
5	.63	.61
6	.60	.57
7	.55	.58
8	.70	.60
9	.63	.61
.	.	.
.	.	.
.	.	.
23	.62	.64

Hypothesis testing: binomial two-sample

If we compute the difference between each pair of estimates, we get the following.

student	$\hat{p}_1$	$\hat{p}_2$	differences $\hat{p}_1 - \hat{p}_2$
1	.62	.57	.05
2	.61	.62	−.01
3	.64	.60	.04
4	.63	.59	.04
5	.63	.61	.02
6	.60	.57	.03
7	.55	.58	−.03
8	.70	.60	.10
9	.63	.61	.02
.	.	.	.
.	.	.	.
.	.	.	.
23	.62	.64	−.02

The distribution of numbers in the last column is called a **distribution of differences** of sample proportions. We will use the symbol $d\hat{p}$ to indicate the quantity $\hat{p}_1 - \hat{p}_2$. Theoretically, if we take all possible pairs of random samples from two populations, the resulting distribution of differences will have three important properties. (The first two properties you may have already guessed.)

1. The distribution of differences will be approximately a normal distribution if $n_1 p_1$, $n_1 q_1$, $n_2 p_2$, and $n_2 q_2$ are all larger than 5.

2. The mean of the differences will equal $p_1 - p_2$. We write

$$\mu_{d\hat{p}} = p_1 - p_2$$

3. The standard deviation of the differences will equal

$$\sqrt{\frac{p_1 q_1}{n_1} + \frac{p_2 q_2}{n_2}}$$

We write

$$\sigma_{d\hat{p}} = \sqrt{\frac{p_1 q_1}{n_1} + \frac{p_2 q_2}{n_2}}$$

where n_1 = number of persons in each sample taken from first population (which population is denoted "first" and which is denoted "second" is arbitrary)

p_1 = true proportion from first population sampled

$q_1 = 1 - p_1$

n_2 = number of persons in each sample taken from second population

p_2 = true proportion from second population sampled

$q_2 = 1 - p_2$

This distribution is shown in Figure 9-1. In the particular case where p_1 and p_2 have the *same* value (say, p), the formula for $\sigma_{d\hat{p}}$ becomes

Understanding statistics

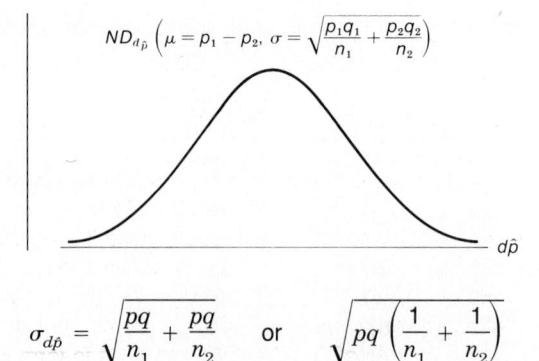

Figure 9-1

$$ND_{d\hat{p}}\left(\mu = p_1 - p_2,\ \sigma = \sqrt{\frac{p_1 q_1}{n_1} + \frac{p_2 q_2}{n_2}}\right)$$

$d\hat{p}$

$$\sigma_{d\hat{p}} = \sqrt{\frac{pq}{n_1} + \frac{pq}{n_2}} \quad \text{or} \quad \sqrt{pq\left(\frac{1}{n_1} + \frac{1}{n_2}\right)}$$

where p is the common value of p_1 and p_2 and $q = 1 - p$.

EXAMPLE 9-1   Leonard has been accepted at Adelphi University and at Hofstra University. He decides that he will go to the one that has the greater percentage of dorm students. A friend tells him that it does not matter because the proportions are about equal. Doubting his friend and unable to obtain the exact information, Leonard decides to perform a two-sample binomial test. His motivated hypothesis is that the percentage of dorm students at Adelphi is different from the percentage at Hofstra. His null hypothesis is that these proportions are the same. If we let $p_1 = P$(a randomly selected student at Adelphi is a dorm student) and $p_2 = P$(a randomly selected student at Hofstra is a dorm student), then symbolically our hypotheses are

H_0: $p_1 = p_2$ or $p_1 - p_2 = 0$
H_a: $p_1 \neq p_2$ or $p_1 - p_2 \neq 0$ (two-tail test)

After Leonard has taken his two random samples he will compute $\hat{p}_1$, $\hat{p}_2$, and their difference $d\hat{p} = \hat{p}_1 - \hat{p}_2$. The difference that Leonard calculates will be one of many possible differences which could have been obtained from other possible pairs of random samples. We will see below that the theoretical distribution of differences is normal. According to Leonard's null hypothesis the mean of this distribution $\mu_{d\hat{p}}$ is zero. Leonard's problem is to decide whether or not his sample difference $d\hat{p} = \hat{p}_1 - \hat{p}_2$ is significantly far from the mean. If this sample difference is far from the mean, it would indicate that the null hypothesis is probably wrong. In this case it would mean that the true population proportions p_1 and p_2 are not equal.

In order to test this null hypothesis, Leonard gathers two random samples, one on each campus. At Adelphi a random sample of 100 students contains 60 dorm students. Therefore, $n_1 = 100$, $\hat{p}_1 = 60/100 = .60$, and $\hat{q}_1 = 1 - \hat{p}_1 = .40$. At Hofstra a random sample of 110 students contains 50 dorm students. Therefore, $n_2 = 110$, $\hat{p}_2 = 50/110 = .45$, and $\hat{q}_2 = 1 - \hat{p}_2 = 1 - .45 = .55$. The sample difference $d\hat{p} = \hat{p}_1 - \hat{p}_2 = .60 - .45 = .15$.

Since we do not know the values of p_1, q_1, p_2, or q_2 we cannot tell if $n_1 p_1$, $n_1 q_1$, $n_2 p_2$, or $n_2 q_2$ are larger than 5. Therefore we use our estimates and check instead to see if the four numbers $n_1 \hat{p}_1$, $n_1 \hat{q}_1$, $n_2 \hat{p}_2$, and $n_2 \hat{q}_2$ are larger than 5.

Hypothesis testing: binomial two-sample

$$n_1\hat{p}_1 = 100\left(\frac{60}{100}\right) = 60 \qquad n_1\hat{q}_1 = 100\left(\frac{40}{100}\right) = 40$$

$$n_2\hat{p}_2 = 110\left(\frac{50}{110}\right) = 50 \qquad n_2\hat{q}_2 = 110\left(\frac{60}{110}\right) = 60$$

(Notice that these four numbers are the same as our four sample outcomes: at Adelphi, 60 dorm and 40 nondorm students; at Hofstra, 50 dorm and 60 nondorm students. This will always be the case.) Since 60, 40, 50, and 60 are all larger than 5, the distribution of the differences will be approximately normal. Under our null hypothesis $\mu_{d\hat{p}} = 0$, but what value can we use to estimate $\sigma_{d\hat{p}}$? According to the null hypothesis, p_1 and p_2 are equal. But what is their actual common numerical value? The best we can do is to form what is called a **pooled estimate** of their value. Altogether, Leonard looked at 210 students ($n_1 + n_2$) and found that 110 were dorm students (60 + 50). So the pooled estimate for the true proportion of dorm students would be

$$\frac{60 + 50}{100 + 110} = \frac{110}{210} = .52$$

We write $\hat{p} = .52$, and use this in the formula for $\sigma_{d\hat{p}}$. The result will be an estimate of $\sigma_{d\hat{p}}$. We call it $\hat{\sigma}_{d\hat{p}}$.

$$\hat{\sigma}_{d\hat{p}} = \sqrt{\frac{\hat{p}\hat{q}}{n_1} + \frac{\hat{p}\hat{q}}{n_2}} \quad \text{or} \quad \sqrt{\hat{p}\hat{q}\left(\frac{1}{n_1} + \frac{1}{n_2}\right)}$$

$$= \sqrt{\frac{.52(.48)}{100} + \frac{.52(.48)}{110}} = \sqrt{.004765} = .069$$

This distribution is shown in Figure 9-2.

Figure 9-2

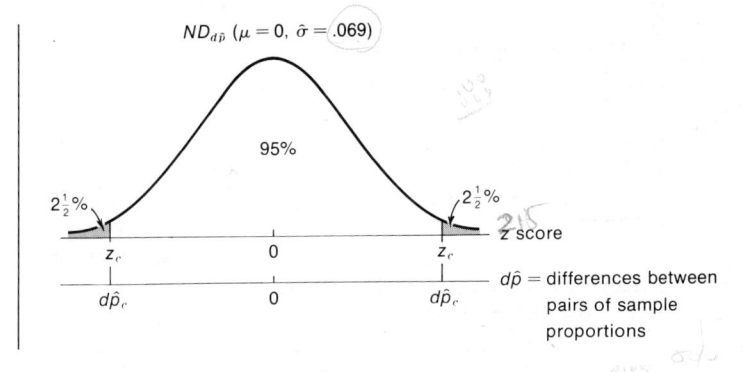

$ND_{d\hat{p}}\ (\mu = 0, \hat{\sigma} = .069)$

95%

$2\frac{1}{2}\%$ ... $2\frac{1}{2}\%$

z_c ... 0 ... z_c — z score

$d\hat{p}_c$ ... 0 ... $d\hat{p}_c$

$d\hat{p}$ = differences between pairs of sample proportions

If Leonard decided to use a significance level of $\alpha = .05$, what would his decision rule be for this experiment? Since we have a two-tail test,

$$z_c = \pm 1.96$$

Now we convert this critical z score to a critical difference $d\hat{p}_c$:

$$d\hat{p}_c = \mu_{d\hat{p}} + \hat{\sigma}_{d\hat{p}} z_c$$

$$= 0 + .069(\pm 1.96) = \pm .14$$

His decision will be to reject the null hypothesis if his sample difference $d\hat{p} = \hat{p}_1 - \hat{p}_2$ is either less than $-.14$ or greater than $+.14$.

Since his sample difference, .15, is greater than .14, he rejects the null hypothesis and claims that the percentage of dorm students at Adelphi is different from the percentage of dorm students at Hofstra. Evidently, there is a higher percentage of dorm students at Adelphi. ∎

SUMMARY OF TWO-SAMPLE BINOMIAL TESTS

The two-sample binomial test is used when we compare two populations. We let p_1 be the true proportion of successes on the first population, that is, $p_1 = P$(success in population 1) and p_2 the true proportion of successes on the second population, that is, $p_2 = P$(success in population 2).

The null hypothesis states that the difference between these true proportions is a fixed number. For most of the problems in this book our null hypothesis is that $p_1 = p_2$, which implies that $p_1 - p_2 = 0$. This means that in the distribution of differences if we took the mean of all the possible sample differences, we would get zero. Thus, we write

$$H_0: \mu_{d\hat{p}} = p_1 - p_2 = 0$$

As before, the alternative will have a $<$, $>$, or $\neq$ symbol, which will tell us whether we are doing a one-tail or a two-tail test.

To compute our decision rule we need the mean and the standard deviation of the distribution of differences. We get the mean from the null hypothesis and we estimate the standard deviation using the formula

$$\hat{\sigma}_{d\hat{p}} = \sqrt{\frac{\hat{p}\hat{q}}{n_1} + \frac{\hat{p}\hat{q}}{n_2}}$$

where $\hat{p}$ is the pooled estimate of the common value of p.

A SECOND EXAMPLE

EXAMPLE 9-2 A researcher is comparing the safety records of two popular automobile models. She is interested in the percentage of accidents in which the driver is killed. She has heard that the Boomer model is more dangerous than its competitor, the Zoomer. Checking recent records, she finds that over the past several months in a large metropolitan area there were 423 accidents involving the Zoomer. In 34 of these the driver was killed. For the Boomer there were 580 accidents, and in 58 the driver was killed. Does this indicate at the .01 significance level, that a driver who gets in an accident is more likely to be killed if he is driving a Boomer?

SOLUTION Let $p_1 = P$(driver of Boomer is killed if he is in an accident), and $p_2 = P$(driver of Zoomer is killed if he is in an accident),

$$H_0: p_1 = p_2$$
$$H_a: p_1 > p_2 \quad \text{or} \quad p_1 - p_2 > 0 \quad \text{(one-tail test on the right)}$$

(If we had let population 1 be the Zoomer population, we would have a one-tail test on the left.)

Hypothesis testing: binomial two-sample

$n_1 = 580 \quad \hat{p}_1 = 58/580 = .10 \quad \hat{q}_1 = .90$
$n_2 = 423 \quad \hat{p}_2 = 34/423 = .08 \quad \hat{q}_2 = .92$

Since $n_1\hat{p}_1$, $n_1\hat{q}_1$, $n_2\hat{p}_2$, and $n_2\hat{q}_2$ are all larger than 5, the distribution of differences is approximately normal. The pooled estimate of p is

$$\hat{p} = \frac{58 + 34}{580 + 423} = \frac{92}{1003} = .09$$

So $\hat{q} = .91$.

Therefore,

$$\hat{\sigma}_{d\hat{p}} = \sqrt{\frac{.09(.91)}{580} + \frac{.09(.91)}{423}} = \sqrt{.09(.91)\left(\frac{1}{580} + \frac{1}{423}\right)}$$

$$= \sqrt{.000335} = .018$$

This distribution is illustrated in Figure 9-3.

Figure 9-3

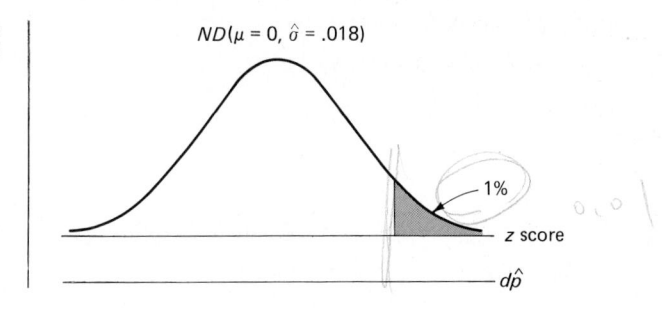

$ND(\mu = 0, \hat{\sigma} = .018)$

1%

z score

$d\hat{p}$

$d\hat{p}_c = \mu + \hat{\sigma}z_c$
$\quad\quad = 0 + .018(2.33) = .04$

We now have Figure 9-4.

Figure 9-4

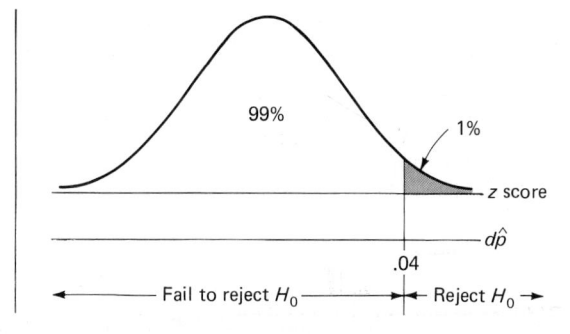

99% 1%

z score

$d\hat{p}$

.04

← Fail to reject H_0 —— Reject H_0 →

Our critical value of $d\hat{p}$ is .04. Our decision rule is that if our sample difference $d\hat{p}$ is to the right of .04, we will reject the null hypothesis.

Our difference is $\hat{p}_1 - \hat{p}_2 = .10 - .08 = .02$, which is not in the rejection region. We have failed to establish at $\alpha = .01$ that the fatality rate is higher in the Boomers. The difference between the two fatality rates is not statistically significant.

Understanding statistics

STUDY AIDS

VOCABULARY

1. Estimate
2. Two-sample hypothesis test
3. Distribution of differences
4. Pooled estimate

SYMBOLS

1. n_1, n_2
2. p_1, p_2, q_1, q_2
3. $\hat{p}_1, \hat{p}_2, \hat{q}_1, \hat{q}_2$
4. $d\hat{p}, d\hat{p}_c$
5. $\mu_{d\hat{p}}$
6. $\hat{p}, \hat{q}$
7. $\sigma_{d\hat{p}}, \hat{\sigma}_{d\hat{p}}$

FORMULAS

1. $\mu_{d\hat{p}} = p_1 - p_2$
2. $\hat{\sigma}_{d\hat{p}} = \sqrt{\dfrac{\hat{p}\hat{q}}{n_1} + \dfrac{\hat{p}\hat{q}}{n_2}}$ or $\sqrt{\hat{p}\hat{q}\left(\dfrac{1}{n_1} + \dfrac{1}{n_2}\right)}$
3. Sample difference $d\hat{p} = \hat{p}_1 - \hat{p}_2$
4. Critical difference $d\hat{p}_c = \mu_{d\hat{p}} + z_c \hat{\sigma}_{d\hat{p}}$

EXERCISES

9-1 Randy Semple, a conscientious statistics student, did a field project for Chapter 8. He read in the newspaper that 28 percent of the families in New York City own a dog. He tested whether 28 percent of the families in his community owned a dog. This of course was a *one-sample* hypothesis test. Now that he has studied Chapter 9, he realizes that the New York City figure must have been derived from a sample also, and he wonders if he should have done a *two-sample* test. Does he have enough data?

9-2 We know that few coins have a probability of coming up heads *exactly* equal to 1/2. If pennies and dimes have $p = P(\text{heads})$ *slightly* different from .5, we wish to test whether they differ significantly from each other. We get 100 newly minted pennies and 50 newly minted dimes. We toss them 10 times each and count the number of heads. For the pennies we have $n_1 = 1000$, and we tossed 490 heads. For the dimes we have $n_2 = 500$, and we tossed 240 heads.
(a) State the hypotheses.
(b) Is this a one-tail or a two-tail test?
(c) Find $\hat{p}_1, \hat{p}_2,$ and $\hat{p}$.
(d) Find $\hat{\sigma}_{d\hat{p}}$.
(e) Using $\alpha = .05$, find the decision rule for this experiment.
(f) Compute $d\hat{p} = \hat{p}_1 - \hat{p}_2$ and state the conclusion reached for this experiment.

9-3 Executive Airlines is analyzing its passenger service. One question is: which of two flights carries a larger percent of people on business trips? Over a period of one month they interview a random sample of passengers. On the first flight 130 out of 200 were on business trips. On the second flight 120 out of 200 were on business trips.
(a) Using $\alpha = .01$, find the decision rule for this test.
(b) Compute $d\hat{p}$ and state your conclusion.

Hypothesis testing: binomial two-sample

9-4 One way of comparing two colleges is to look at the percentage of students in each who hold part-time jobs during the school year. A survey was taken at two colleges with these results. At each college 100 students were interviewed. At Cardinal College 70 percent of those interviewed held jobs. At Blank College 75 percent held jobs.
(a) Using $\alpha = .05$, find the decision rule for this test.
(b) Compute $d\hat{p}$ and state your conclusion.

9-5 In certain medical procedures a small plastic tube must be permanently attached to a patient's vein. This is a potential source of blood clots. A clinical trial could be done to see if putting such patients on a daily low dose of aspirin would reduce the occurrence of blood clots. In one such experiment the results were that 6 out of 19 patients on aspirin developed clots, while 18 out of 25 patients on placebo developed clots. Is this evidence at $\alpha = .01$ in favor of using the aspirin?

9-6 A teacher reads of a new approach to teaching a difficult idea. She has two classes of equal background, intelligence, and ability. She teaches one class of 20 students the traditional way, and after testing finds that 12 have grasped the topic. She teaches the second class of 25 students by the new technique, and 16 are found to have grasped the topic. Test the theory that the new technique is significantly better at the .01 significance level.

9-7 To test the hypothesis that the attitude toward birth control of Roman Catholics and Orthodox Jews is significantly different, we gather two random samples. Among 60 Catholics questioned, 42 opposed birth control, while among 60 Jews, 29 opposed birth control. Use $\alpha = .05$.

9-8 The manager of a paint store wants to determine the effect advertising has on her sales. She has a paint which ordinarily sells for $10.98. She decides to run an ad calling this paint a special item on sale for $10.98, on Friday, Saturday, and Sunday only. To find out whether or not the advertisement is effective in increasing the percentage of sale of this item, she gathers two random samples from her customers. On the weekend prior to the ad, she found that 12 out of 100 randomly selected customers bought the paint. On the sale weekend she found that 21 out of 110 randomly selected customers bought the paint. Decide at the .05 significance level if the ad was effective.

9-9 A dietitian in a large university suspects that the proportion of male college students who eat 3 meals a day is greater than the proportion of female college students who eat 3 meals a day. Two random samples are gathered. Of 500 men interviewed, 432 said they usually eat 3 meals a day, while of 500 women interviewed, 401 said they eat 3 meals a day. Determine if this is a significant difference at the .01 significance level.

9-10 Oscar, a Ph.D. candidate, is doing a survey concerning the children's television program "Sesame Street." He wants to know if there is a difference between the proportion of suburban children who regularly watch the program and the proportion of inner city children who regularly watch the program. In two random samples, he finds 24 out of 30 suburban children who watch it and 19 out of 25 inner city children who watch it. Test with $\alpha = .05$.

9-11 In a check of records of deaths in a veterans' hospital, it was found that of 50 nonsmokers, 6 had lung cancer, while of 60 smokers, 15 had lung cancer. Using a one-tail test, decide if this difference is significant at the .05 level.

9-12 Dustin is attending his graduation party. Since the guests are primarily his parents' friends, to while away the time he decides to do a two-sample

Understanding statistics

binomial hypothesis test. He notices that there is a difference between men and women! He thinks that a greater proportion of men extinguish their matches by blowing them out. He gathers two random samples and finds that 7 of 13 men blew out their matches while only 6 of 14 women blew out their matches. Test if there is a difference at the .01 significance level.

9-13 Marc and Anthony are arguing about which of them has more ESP. They split a pair of honest dice between themselves and alternately toss one die each. After Marc rolls his die and looks at the result, Anthony attempts, without looking, to announce the correct result. Then Anthony rolls his die and Marc tries to determine the result. After 60 tosses each, Marc was correct 20 of 60 times, while Tony was right 15 of 60 times.
(a) Does this indicate any difference in their ESP at the .05 significance level?
(b) Did Marc show any extraordinary ESP at the .01 significance level?
(c) Did Tony?

9-14 Prof. Laura Hardy is studying whether primitive man could have had webbed feet. In her work she examined the toes of 1000 school children. Of the boys examined, 45 of 500 had webbing between their second and third toes. Of the girls, 33 of 500 did. In some children the webbing was present between all the toes, and Prof. Hardy ignored these cases, as this phenomenon is unknown among other primates. Do these figures indicate a difference between the percentage of boys and the percentage of girls with webbing between their second and third toes? Use $\alpha = .01$.

9-15 Recently 1457 persons between 18 and 64 years of age were surveyed. Similarly, 2797 people over 65 were surveyed. 50 percent of the younger group reported infirmities, yet only 23 percent of the senior citizens reported infirmities. Perform a one-tail hypothesis test on these data at the .01 significance level. What exactly are you testing?

9-16 An oil company has two methods of deciding where to drill for oil. For method A they have been successful 10 of the last 30 tries. For method B they have been successful 7 of 27 times. Does this indicate at $\alpha = .05$ that method A is superior?

9-17 In a recent National Mathematics Aptitude Examination, a sample of 500 adults (ages 26 to 35) and a sample of 500 teenagers (age 17) were given the following problem. A fruit punch is to be made with equal amounts of lemonade, limeade, orange juice, and ginger ale. You want to make 2 gallons of punch. How much ginger ale should you use? It was found that 30 percent of the teenagers got the correct answer and 36 percent of the adults got the correct answer. Does this support the idea that the older people are better at this type of question? Use $\alpha = .01$. By the way, what is the correct answer?

9-18 In a study of who was at fault in bicycle-automobile collisions, a survey revealed that cyclists 12 years old or younger were probably responsible for 92 percent of the accidents in which they were involved. The ratio dropped to 43 percent for cyclists 20 to 24 years old, and to 34 percent for those over 25.
(a) If the survey included 500 bicycle-auto accidents involving cyclists 12 or younger, 300 bicycle-automobile accidents involving 20-to-24-year-olds, and 400 accidents involving those over 25, test the hypothesis that the percentage of 20-to-24-year-olds who are responsible for collisions is greater than the percentage of over-25 cyclists who are responsible. Use the .05 significance level.

Hypothesis testing: binomial two-sample

(b) In a certain city an intensive bike safety program was initiated in the elementary schools. After 1 year, the reports detailing responsibility for accidents of this type were summarized as follows.

age group	responsible for accident
under 12	240 of 300
20 to 24	40 of 100
over 25	25 of 76

Do these results indicate that this kind of program reduces the responsibility for accidents in the under-12 age group? Use $\alpha = .05$.

9-19 A government-funded study of 2100 adults (ages 26 to 35) and 1700 teenagers (age 17) whose instruction was with the "new" math showed that some consumers lose hundreds of dollars annually because they can't apply math to everyday purchases. Dr. Noah Progress, a member of the commission doing the survey, speculated that the nation has lost a generation of students who were taught how math works rather than how to use it. "They can't apply math to everyday problems" he said. Are his conclusions valid at the .01 significance level based on the evidence that only 40 percent of the 17-year-olds and 45 percent of the adults could calculate the lowest price per ounce for a box of rice?

9-20 A physician was studying the use of anticoagulants in treating acute heart attacks. She found that of 1104 patients who received anticoagulants, 8.3 percent died within 21 days of the attack. Of 1226 patients who did not receive this treatment, 27.3 percent died within 21 days. Show that this difference is significant at the .01 significance level.

9-21 In a random sample of the visiting records at the Sea View retirement home, the following data were gathered:

	patient is visited regularly	patient is not visited regularly	
patient has grandchildren	20	40	60
patient does not have grandchildren	10	30	40
	30	70	

A typical conclusion often drawn is that since 20/30, or 2/3, of those who are regularly visited have grandchildren, someone who has grandchildren is more likely to be regularly visited. But this is *false*, because 20/60, or 1/3 (which is less than 1/2), of the grandparents receive visitors. We can pose two *different* questions here.

(a) What is the probability that a person who is visited regularly is a grandparent? What is the probability that a person who is not visited regularly is a grandparent? Let population 1 be those patients who are visited regularly. Let population 2 be those who are not visited regularly. At $\alpha = .01$ do the data indicate that there is a different percentage of grandparents in population 1 than in population 2?

(b) What is the probability that a grandparent is visited regularly? What is the probability that a nongrandparent is visited regularly? Designate population 1 the grandparents and population 2 the nongrandparents. At $\alpha = .01$ do the data indicate that there is a different percentage of patients who receive regular visits in population 1 than in population 2?

9-22 Donald Wahn was doing a survey on attitudes toward school bussing. He interviewed 400 students and found that 100 approved of school bussing. He interviewed 100 teachers, and 10 approved of school bussing. Show that this indicates a difference in attitude at the .01 significance level.

9-23 (a) Winny rolls a die. She considers rolling an odd number as winning, and an even number as losing. About what percentage of games should Winny win?

(b) Lucy draws a card from an ordinary 52-card deck. She considers drawing a heart as losing. About what percentage of games should Lucy lose?

(c) Winny rolls the die 20 times and records the percentage of times she wins as $\hat{p}_1$; Lucy draws 50 cards and records the percentage of times that she loses as $\hat{p}_2$. Drew subtracts $\hat{p}_1 - \hat{p}_2$ and records the differences as $d\hat{p}$. They repeat this procedure many times, obtaining a large distribution of $d\hat{p}$'s. Estimate the mean of this distribution.

(d) Normally, Drew draws a graph of the $d\hat{p}$'s. Explain why the graph that Drew drew should be approximately a normal curve.

(e) Since $p_1 \neq p_2$, the standard deviation of the $d\hat{p}$'s is found by a different formula:

$$\sigma_{d\hat{p}} = \sqrt{\frac{p_1 q_1}{n_1} + \frac{p_2 q_2}{n_2}}$$

$$= \sqrt{\frac{.5(.5)}{20} + \frac{.25(.75)}{50}} = \sqrt{.0125} = .11$$

Let D be the number that separates the bottom 95 percent of Drew's outcomes from the top 5 percent. Find D.

9-24 Instead of testing a hypothesis that $p_1 = p_2$, we can test that p_1 is 20 percent more than p_2. $\mu_{d\hat{p}}$ is still $p_1 - p_2$, but we do not pool the experimental results, and

$$\hat{\sigma}_{d\hat{p}} = \sqrt{\frac{\hat{p}_1 \hat{q}_1}{n_1} + \frac{\hat{p}_2 \hat{q}_2}{n_2}}$$

Using this idea, test the hypothesis that 20 percent more young females than young males smoke. Use the data that of 200 young females interviewed, 120 smoked, and of 400 young males interviewed, 150 smoked. Let $\alpha = .05$.

CLASS SURVEY

Assuming that the breakdown of hair color in your class is representative of hair color in the whole school, perform the following experiment. Test whether the percentage of light-haired females in the school is the same as the percentage of light-haired males.

FIELD PROJECTS

Review the general instructions for special projects given at the end of Chapter 8, and then select one of the following.

Hypothesis testing: binomial two-sample

1. Perform an experiment similar to Exercise 9.2.

2. Perform an experiment to decide whether or not there is a significant difference between the percentage of males who are left-handed and the percentage of females who are left-handed.

3. Perform a two-sample binomial test of your own choosing. *Hints:* Differences between age groups, political groups, sexes, religions, races, etc., on fashions, politics, preferences for food, literature, music, etc.

EXAMPLE OF A FIELD PROJECT

Proposal Is there any difference between the percentage of Washington College women who attend religious services regularly and the percentage of local noncollege women who attend religious services regularly?
I will ask 30 women at Washington College the following questions.

1. I am doing a survey for my college statistics class. Would you answer two questions for me with a yes or no answer?

2. Are you a student at Washington College?

3. Do you attend religious services at least twice a month?

I will continue this survey until I get 30 women who answer yes to questions 1 and 2 and either yes or no to question 3.
I will ask 30 women at the Roosevelt Field shopping center the following questions.

1. I am doing a survey for my college statistics class. Would you answer two questions for me with a yes or no answer?

2. Did you attend college?

3. Do you attend religious services at least twice a month?

I will continue this survey until I get 30 women who answer yes to question 1, no to question 2, and either yes or no to question 3.
I can then formulate the following hypotheses:

H_a: The percentage of Washington College women who attend religious services is different from the percentage of local noncollege women.
H_0: The two percentages are the same.

Let $p_1 = P$(a Washington College woman attends religious services regularly) and let $p_2 = P$(a noncollege woman attends religious services regularly).

Report On Tuesday, April 27, from 3:15 to 4:30 P.M. I questioned 47 women in the Washington College cafeteria. Of these, 4 would not answer question 1, 3 were not students, 8 would not answer question 3, and 2 answers were vague, neither yes nor no. Of the 30 who answered, 13 said yes. Therefore, $\hat{p}_1 = 13/30 = .43$ and $n_1 = 30$.

On Wednesday, April 28, from 1:20 to 3:05 P.M. I questioned 56 women at Roosevelt Field. Of these, 10 would not answer question 1, and 16 had attended college. Of the 30 who responded, 15 said yes. Therefore, $n_2 = 30$ and $\hat{p}_2 = 15/30 = .50$. Since $n_1\hat{p}_1$, $n_1\hat{q}_1$, $n_2\hat{p}_2$, and $n_2\hat{q}_2$ are 13, 17, 15, and 15, and are all larger than 5, we can use the normal distribution. $\alpha = .05$. This is a two-tailed test. Therefore, $z_c = \pm 1.96$.

$$\hat{p} = \frac{13 + 15}{30 + 30} = \frac{28}{60} = .47$$

$$\hat{\sigma}_{d\hat{p}} = \sqrt{\frac{\hat{p}\hat{q}}{n_1} + \frac{\hat{p}\hat{q}}{n_2}}$$

$$= \sqrt{\frac{.47(.53)}{30} + \frac{.47(.53)}{30}} = \sqrt{0.166} = .13$$

This distribution is represented in Figure 9.5. The critical difference is then

Figure 9-5

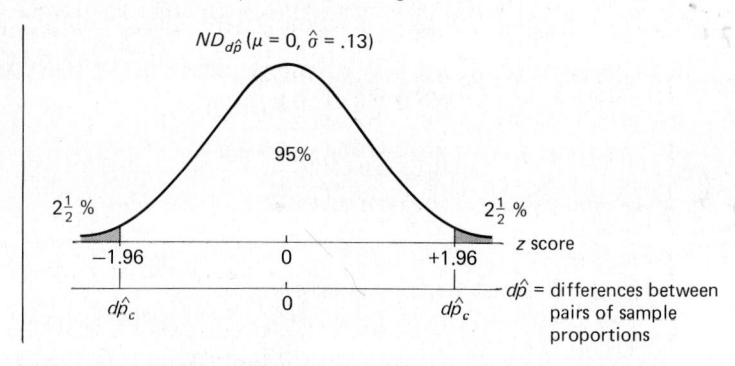

$ND_{d\hat{p}}(\mu = 0, \hat{\sigma} = .13)$

95%

$2\frac{1}{2}$% $2\frac{1}{2}$%

−1.96   0   +1.96   z score

$d\hat{p}_c$ 0 $d\hat{p}_c$ $d\hat{p}$ = differences between pairs of sample proportions

$$d\hat{p}_c = \mu_{d\hat{p}} + z_c \hat{\sigma}_{d\hat{p}}$$
$$= 0 + (\pm 1.96)(.13) = \pm .25$$

My decision rule will be to reject H_0 if the sample difference is less than $-.25$ or greater than .25.

The sample difference $d\hat{p} = \hat{p}_1 - \hat{p}_2 = .43 - .50 = -.07$. Therefore, I have failed to reject the null hypothesis, that is, I have failed to prove that there is any significant difference between the attendance at religious services and attendance at Washington College (at the .05 significance level). I feel that it is unlikely that those who did not respond would have changed the outcome significantly.

Hypothesis Testing with Sample Means: Large Samples

10

A young man being offered a job as a secretary in a large company asked the personnel director what the average age of the secretaries in the company was. She replied that the company employed several hundred secretaries, and she did not know the correct answer. But looking around the personnel office at the 38 secretaries there, she said that the average age in her office was about 20 and that the secretaries had been selected at random from the secretarial pool. The young man figured that the average for the whole company could not be much different from that, and so he agreed to work for the firm.

The young man has informally performed a type of hypothesis test which is very commonly done by statisticians. He came to a conclusion about a *population mean,* denoted by μ_{pop}, on the basis of what he knew about the *mean of a sample* taken from that population. Such a mean is referred to briefly as a **sample mean,** denoted m. In this chapter we will explain formally how to conduct hypothesis tests with sample means, but the basic ideas are exactly the same as those of the hypothesis-testing procedures you already know. Precisely the same seven steps are carried out. The only differences will be these:

1. The hypotheses will be statements about μ instead of p.

2. The formulas for computing the mean and the standard deviation of the normal curve will be changed.

THEORETICAL DISTRIBUTION OF SAMPLE MEANS

Imagine that the young man of our example goes from office to office and *for each office separately* computes the mean age of the secretaries. By the end

Understanding statistics

of the day he would have a long list of sample means, and these averages would certainly vary. Such a list of averages is called a **distribution of sample means.** Theoretically, you can imagine that the ages of all the secretaries in the company could be written, one per slip of paper, and put into a giant drum from which they could be picked at random. Suppose the young man picked 38 slips at a time, computed the mean, wrote it down, and returned the slips to the drum to be remixed. You can see that he could do this hour after hour, getting a longer and longer distribution of sample means. A typical distribution of sample means could be listed as shown in Table 10-1.

Table 10-1 Distribution of sample means

sample number (each sample composed of 38 ages)	sample mean, m (average age of the 38 secretaries sampled)
1	26
2	22
3	19
4	21
5	29
6	20
7	20
8	19
9	18
10	26
11	21
12	20
13	20
14	19
15	22
16	21
17	20
.	.
.	.
.	.

The mean of the distribution of sample means is denoted by

μ_m or $\mu_{\bar{x}}$

The standard deviation of the distribution of sample means is denoted by

σ_m or $\sigma_{\bar{x}}$

Mathematicians have analyzed this type of distribution and have learned some useful facts about what would appear after many, many sample means have been computed. The situation is summarized by the **central limit theorem.**

CENTRAL LIMIT THEOREM

Under most circumstances the distribution of means of **large samples** taken from a population has, in theory, three characteristics.

Hypothesis testing with sample means: large samples

1. The shape is *normal*.

2. The mean μ_m is the *same* as the mean of the original population,

$$\mu_m = \mu_{pop}$$

3. The standard deviation is *smaller* than the standard deviation of the original population. How much smaller depends on the sample size,

$$\sigma_m = \frac{\sigma_{pop}}{\sqrt{n}}$$

How accurately the theorem holds in any given application depends mainly on two factors.

1. The size n of each of the many samples. In theory the larger the sample size, the more closely the distribution is normal in shape. For most applications, a sample where n is *larger than 30* will lead to a distribution of means close enough to normal so that calculations based on the normal curve table will be reasonable.

2. The "shape" of the original population. The closer the original population is to normal, the smaller the samples we can use and expect our sample means to have approximately a normal distribution. In most statistical applications the populations are such that you can assume that the theorem holds. We will assume throughout the text that there is no problem on this account.

This theorem is really quite useful because it gives the properties of distributions of sample means. This in turn allows you in an experiment to estimate how far the population mean is likely to be from the experimental sample mean. And since in many experiments the main question concerns the population mean, the theorem allows you to "connect" your sample evidence to the main question of interest.

For example, let us see how the theorem can help with the secretary problem. We can indicate in a sketch (Figure 10-1) the relation between the distribution of the ages of the individual secretaries and the distribution of sample

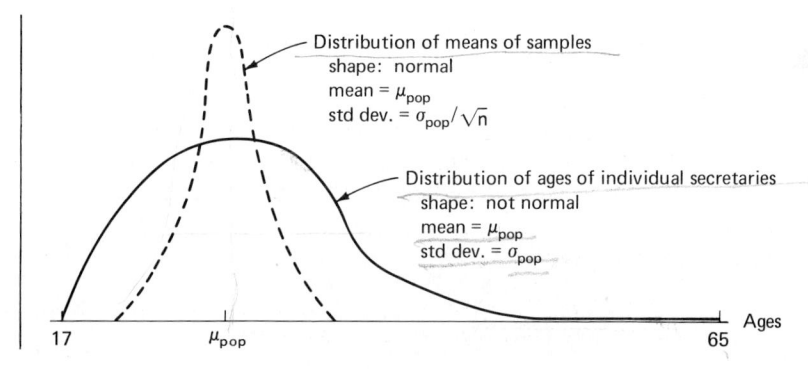

Figure 10-1

Distribution of means of samples
shape: normal
mean = μ_{pop}
std dev. = $\sigma_{pop}/\sqrt{n}$

Distribution of ages of individual secretaries
shape: not normal
mean = μ_{pop}
std dev. = σ_{pop}

17   μ_{pop} 65   Ages

Understanding statistics

means as given by the central limit theorem. We would know, for example, that most of the sample means cluster near the population mean. More precisely, since the distribution is normal, we know that only 5 percent of the sample means are more than 1.96 standard deviations away from the population mean. Another way of looking at this is as follows. If we select one sample at random, the probability that the mean of this sample will be more than 1.96 standard deviations from the population mean is less than .05.

This information can be very helpful. For example, suppose the claim were made that the mean age of secretaries in the company was 20, but when the young man took his sample of 38 secretaries, he found that his sample mean was 47. He would think in this situation, "This sample mean is so far away from the *supposed* true mean that the probability of my having obtained this result is extremely small. Either a fantastically unlikely event has happened, or somebody lied to me about the population mean." He would make his job decision according to how he felt about these two alternatives. In formal hypothesis tests, when the sample mean is so far from the claimed population mean that the probability of its occurrence is less than some specified small number α, statisticians do *not* decide that they have just witnessed a fantastically unlikely event. They reason that what they witnessed *was* likely, and on the basis reject the claim about the population mean.

EXAMPLE 10-1 Suppose we take as our population the weights of all U.S. Army lieutenants. Also, suppose it happens to be true that the mean weight μ_{pop} equals 159 pounds, and the standard deviation σ_{pop} equals 24 pounds. Describe the theoretical distribution of sample means that you would get by taking many, many random samples of size 36.

SOLUTION (a) The distribution of sample means will be *normal*, since $n = 36$ is bigger than 30.

(b) $\mu_m = \mu_{pop} = 159$ pounds

(c) $\sigma_m = \dfrac{\sigma_{pop}}{\sqrt{n}} = \dfrac{24}{\sqrt{36}} = \dfrac{24}{6} = 4$ pounds

Figure 10-2

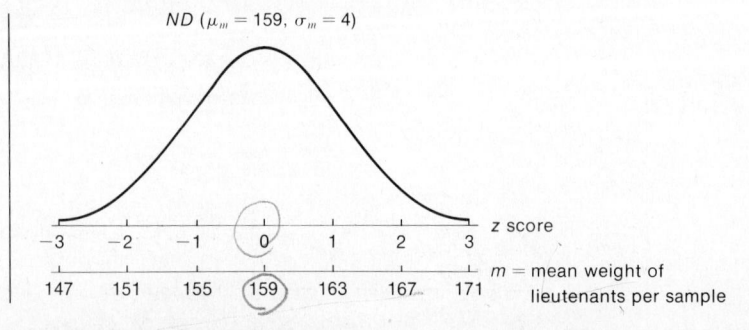

ND ($\mu_m = 159$, $\sigma_m = 4$)

z score:	-3	-2	-1	0	1	2	3
m = mean weight of lieutenants per sample:	147	151	155	159	163	167	171

Recall that the standard deviation measures the variability in a distribution. The standard deviation of the population σ_{pop} reflects the variability among *individual weights*. These weights might range from 120 to 200. The standard deviation of the sample means σ_m reflects the variability among the means of samples of 36 weights, and it would be quite unlikely to have a *mean* of 120 pounds

Hypothesis testing with sample means: large samples

in a random sample of 36 lieutenants. That is because in each sample there will probably be both light and heavy people whose weights will tend to balance one another. These means tend to vary much less, and hence σ_m is smaller than σ_{pop}. This distribution is represented in Figure 10-2.

ESTIMATING THE STANDARD DEVIATION OF A DISTRIBUTION OF SAMPLE MEANS

In hypothesis tests based on the central limit theorem it is necessary to know the standard deviation of the population in order to calculate the standard deviation of the distribution of sample means. Recall that

$$\sigma_m = \frac{\sigma_{pop}}{\sqrt{n}}$$

Often, however, statisticians wish to use the theorem when they do not know σ_{pop}. They must then do two things.

1. Find an estimate for σ_{pop}. You will recall that we denote this by s.
2. Use s to get an estimate for σ_m, denoted by s_m, that is,

$$s_m = \frac{s}{\sqrt{n}}$$

EXAMPLE 10-2 A claim is made that the American family, on the average, produces 5.2 pounds of organic garbage per day. A public health officer feels that the figure is probably incorrect. To test this, an experiment is set up to be analyzed at the .05 significance level. 40 families are chosen at random and their organic garbage for 1 day is weighed. The results are shown in Table 10-2.

SOLUTION From the data the health officer computes:

$n = 40$
$\Sigma X = 180.3$
$(\Sigma X)^2 = 32{,}508.09$
$\Sigma(X^2) = 883.65$

Step 1 H_a: the claim made about the organic garbage produced by American families is false, $\mu_{pop} \neq 5.2$.

Step 2 H_0: $\mu_{pop} = 5.2$.

Step 3 $\alpha = .05$.

Step 4 $n = 40 > 30$. Therefore the distribution of sample means will be approximately normal.

Step 5 Since H_a involves $\neq$, this is a two-tail test with .025 in each tail, so $z_c = \pm 1.96$. Under the assumption of the null hypothesis,

$\mu_m = \mu_{pop} = 5.2$

$$\sigma_m = \frac{\sigma_{pop}}{\sqrt{n}}$$

Table 10-2 Results of Garbage-Weighing Experiment for 1 Random Sample of 40 Families

family number	X, pounds of garbage
1	2.6
2	4.8
3	5.0
4	7.3
5	2.2
6	3.4
7	4.6
8	5.8
9	5.0
10	4.0
11	3.1
12	2.2
13	5.1
14	4.7
15	4.8
16	3.0
17	7.3
18	7.1
19	6.2
20	6.0
21	4.3
22	4.2
23	4.1
24	4.0
25	3.6
26	3.8
27	7.0
28	6.2
29	5.5
30	4.3
31	4.2
32	3.2
33	2.7
34	4.0
35	4.0
36	3.2
37	4.1
38	4.0
39	4.2
40	5.5

$n = 40$ $\Sigma X = 180.3$

Since σ_{pop} is unknown, he must compute s:

$$s = \sqrt{\frac{\Sigma(X^2) - \frac{(\Sigma X)^2}{n}}{n - 1}}$$

Hypothesis testing with sample means: large samples

$$= \sqrt{\frac{883.65 - \frac{32{,}508.09}{40}}{40 - 1}} = \sqrt{\frac{883.65 - 812.70}{39}} = \sqrt{\frac{70.95}{39}}$$

$$= \sqrt{1.82} = 1.35$$

Since σ_{pop} is unknown and is estimated by s, we estimate σ_m by s_m. Therefore,

$$s_m = \frac{s}{\sqrt{n}} = \frac{1.35}{\sqrt{40}} = \frac{1.35}{6.32} = .21$$

This distribution is illustrated in Figure 10-3.

Figure 10-3

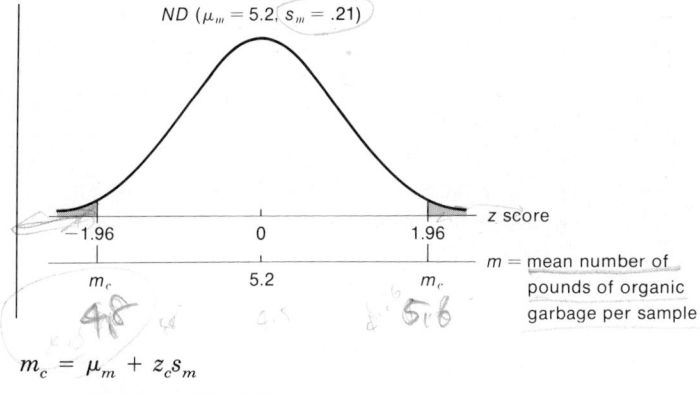

ND ($\mu_m = 5.2$, $s_m = .21$)

z score

m = mean number of pounds of organic garbage per sample

$m_c = \mu_m + z_c s_m$
$= 5.2 + (\pm 1.96)(.21)$
$= 5.2 \pm .41 = 4.8$ and 5.6

The decision rule is that a sample mean outside the range 4.8 to 5.6 will lead to rejection of the null hypothesis that the mean of the population is 5.2 pounds of garbage.

Step 6 *Outcome of Experiment* The sample mean is

$$m = \frac{\Sigma X}{n} = \frac{180.3}{40} = 4.5 \text{ pounds of garbage}$$

Step 7 *Conclusion* The outcome 4.5 is outside the range 4.8 to 5.6. Based on this evidence, the health officer rejects the null hypothesis that the average amount of organic garbage in the population is 5.2 pounds. Evidently, it is less.

EXERCISES

10-1 In the class survey you took in Chapter 1 you found the ages of this class. If you were to find the mean, the range, and the standard deviation of these ages (you may have done this in Chapter 2, but you can answer this question whether or not you know the numerical values), how would they compare with the mean, range, and standard deviation of the ages of the entire school? *Hint:* do you think that this class contains the oldest student on campus? the youngest?

Understanding statistics

10-2 A commuter buys peanuts from a vending machine each evening on his way home from work. On the last 40 purchases he received the following numbers of peanuts per purchase: 12, 10, 0, 5, 15, 16, 20, 3, 12, 0, 12, 10, 9, 11, 8, 13, 15, 20, 18, 19, 20, 0, 14, 13, 15, 16, 15, 19, 11, 10, 10, 10, 3, 8, 2, 0, 0, 20, 12, 12.
(a) What is the population being sampled?
Considering these 40 purchases as one sample of size 40 from the population:
(b) Compute the mean of the sample.
(c) Estimate the mean of the population.
(d) Estimate the standard deviation of the population.
(e) Describe the distribution of sample means.
(f) Estimate the mean of the distribution of sample means.
(g) Estimate the standard deviation of the distribution of sample means.

10-3 A random sample of the contributions of physicians to the United Fund was taken; 50 doctors were sampled. The results in dollars were as follows: 100, 95, 92, 92, 91, 90, 86, 85, 81, 80, 76, 76, 73, 73, 70, 70, 69, 69, 67, 66, 65, 61, 57, 52, 50, 49, 48, 47, 45, 39, 35, 35, 35, 35, 35, 30, 30, 30, 25, 25, 20, 20, 15, 15, 10, 10, 9, 5, 5, 0. Answer parts (a) to (g) from Exercise 10-2.

10-4 A study was done in Europe recently to investigate the health hazards of working long hours in front of computer or word processor video displays. It found that it took an average of 2.6 hours before a certain symptom of eye strain developed. If a similar experiment in the United States using $n = 100$ showed $m = 2.8$ hours, with $s = .5$ hour, would this indicate that the American results are in conflict with the European results? $\alpha = .05$. Would you think that the difference between 2.8 and 2.6 hours is important in any practical way?

10-5 Brad Brandt bands brants. As a government ecologist, he has put leg bands on thousands of brants in order to study their migratory habits. He bands, on the average, 50 brants per week. This distribution of the number of brants banded is approximately normal with a standard deviation of 7 brants banded. His supervisor periodically checks up on him and the number of brants banded.
(a) What is the probability that the supervisor randomly picks a week in which Brad has banded less than 40 birds? *Hint:* The number of bandings is a whole number.
(b) If the supervisor randomly picks 36 weeks of the last 3 years' work, what is the probability that she gets an average of less than 45 birds banded per week? *Hint:* Consider the averages to be continuous data.

10-6 Ms. Kupp, the owner of Mae's cosmetic firm, commissioned a study of her customers' buying habits. Among the results, Mae learned that her customers bought on the average 7.3 tubes of *Passion Flower* lipstick per annum. This distribution was approximately normally distributed with a standard deviation of 2.3 tubes of lipstick.
(a) What is the probability that a customer picked at random bought 10 or more tubes of *Passion Flower*? *Hint:* The number of tubes purchased is a whole number.
(b) If a random sample of 100 customers is taken, what is the probability that the sample average is more than 8 tubes of *Passion Flower*? *Hint:* Consider the averages to be continuous data.

10-7 A college admissions officer believes that this year's freshman class is

Hypothesis testing with sample means: large samples

superior in math aptitude to previous freshman classes. The mean score on the math aptitude test for previous classes was 470 with $\sigma = 120$.

(a) Assuming that σ is the same for this year's class, if you test at the .01 significance level a random sample of 400 freshmen, what is the critical aptitude-test score to support the officer's belief?

(b) If the sample of this year's freshmen has a mean score $m = 490$, would this support the officer's belief?

10-8 A newspaper article states that college students at a large state university campus spend on the average $56 a year on illegal drugs. A student investigator wishes to test this hypothesis at the .05 significance level. He gets a random sample of 144 students and finds that $m = \$70$ and $s = \$54$ for this sample.

(a) What are the critical amounts of money for the test?
(b) Does the evidence support the claim of the newspaper article?

10-9 In an inspection of a chemical plant, the inspector, John Smith, takes a specified quantity of a particular chemical, puts it on a microscope slide, and counts the number of microorganisms in it. The mean number of microorganisms in this quantity of chemical is supposed to be 1000. In an inspection one day, John finds for a random sample of 36 slides that $m = 1030$ and $s = 180$. If the mean truly is 1000, find the probability that Mr. Smith would get a sample mean of 1030 or more.

10-10 A claim is published that in a certain area of high unemployment, $95 is the average amount spent on food per week by a family of four. A home economist wants to test this claim against the suspicion that the true average is lower than $95. She surveys a random sample of 36 families from the locality and finds $m = \$93.20$ and $s = \$4.80$. Testing at $\alpha = .05$, what should be her conclusion?

10-11 The mean grade-point average of graduating college seniors who have been admitted to graduate school is 3.1, where an A is given 4 points. At Ivy University a random sample of 36 incoming graduate students yielded a grade-point average of 3.2 with a standard deviation of .24. Can we claim that the students going to Ivy have better grades than the national average, using the .01 significance level?

10-12 A normal distribution of the diameters of marbles has a mean equal to .52 inch and a standard deviation equal to .04 inch. If we took a random sample of 50 marbles, what is the probability that our sample mean will lie between .53 and .55 inch?

TWO-SAMPLE TESTS OF MEANS

Very often a statistician wishes to compare the means of *two populations*. One way to do this is to examine the **difference between the means of samples** taken from each of the populations.

EXAMPLE 10-3 A spokeswoman for a women's group wishes to present evidence to support the claim that in their first year of employment, male scientists in industry are paid more than female scientists doing the same work. She gathers data from two random samples as shown in the following table.

Understanding statistics

	male	female
n sample size	100	86
m sample mean	$22,400	$21,300
s estimate of population standard deviation based on sample data	$1200	$1000

It is clear that for the 186 scientists sampled the men are better paid on the average. The difference between the two sample means is $22,400 − $21,300 = $1100. The question for the statistician is: is this a statistically significant difference? Should we infer from it that the mean of the *entire population* of first-year male scientists is higher than the mean of the *entire population* of first-year female scientists? Or is it possible that the means of the two populations are equal, and this outcome happened just by chance—just because of the 186 people we happened to select?

■ When we are faced with the question: is the difference between two observed sample means statistically significant?, then the situation has to be analyzed in a manner similar to the one you used in Chapter 9, where you tested the difference between two sample *proportions*. In theory we can think of the sampling of male and female scientists being repeated many, many times with pairs of samples of 100 men and 86 women. If we actually did this, our results might look like those listed in Table 10-3.

Table 10-3

	mean salary for men m_1	mean salary for women m_2	differences of means $dm = m_1 - m_2$
results of first sampling	$22,400	$21,300	$1100
results of second sampling	22,200	21,700	500
results of third sampling	22,100	22,300	−200
results of fourth sampling	22,104	22,102	2
.	.	.	.
.	.	.	.
.	.	.	.

When both samples are large, that is, both n_1 and n_2 are bigger than 30, then the distribution of numbers in the third column, the differences of sample means, is approximately normal. The mean of these dm's, μ_{dm}, is equal to $\mu_{\text{pop 1}} - \mu_{\text{pop 2}}$.

Notice that if the men actually had the larger salaries, then the differences would tend to be positive, and so μ_{dm} would be greater than zero. If the women actually had the larger salaries, then the differences would tend to be negative, so that μ_{dm} would be less than zero. If there were really no differences between the male and female salaries, then some differences would be positive and some negative, and they would tend to cancel each other out, so that μ_{dm} would be zero. Note that our interpretation of μ_{dm} being positive or negative depended upon which population was called population 1. We summarize these ideas with the following theorem.

Hypothesis testing with sample means: large samples

THEOREM ABOUT THE DIFFERENCE BETWEEN MEANS OF TWO LARGE SAMPLES

1. Suppose there are two populations, call them population 1 and population 2.
2. A large random sample of size n_1 is picked from population 1, and a separate and independent large random sample of size n_2 is picked from population 2.†
3. The *mean* of each sample is computed.
4. The *difference between these two means* is written down.
5. Steps 2, 3, and 4 are repeated many times (in theory an infinite number of times), giving a long list of differences. This list of differences is called the **distribution of differences of means** (denoted by dm).

CONCLUSION

The following statements are true about the distribution of differences of sample means.

1. The distribution of differences is approximately *normal*.
2. The mean of the differences $\mu_{dm} = \mu_{\text{pop 1}} - \mu_{\text{pop 2}}$. If we are assuming for our null hypothesis that $\mu_{\text{pop 1}} = \mu_{\text{pop 2}}$, then $\mu_{dm} = 0$.
3. The standard deviation of the differences σ_{dm} is given by the formula

$$\sigma_{dm} = \sqrt{\frac{\sigma_{\text{pop 1}}^2}{n_1} + \frac{\sigma_{\text{pop 2}}^2}{n_2}}$$

Usually the experimenter does not know $\sigma_{\text{pop 1}}$ and $\sigma_{\text{pop 2}}$, in which case they can be estimated by s_1 and s_2. Using s_1 and s_2, we get an estimate for σ_{dm}:

$$s_{dm} = \sqrt{\frac{s_1^2}{n_1} + \frac{s_2^2}{n_2}}$$

■

APPLICATION OF THE THEOREM

Let us apply the theorem to the problem of whether male scientists are paid more than female scientists during their first year of employment in industry. Recall the given information.

	male (population 1)	female (population 2)
n	100	86
m	$22,400	$21,300
s	$1200	$1000

†For good results n_1 and n_2 should each be larger than 30. In general, the larger the samples, the closer the distribution of differences is to normal.

We conduct a hypothesis test at the .05 significance level. We call the males population 1 and the females population 2.

Step 1 H_a: the mean salary for male scientists is larger than the mean salary for female scientists, $\mu_{dm} = \mu_1 - \mu_2 > 0$.

Step 2 H_0: the mean salaries are equal, $\mu_{dm} = \mu_1 - \mu_2 = 0$.

Step 3 $\alpha = .05$.

Step 4 $n_1 = 100 > 30$ and $n_2 = 86 > 30$. Therefore the distribution of differences is approximately normal.

Step 5 Since H_a involves $>$, this will be a one-tail test. It will be on the right because we chose the males to be population 1. So, $z_c = 1.65$.

$$\mu_{dm} = \mu_1 - \mu_2 = 0 \quad \text{(based on } H_0\text{)}$$

$$\sigma_{dm} = \sqrt{\frac{\sigma_1^2}{n_1} + \frac{\sigma_2^2}{n_2}}$$

Since σ_1 and σ_2 are unknown, we estimate them by s_1 and s_2. Thus we estimate σ_{dm} by s_{dm}:

$$s_{dm} = \sqrt{\frac{s_1^2}{n_1} + \frac{s_2^2}{n_2}}$$

$$= \sqrt{\frac{(1200)^2}{100} + \frac{(1000)^2}{86}} = \sqrt{\frac{1,440,000}{100} + \frac{1,000,000}{86}}$$

$$= \sqrt{14,400 + 11,627.9} = \sqrt{26,027.9} = 161.33$$

This distribution is illustrated in Figure 10-4.

Figure 10-4

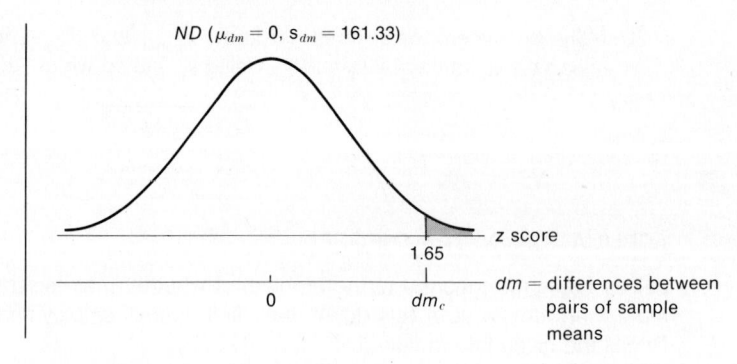

ND ($\mu_{dm} = 0$, $s_{dm} = 161.33$)

z score

dm = differences between pairs of sample means

$$dm_c = \mu_{dm} + z_c s_{dm}$$
$$= 0 + 1.65(161.33) = 266.19$$

The decision rule then is that if the difference between the two sample means is more than $266.19, we will say that the men are paid higher salaries than the women.

Step 6 The experimental results are

$$m_1 - m_2 = \$22,400 - \$21,300 = \$1100$$

Hypothesis testing with sample means: large samples

Step 7 The difference between the sample means is $1100, which is more than the critical difference, $266.19. Therefore, we claim that the average salary for the males is higher than the average salary for the females.

Under the assumption that the two populations have the same mean, we ask: what is the probability that the difference between the mean of a randomly picked sample from population 1 and the mean of a randomly picked sample from population 2 will be as large as $1100? Our analysis shows that the probability of this happening is less than .05. We conclude that since we did get a difference of $1100, then the two populations probably do not have the same mean, and the spokeswoman is correct.

STUDY AIDS

VOCABULARY

1. Sample mean
2. Distribution of sample means
3. Central limit theorem
4. Difference between two sample means
5. Large samples
6. Theorem about the difference between means of two samples
7. Distribution of differences of sample means

SYMBOLS

1. μ_{pop}
2. σ_{pop}
3. s
4. m
5. μ_m
6. σ_m
7. s_m
8. dm
9. μ_{dm}
10. σ_{dm}
11. s_{dm}

FORMULAS

One-Sample Tests

1. $\mu_m = \mu_{pop}$

2. $s = \sqrt{\dfrac{\Sigma X^2 - \dfrac{(\Sigma X)^2}{n}}{n-1}}$

3. $s_m = \dfrac{s}{\sqrt{n}}$

4. $m_c = \mu_m + z_c s_m$

5. Experimental outcome, $m = \dfrac{\Sigma X}{n}$

Two-Sample Tests

6. $\mu_{dm} = \mu_1 - \mu_2$ (if H_0 states that $\mu_1 = \mu_2$, then $\mu_1 - \mu_2 = 0$)

7. $s_{dm} = \sqrt{\dfrac{s_1^2}{n_1} + \dfrac{s_2^2}{n_2}}$

8. $dm_c = \mu_{dm} + z_c s_{dm}$

9. Experimental outcome, $dm = m_1 - m_2$

EXERCISES

10-13 All the high school students in the huge metropolis of Futura were given an aptitude test to measure their fitness as potential colonists for a space station. A guidance counselor considered each of the many schools as a random sample from the population of all the high school students in Futura. The counselor found the difference between the average score of the boys and the average score of the girls in each school. Should the graph of these differences be approximately normal? Why?

10-14 A design engineer, Wilbur Orville, wants to compare two mechanisms for use in pilot ejection seats. He makes 50 of each and subjects them to a stress test. He records the amount of stress (in pounds) that causes each to malfunction. Here are the results. Should this evidence be conclusive in favor of model 1? Let $\alpha = .01$.

model 1	model 2
$m = 600$ pounds	$m = 550$ pounds
$s = 75$ pounds	$s = 75$ pounds

1 tail test

10-15 A teacher used two different teaching methods in two similar statistics classes of 35 students each. Then each class took the same exam. In one class we get $m = 82$ and $s = 4$. In the other class we have $m = 77$ and $s = 7$. Test to see if we have evidence that one method is significantly better than the other. Use $\alpha = .05$.

10-16 In an unusual experiment, Professor Stever had some students take an exam while hanging upside down, and another group of students while lying on the floor. The results were as follows. Hanging group: $m = 52$, $s = 10$, $n = 36$. Lying group: $m = 60$, $s = 7$, $n = 36$. Does this indicate a significant difference in performance? Use $\alpha = .01$.

10-17 Have the high school averages of a college's entering freshman class gone up if one year the mean high school average of 80 freshmen picked at random is 82.5 with $s = 2.5$, while the next year the mean high school average of 84 freshmen picked at random is 83.1 with $s = 2.6$? Use $\alpha = .05$.

10-18 It is well known that for most people the older they get, the poorer their hearing becomes. A hearing test was given to a group of 40 boys (age 10) and a group of 40 men (age 50). A high score on the test means that the person could hear high-pitched sounds. The mean score for the boys was 200 with $s = 20$. The mean score for the men was 170 with $s = 20$. Show that this is statistically significant at the .05 significance level. (Recall that the phrase "statistically significant" means that the difference is large enough to call for rejection of the null hypothesis.)

10-19 In a comparison of buying habits, the following data were obtained from two samples, each consisting of 64 nuns: 10 years ago nuns bought an average of 120 habits per year with $s = 8$. Today the average is 30 habits per year with $s = 12$. Using $\alpha = .01$, does this indicate a change in the buying habits of nuns buying habits?

10-20 A school psychologist in California administered a standardized aptitude test in arithmetic to a group of 75 randomly picked sixth-grade students who had come to California from Vietnam the previous year. She gave the same test to 75 randomly picked sixth-graders who had attended California elementary schools from first grade. The mean score for the Vietnamese was 150 with $s = 25$. The California students had a mean score of 100 with $s = 40$. Show that the Vietnamese scores are significantly higher at the .01 significance level.

Hypothesis testing with sample means: large samples

10-21 A group of 40 left-handed people were asked to pick up 10 pennies quickly with their right hands. Then a group of 80 right-handed people were asked to pick up 10 pennies quickly with their left hands. The length of time each person took was recorded. The following information was gathered:

$n_1 = 40$, $m_1 = 2.8$ seconds, $s_1 = 1.0$ second
$n_2 = 80$, $m_2 = 3.2$ seconds, $s_2 = 2.0$ seconds

Is this difference significant at $\alpha = .05$?

10-22 In a carefully controlled experiment, Etherea raised 35 sunflower plants by reciting a tender poem by Kahlil Gibran to each plant whenever she fed and watered it. She also raised 35 other sunflower plants without talking to them at all. After 1 month the results were as follows. For the talked-to plants the mean growth was 10.1 inches with $s = 1$ inch. For the others, the mean growth was 9.8 inches with $s = 1$ inch. Does this indicate at $\alpha = .05$ that reciting the poem is associated with superior growth?

10-23 Professor Signo Diferens wants to see if there is any significant difference between the average grades of students who hand in their test papers early and those who hand them in later. In a recent test of 80 students, the first 40 papers had an average grade of 83 with $s = 10$. The last 40 papers had an average grade of 78 with $s = 6$. Does this information indicate a significant difference? Use $\alpha = .05$.

CLASS SURVEY

If you have data for more than 30 students, do a hypothesis test to see whether or not the average of the last digit of all social security numbers is 4½. What about the average of the fifth digit?

FIELD PROJECTS

Select one of the following as a special project.

1. To test whether or not people can pick a number at random, ask 100 people to pick a number from 1 to 10. The null hypothesis is that the mean of the 100 replies should be 5.5. Test at the 10 percent significance level.

2. Test whether male and female students on your campus carry the same average amount of coin change with them.

3. Perform a large-sample hypothesis test of your own choosing. Do either a one- or a two-sample test. Some projects which students have performed include the following.
    (a) Is there a difference between the average number of cigarettes smoked per day by the sales persons at Brandt's Department Store and the average number of cigarettes smoked per day by the stockroom employees at Brandt's?

    (b) Testing the difference between the average test grades for two different teachers.

    (c) Testing the difference between the average age of teaching faculty and the average age of the rest of the college staff.

(d) A newspaper claimed that people averaged a certain amount of money spent on entertainment each week. Test to see if the average in your neighborhood is different.

(e) A report claimed that people averaged a certain amount of time watching TV each week. Is the average at your place of employment different?

Hypothesis Testing with Sample Means: Small Samples

11

STUDENT'S t DISTRIBUTION

In our earlier discussion of the central limit theorem (Chapter 10) we said that the distribution of means of large samples from *any* population is normal in shape with

$$\mu_m = \mu_{\text{pop}} \quad \text{and} \quad \sigma_m = \frac{\sigma_{\text{pop}}}{\sqrt{n}}$$

In any given hypothesis test, of course, we construct a *particular* normal curve, usually taking for μ_{pop} a value given in our null hypothesis, and taking s from our sample data as our estimate of σ_{pop}. This procedure can be shown to produce good results both in theory and in practice. That is, when we set $\alpha = .05$, for example, in many repeated tests we actually do reject a true hypothesis about 5 percent of the time.

W. S. Gossett was a British statistician who worked in a brewery where he took small samples. In 1908 he wrote a paper under the pseudonym "A. Student," showing that if you use this same procedure when you have **small samples,** Type I errors will be made *more* than 5 percent of the time. Basically, this is due to the fact that in repeated experiments with *small* samples, the values of s tend to be quite variable, so that if in any *one* experiment you take your current value of s and use it in the formula

$$m_c = \mu + \frac{s}{\sqrt{n}} z_c$$

you really run *more* of a chance of a Type I error than z_c would indicate. To solve this problem, what Gossett did, from our point of view, was to come up with *different sets* of critical scores, called **Student's *t* scores,** to be used in place of the critical z scores, depending on how big the sample actually is. These scores are also called critical values of t. They are printed in Table C-5 for two-tail tests and in Table C-6 for one-tail tests. Gossett showed that even these t scores are *not always* reliable. They *are* reliable, however, in the case where the *original* population from which we are taking our sample is near normal to begin with.

For example, using t scores to analyze small samples of weights of loaves of bread made by Acame bakery would probably be all right, since these weights are probably distributed close to normal. But an experiment based on small samples from a nonnormal variable should not be analyzed using t scores. For example, an experiment to test a claim about the mean annual income of faculty at a certain university should not be analyzed with t scores if we can only sample, say, 20 faculty members, because incomes are known to be usually distributed in a nonnormal pattern.

In summary, the procedure we have been using with the normal curve and z scores is correct for large-sample testing from *any* population. Student's t scores are correct for small-sample testing from *normal* populations. There is no simple general approach for small samples from nonnormal populations. Certain tests, called nonparametric tests, may be useful in such cases. See Chapter 16 for a brief introduction to this topic.

When you draw your curve for a particular experiment using a small sample, you will be using a particular set of critical t scores corresponding to your sample size. A curve which corresponds to a set of t scores is called a ***t* curve.** There are many t curves, and for each sample size there is a slightly different curve. The t curves are similar to the normal curve in that they are symmetrical and bell-shaped. For your rough sketches there is no need to distinguish between the shapes. When drawn precisely, however, it can be seen that the t curves are somewhat flatter than the normal curve, and that the smaller the sample size, the flatter is the curve. Conversely, the larger the sample size, the more closely the t curve resembles the normal curve. Once the sample size is larger than 30, the t curve and the normal curve are practically the same. Many experimenters, therefore, use t curves for samples of size 30 or less, and the normal curve for samples larger than 30, although it is always correct to use t curves.

In small-sample testing, the critical t scores are numerically larger than the critical z scores from large-sample testing, and the smaller the sample, the larger must be the critical t values. You will see that this is the case by looking at the t tables (Tables C-5 and C-6). This should make sense, because in using only a small sample we should require more conclusive evidence to reject a null hypothesis.

A mean from a small sample should really be *quite different* from that predicted by the null hypothesis if we are going to use it to reject the null hypothesis. By using larger critical t values we are, in practice, requiring stronger evidence to reject a null hypothesis.

DEGREES OF FREEDOM AND USING THE *t* TABLE

To use the t table, you must find the entry that corresponds to your particular sample size. You would therefore expect the table to have a column labeled

Hypothesis testing with sample means: small samples

n for sample size. However, it turns out that this same table can be used in other problems (for example, problems involving two different samples of different sizes) where that label would not make sense. It is usual, instead, to have a column labeled **degrees of freedom**. It is not obvious why this name should be used. Technically it is used because the t curve is related to another curve we will study later (called the chi-square curve) for which the phrase "degrees of freedom" makes more intuitive sense. We will not go into this relationship in this chapter. For our purposes it is sufficient to know that in a hypothesis test about a population mean where we are working with a single small sample of size n, the correct numerical value of the degrees of freedom for t is $n - 1$.

EXAMPLES OF DEGREES OF FREEDOM

The phrase "degrees of freedom" comes up often in statistical work. Here are some examples of contexts in which it may appear.

EXAMPLE 11-1 Suppose that somebody selects two angles of a triangle as 40° and 50°. Furthermore, he also informs you that the third angle is 90°. He was free to pick different values for the first two angles, but once they were selected he was *not* free in his choice of the third angle, because the three angles must add up to 180°. We say that in the selection of the values for the three angles of a triangle you have only 2 degrees of freedom.

EXAMPLE 11-2 Again, consider a teacher who wishes to pick 5 numbers for an example. She desires that the mean of the numbers be 10. She is free to pick any 4 numbers she wishes, but the fifth number is decided. We say that in selecting 5 numbers whose mean is 10, we have only 4 degrees of freedom.

EXAMPLE 11-3 Suppose a statistician wanted to pick 200 persons so that 100 were male and 100 were female, including 50 Democrats and 150 Republicans. He would need four groups, namely, male Democrats, female Democrats, male Republicans, and female Republicans. However, for the sizes of these four groups, he has only 1 degree of freedom.

Table 11-1

	Democrats	Republicans	
male	20		100
female			100
	50	150	200

As you can see from Table 11-1, if he selects 20 male Democrats he must pick 30 female Democrats, 80 male Republicans, and 70 female Republicans.

EXAMPLE 11-4 In computing the standard deviation of 10 numbers, we use the deviations from the mean. These deviations must always add up to 0. Therefore, when we compute the standard deviation of 10 numbers, there are only 9 degrees of freedom.

Understanding statistics

In general, when we are trying to estimate the standard deviation of a population by using a sample of size n we have $n - 1$ degrees of freedom.

ILLUSTRATION OF A t TEST

EXAMPLE 11-5 A claim is made that Eagle Scouts have a mean age of 14 years. Doubting that this is true, we take a random sample of 16 Eagle Scouts. The mean age of the sample is found to be 12 years. The estimate s of the standard deviation of the population was calculated to be 4. Test at the .05 significance level, assuming that the ages of all Eagle Scouts are in fact distributed in a pattern not too different from normal.

SOLUTION
H_0: $\mu_{pop} = 14$
H_a: $\mu_{pop} \neq 14$ (Therefore, we are doing a two-tail test.)
$m = 12$
$s = 4$
$n = 16 \leq 30$ (Therefore, we will use the t distribution.)

The number of degrees of freedom is given by $n - 1$, or $16 - 1 = 15$.

$\mu_m = \mu_{pop} = 14$

$s_m = \dfrac{s}{\sqrt{n}} = \dfrac{4}{\sqrt{16}} = \dfrac{4}{4} = 1$

For this t distribution we write $tD(\mu_m = 14, s_m = 1)$. Looking up .025 and .975 at 15 degrees of freedom in Table C-5, we find our critical t scores, $t_c = \pm 2.13$. Converting these to raw scores, we get

$m_c = \mu_m + t_c s_m$
$= 14 + (\pm 2.13)(1) = 14 \pm 2.13 = 11.87$ and 16.13

Therefore, our decision rule is to reject the null hypothesis if we get a sample mean less than 11.87 or greater than 16.13. Since our outcome was 12, we have failed to establish that the mean age of Eagle Scouts is different from 14 years.

You will notice that the only difference between this problem and one with n larger than 30 is that we used $t_c = \pm 2.13$ rather than $z_c = \pm 1.96$.

EXERCISES

11-1 If you did not know any better, and you performed a z test instead of a t test on data from a small sample, would you increase or decrease α? What about β?

11-2 Which one of the following statements is true?
(a) Every normal curve is bell-shaped.
(b) Every bell-shaped curve is a normal curve.

11-3 It is claimed that a large corporation discriminates against its women employees in its promotion practices. Over many years, the mean time before first promotion for its male employees has been 3 years. A random sample of

Hypothesis testing with sample means: small samples

20 females who had worked there many years showed a mean time of 3.8 years before first promotion, with $s = 1.2$ years. Using $\alpha = .05$, see if the data support the claim.

11-4 A species of peccary formerly thought to be extinct was recently discovered living in Paraguay. The basis for this claim is that several different skull measurements are very near to the measurements of the ancient fossils. For example, a sample of 13 Paraguayan peccaries had a mean width across the canines of 60.8 mm with $s = 1.8$ mm. The extinct peccary has a mean width of 60 mm across the canines. Assuming that these measurements are from a normal distribution, show at $\alpha = .05$ that you cannot reject the hypothesis that both species have the same average width across the canines.

11-5 The average IQ of Martians is quoted to be 260. Thinking that this figure is too large, an earthling tested the crew of the first Martian spaceship to land in New Jersey. The average IQ of the 8 crew members was 250 with $s = 8$. Test at $\alpha = .01$, assuming that these IQs are distributed near normal.

11-6 The secretary of an association of professional landscape gardeners claims that the average cost of services to customers is $90 per month. Feeling that this figure is too low, we question a random sample of 10 customers. Our sample yields a mean cost of $125 with $s = $20. Test at the .05 significance level. Assume that such costs are normally distributed.

11-7 A claim is made that the mean height of police officers is 5 feet 11 inches. To test whether this claim is true or not, a random sample of 25 police officers was gathered. This sample had a mean height of 5 feet 10 inches with $s = 3$ inches. Test at the .05 significance level. Assume that heights are normally distributed.

11-8 The mean score on a standardized psychology test is supposed to be 50. Believing that a group of psychologists will score higher, we test a random sample of 11 psychologists. Their mean score is 45 with $s = 3$. Test at $\alpha = .01$. Assume that such scores are distributed near normal.

11-9 Does the average lecture consist of 3000 words if a sample of the lectures of 16 professors had a mean of 3472 words with $s = 500$ words? Use $\alpha = .01$. Assume that lecture lengths are approximately normally distributed.

11-10 An astronomer is testing the claim that the mean brightness of a certain star is now more than 30 units. She is able to get 6 readings on the star during her experiment. Using $\alpha = .01$, does this indicate a new mean brightness of more than 30?

reading	brightness
1	30
2	30
3	29
4	31
5	32
6	33

Assume that such readings are normally distributed.

11-11 Count Fatchula supervises students who are training to be hematologists. For one project his 8 students had to count certain cell types in blood samples. Their counts were 103, 75, 82, 109, 63, 240, 81, and 72. Does this support the hypothesis that the mean count is 100? Use $\alpha = .05$. Assume that the cell type is normally distributed.

Understanding statistics

11-12 If a population is *known* to be normally distributed, and the value of σ is *known* also, then the z scores may be used to test hypotheses regardless of the size of n. Repeat Exercise 11-7 with $\sigma = 3$ instead of $s = 3$.

11-13 Skippy, the skeptic, doubts his statistics teacher's claim that 95 percent of the means of samples of size 9 will have a t score less than 1.86. On his computer terminal a subroutine is available that picks a number randomly from a normal distribution with mean equal to zero. He programs the computer to pick 9 of these random numbers. Then he computes m, s, and s_m. Finally, using

$$t = \frac{m - \mu}{s_m}$$

he computes the t score for his sample mean. Using a loop, he has the computer repeat this experiment 500 times; 468 times he gets a t score less than 1.86! Therefore he claims that his teacher was wrong. Test using $\alpha = .05$.

TWO-SAMPLE *t* TESTS

The problem of testing the difference between two means when one or both samples are small (less than 30) is simplest when we can make two additional assumptions.

1. Both populations are approximately normal.

2. The variances of the two populations are approximately equal.

When these assumptions are reasonable, it follows that the distribution of the differences of sample means can be related to the t distribution. To be able to use the t table, we will need to know the appropriate degrees of freedom. To calculate the critical difference we will need a formula for s_{dm}, the standard deviation of the sample differences. We give both of these results:

Degrees of freedom $= (n_1 - 1) + (n_2 - 1) = n_1 + n_2 - 2$

$$s_{dm} = \sqrt{\frac{s_p^2}{n_1} + \frac{s_p^2}{n_2}} = \sqrt{s_p^2\left(\frac{1}{n_1} + \frac{1}{n_2}\right)}$$

where s_p^2 is the pooled estimate of the variance

$$s_p^2 = \frac{(n_1 - 1)s_1^2 + (n_2 - 1)s_2^2}{n_1 + n_2 - 2}$$

Note that s_p^2 is a weighted average of the two variance estimates. According to assumption 2, both sample variances are estimating the same true variance. So it makes sense to combine both estimates. The weighting simply assigns more weight to the estimate from the bigger sample. If both samples happen to be the same size, then

$$s_p^2 = \frac{s_1^2 + s_2^2}{2}$$

Note: If the number of degrees of freedom given by $n_1 + n_2 - 2$ is not in our table, we will approximate it by the nearest value in the table. If the number

Hypothesis testing with sample means: small samples

EXAMPLE 11-6 A claim is made that Harvard graduates have higher IQs than Havemor graduates. In an attempt to establish this claim at the .01 significance level, we select a random sample from each population. (Assume that both populations are normal.) In a sample of 15 Harvard graduates, the mean IQ was 120 with $s_1 = 10$, while in a sample of 18 Havemor graduates, the mean was 110 with $s_2 = 5$.

SOLUTION $n_1 = 15 \qquad n_2 = 18$
$m_1 = 120 \qquad m_2 = 110$
$s_1 = 10 \qquad s_2 = 8$

H_0: mean IQ for both groups of graduates is the same, $\mu_{dm} = \mu_1 - \mu_2 = 0$
H_a: mean IQ of Harvard graduates is higher than mean IQ of Havemor graduates, $\mu_{dm} = \mu_1 - \mu_2 > 0$ (Therefore we have a one-tail test to the right.)

Degrees of freedom = $15 + 18 - 2 = 31 \qquad \alpha = .01$

Since the number of degrees of freedom, 31, is not in our table, we use the nearest value, namely, degrees of freedom = 30. Looking up the critical t score, we find $t_c = 2.46$.

We find the pooled value of s^2:

$$s_p^2 = \frac{(n_1 - 1)s_1^2 + (n_2 - 1)s_2^2}{n_1 + n_2 - 2}$$

$$= \frac{14(100) + 17(64)}{15 + 18 - 2} = \frac{2488}{31} = 80.26$$

$$s_{dm} = \sqrt{s_p^2 \left(\frac{1}{n_1} + \frac{1}{n_2}\right)}$$

$$= \sqrt{80.26\left(\frac{1}{15} + \frac{1}{18}\right)} = \sqrt{9.81} = 3.13$$

We have $tD(\mu_{dm} = 0, s_{dm} = 3.13)$, so that

$dm_c = \mu_{dm} + t_c s_{dm}$
$= 0 + 2.46(3.13) = 7.70$

Therefore, our decision rule is to reject the null hypothesis if we get a difference greater than 7.70. Our experimental outcome, $dm = 120 - 110 = 10$, is greater than 7.70. Therefore, we reject the null hypothesis at the .01 significance level and claim that Harvard graduates have higher IQs than Havemor graduates.

EXERCISES

In these exercises assume that the populations being sampled are approximately normal, and that their variances are approximately equal.

11-14 Goodbenefits Manufacturing Company replaced the Ratari game machine in the employees' lunchroom because of the number of wrist injuries. Dr. Leech, the company doctor, notes that with the new Satari game machine there have been an average of 38 wrist injuries per day over the past 14 working days, with a standard deviation of 4 wrist injuries. Checking back in his previous records, he discovers that with the old Ratari machine there were an average of 40 injuries over the last 10 days they had the machine. The standard deviation was also 4. If these averages are significantly different ($\alpha = .05$), Dr. Leech thinks he can write an article, which will be published in the local medical journal. Are they?

11-15 To test whether or not there is a difference between the grade-point averages of male math majors and female math majors at State U, two random samples were gathered. For the 35 males we find that $m_1 = 3.1$ with $s_1 = .2$. For the 20 females, $m_2 = 3.2$ with $s_2 = .15$. Test at $\alpha = .01$.

11-16 To find out whether fluoride treatment is effective in reducing the occurrence of cavities, 15 children who had fluoride treatment were compared with 15 who did not. Of those who used fluoride, the mean number of cavities m_1 was .8 with $s_1 = .7$. For those who did not use fluoride, m_2 was 1.4 with $s_2 = .9$. Test at the .05 significance level.

11-17 Exemptions from final exams may improve the grades of students. To test this idea, 40 students of similar background and ability in French were picked at random and placed in two classes of 20 students each. In one class the students were told that if they averaged 90 or more on all classwork during the term they would be exempt from the final exam. In the second class the students were told that no exemptions would be given. The average mark on classwork in the first class was 83.3 with $s_1 = 9.4$. The average in the second class was 80.7 with $s_2 = 11.1$. Test using $\alpha = .05$.

11-18 An inventor states that his new method for packing eggs is safer than the old method. Two shipments of eggs are packed, one by the former method and one by this new method. After the eggs are delivered, a sample of 10 gross from each shipment was inspected, and the number of cracked eggs in each gross was counted. In the first sample, the average number of cracked eggs per gross m_1 was 7.2 with $s_1 = 3.2$. In the second sample, m_2 was only 4.1 with $s_2 = 3.0$. Test at the .01 significance level. *Hint:* $n_1 = 10$ and $n_2 = 10$.

11-19 To see whether there is a statistically significant difference between the age of owners of convertibles and the age of owners of sedans, a Madison Avenue advertising agency gathered two random samples. It was found that 15 convertible owners had an average age of 29.2 years with $s_1 = 5$, and 35 sedan owners had an average age of 24.8 years with $s_2 = 8$. Test at the .05 significance level.

11-20 A study was conducted recently of the temperatures of spark plugs when cars were driven at 50 mph and at 55 mph. The study found that of 22 spark plugs tested at 50 mph, their temperatures in °F were as follows: 900°, 920°, 860°, 890°, 910°, 820°, 950°, 880°, 930°, 870°, 900°, 850°, 910°, 830°, 930°, 950°, 840°, 860°, 900°, 900°, 960°, 890°. The 16 spark plugs tested at 55 mph had temperatures of 1000°, 930°, 900°, 1100°, 950°, 1070°, 890°, 1050°, 910°, 1090°, 870°, 1130°, 1000°, 1110°, 1050°, 950°. Do these data suggest that the average temperature is different at 50 mph? Use the .01 significance level.

Hypothesis testing with sample means: small samples

11-21 Amy and Josh go outside in the evenings to collect toads. In June they caught 50 toads with a mean weight of 1.1 ounces and with $s = .05$ ounce. In August they caught 25 toads with a mean weight of 2.1 ounces and $s = .07$ ounce. Does this indicate at the .01 significance level that August toads are heavier than June toads?

11-22 People were asked how long after a meal they became hungry. A random sample of 10 patrons of How Long's restaurant yielded a mean of 1.3 hours and a standard deviation of .5 hour. A sample of 15 patrons of Sum Fun's restaurant had a mean of 1.7 hours and a standard deviation of .2 hour. Is this difference significant at $\alpha = .05$?

11-23 Is there a significant difference between the bowling averages of male nurses and those of businessmen, if last year's league play produced the following data: 10 male nurses averaged 180 with $s = 10$, and 15 businessmen averaged 170 with $s = 10$. Use $\alpha = .05$.

11-24 Casey is batting against a baseball pitching machine. The machine pitches either fast or very fast. The distances of all of Casey's fly balls are recorded. Test at the .05 significance level to see if the very fast pitches tend to be hit further. The data were as follows.

| fast | 250 | 300 | 275 | 280 | 350 | 350 |
| very fast | 300 | 290 | 400 | 325 | 330 | 400 |

11-25 Are Krispy toasters quicker than No-Burn toasters, at the .01 significance level, if the times it took to toast a slice of bread were as follows.

| No-Burn | 60, 55, 70, 65, 80 seconds |
| Krispy | 70, 65, 65, 60, 60 seconds |

11-26 The two formulas for s_p^2 are:

$$\frac{(n_1 - 1)s_1^2 + (n_2 - 1)s_2^2}{n_1 + n_2 - 2} \quad \text{and} \quad \frac{s_1^2 + s_2^2}{2}$$

(a) How different are they?
(b) Calculate s_p^2 by both formulas, using various values of s_1, s_2, n_1, and n_2, such as the following.

s_1	s_2	n_1	n_2
4	6	100	101
50	60	70	80
10	10	50	500
.	.	.	.

If you have access to a computer, you can easily compare many values.

A t TEST FOR PAIRED DIFFERENCES

EXAMPLE 11-7 In Exercise 11-17 we looked at an experiment designed to see if exemptions from final exams improve grades. The two-sample t test showed "no significant difference." For that experiment we randomly placed 40 students in two classes of 20 students each. We say that the two samples were chosen independently of one another because there is no particular connection between the students in one class and those in the other. There are other ways of deciding which students go into which class. For example, suppose we first rank the students from 1 to 40 in terms of their past grades in French. We then pair them off as shown in Table 11-2.

Table 11-2

student numbers	student numbers
1 and 2	21 and 22
3 and 4	23 and 24
5 and 6	25 and 26
7 and 8	27 and 28
9 and 10	29 and 30
11 and 12	31 and 32
13 and 14	33 and 34
15 and 16	35 and 36
17 and 18	37 and 38
19 and 20	39 and 40

This puts two students of *similar ability* in each pair. One student has an odd number and one has an even number. Now we decide which student in each pair goes into the first class and which one goes into the second class. We toss a coin 20 times, once for each pair. We let heads mean that the odd-numbered student goes into the first class. Tails means that the odd-numbered student goes in the second class. If our 20 tosses result in

H T T T T   H T T H T   T H T H T   H H H T H

then our two classes would be as shown in Table 11-3.

Table 11-3

pair number	students in first class	students in second class
1	1	2
2	4	3
3	6	5
4	8	7
5	10	9
6	11	12
7	14	13
8	16	15
9	17	18
10	20	19
11	22	21
12	23	24
13	26	25
14	27	28
15	30	29
16	31	32
17	33	34
18	35	36
19	38	37
20	39	40

Hypothesis testing with sample means: small samples

The experiment has been substantially changed because each student in one class has been *paired off* with a student of similar ability in the second class. Statisticians say that this is a new *experimental design*. For this type of design it is better not to analyze these data as a two-sample t test on the difference of the sample means, because that method ignores the fact that the pairs are matched. In the case of matched pairs, the ordinary two-sample t test is a weaker approach in that it is not as likely to pick up a small but real difference between the two populations. The two-sample t test would not notice, for example, the situation where every student in the first class did better than his or her "partner" in the second class, if the first class students were only a *little* better than their partners. For a somewhat less obvious illustration let us suppose that the marks for classwork were paired as shown in Table 11-4.

Table 11-4

pair number	mark for student in first class (exemptions available)	mark for student in second class (no exemptions allowed)
1	100	98
2	96	100
3	97	95
4	92	90
5	91	91
6	93	88
7	79	80
8	79	83
9	81	71
10	86	85
11	90	88
12	80	77
13	80	74
14	82	78
15	82	80
16	75	71
17	71	60
18	73	69
19	65	72
20	75	65

We would then have a sample of 20 differences. We can consider them to be a sample of size 20 from the population of all such differences. We denote these differences by the letter d. As usual, we denote the mean of a sample (of d's) by the symbol m. Assuming that the population of all d's is approximately normal, then our experiment reduces to a *one-sample* hypothesis test using sample means, where our raw score is now d instead of X.

SOLUTION H_0: exemptions make no difference, $\mu_{pop} = 0$

H_a: exemptions tend to raise grades, $\mu_{pop} > 0$ (one-tail test)

$n = 20$ (Therefore, we have a t test with 19 degrees of freedom.)

$\mu_m = \mu_{pop} = 0$

We need to compute s_{pop} and s_m. Going back to the above data we would now have data as shown in Table 11-5.

Table 11-5

pair number	difference, d	d^2
1	+2	4
2	−4	16
3	+2	4
4	+2	4
5	0	0
6	+5	25
7	−1	1
8	−4	16
9	+10	100
10	+1	1
11	+2	4
12	+3	9
13	+6	36
14	+4	16
15	+2	4
16	+4	16
17	+11	121
18	+4	16
19	−7	49
20	+10	100
	52	542

$$m = \frac{52}{20} = 2.6$$

$$s_{\text{pop}} = \sqrt{\frac{542 - \frac{(52)^2}{20}}{19}} = 4.63$$

$$s_m = \frac{4.63}{\sqrt{20}} = 1.03$$

Therefore, we have $tD(\mu_m = 0, s_m = 1.03)$, and so

Degrees of freedom = 19

$\alpha = .05$

$t_c = 1.73$

$m_c = \mu_m + t_c s_m$

$= 0 + 1.73(1.03) = 1.8$

■ Our decision rule is then to reject the null hypothesis if the outcome is greater than 1.8. Since our outcome, $m = 2.6$, is greater than 1.8, we reject the null hypothesis. Exemptions apparently raise grades.

This type of experiment is referred to as a **matched-pair** experiment, and the d's are called the **paired differences.** This is the usual type of experiment found when we do "before-and-after" comparisons, or when we compare siblings. If the subjects in an experiment can be "matched up" in some reasonable

Hypothesis testing with sample means: small samples

way, this type of one-sample test has *greater power* than the ordinary two-sample t test. With it we are more likely to detect a difference if there is one.

STUDY AIDS

VOCABULARY

1. Small sample
2. Student's t score
3. t curve
4. Degrees of freedom
5. Matched pair
6. Paired differences

SYMBOLS

1. t
2. t_c
3. d
4. $tD(\mu_m = 3, s_m = 1)$

FORMULAS

One-Sample Tests

1. Degrees of freedom $= n - 1$
2. $\mu_m = \mu_{pop}$
3. $s_m = \dfrac{s}{\sqrt{n}}$
4. $m_c = \mu_m + t_c s_m$
5. Experimental outcome, $m = \dfrac{\Sigma X}{n}$

Two-Sample Tests

1. Degrees of freedom $= n_1 + n_2 - 2$
2. $\mu_{dm} = \mu_1 - \mu_2$. If H_0 states that $\mu_1 = \mu_2$, then $\mu_1 - \mu_2 = 0$.
3. $s_{dm} = \sqrt{s_p^2 \left(\dfrac{1}{n_1} + \dfrac{1}{n_2}\right)}$
4. $dm_c = \mu_{dm} + t_c s_{dm}$

with

$$s_p^2 = \dfrac{(n_1 - 1)s_1^2 + (n_2 - 1)s_2^2}{n_1 + n_2 - 2}$$

5. Experimental outcome, $dm = m_1 - m_2$

EXERCISES

11-27 The same data were used in two hypothesis tests in this chapter. In Exercise 11-17, using a t test on the difference between two sample means, we failed to show that granting exemptions affected grades. However, in Example 11-7, using a t test for paired differences, we proved that exemptions did affect the grades. How can the same data lead to different conclusions?

11-28 In a study of the effects of cigarette smoking on blood conditions 11 people had their blood sampled before and after they smoked a single cigarette. Each blood sample then had a chemical added to it which causes blood clots. A numerical value was given to describe the amount of clotting. Here are the results.

Understanding statistics

person	difference in clotting measure, d
1	2
2	4
3	10
4	12
5	16
6	15
7	4
8	27
9	9
10	−1
11	15

You will find that $m = 10.3$ and $s_{pop} = 8.0$. Using $\alpha = .01$, show that this is a statistically significant effect of smoking.†

11-29 (a) In order to test which brand of gasoline gives better mileage, the Acme Cab Company tried Axon gas in 10 cabs and Flug in 10 similar cabs. The cabs using Axon averaged 16.3 miles per gallon with $s_1 = 4.2$ miles per gallon. The cabs using Flug averaged 16.9 miles per gallon with $s_2 = 4.0$ miles per gallon. Test at the .01 significance level for any difference between the two brands.

(b) If Axon were used in 10 cabs for 1 week and then Flug were used in the *same* 10 cabs, we would have a paired-differences test. If $m = -.6$ with $s_{pop} = .39$, test at $\alpha = .01$.

11-30 (a) Assume that X represents the number of children that 7 randomly selected first-born males produced, and Y represents the number of children that 7 randomly selected first-born females produced.

X	3	2	4	1	2	0	5
Y	4	0	6	1	0	3	2

Test to see if $\mu_X > \mu_Y$ at $\alpha = .05$.

(b) Now assume that X represents the number of children of the male sibling and Y represents the number of children of the female sibling in a random sample of pairs consisting of a brother and a sister. Test $\mu_X > \mu_Y$ again at $\alpha = .05$.

11-31 Data were collected on 10 adults who enrolled in the Looze-A-Pound weight reduction program. Their weights were recorded before and after the program. For the 10 pairs of differences m was 5.8 pounds lost with $s = 5$. Using $\alpha = .05$, is this evidence that the program works?

11-32 Dr. Quack has invented a new way to teach reading. 40 sixth-graders were paired by reading ability and randomly divided into 2 sections. 20 were taught by Dr. Quack and 20 were taught in the ordinary manner. At the end of the course, the differences between the reading abilities of the pairs of students were computed, letting d equal the Quack score minus the ordinary score. The average difference was -3.6 with $s_{pop} = 4.0$. Is Dr. Quack's method significantly different? Use $\alpha = .05$.

†Problem based on discussion in Glantz, *Primer of Biostatistics*, McGraw-Hill Book Company, New York, 1981.

Hypothesis testing with sample means: small samples

11-33 An analysis of tennis included a study by Commissioner P. Recka of certain errors. He studied the play of 10 professionals, recording how often they missed on a backhand return and how often they missed on a forehand return, to see if there was any significant difference. Letting d equal the backhand errors minus the forehand errors, he found for these 10 players (over the course of 100 errors) a mean value of $d = 10$ with $s = 6$. Does this indicate, at $\alpha = .05$, that in general more errors are committed on backhand shots?

11-34 The difference between the length of the left arm and the length of the right arm (d equals left length minus right length) was measured on 200 right-handed marines. If $m = -.05$ inch with $s_{pop} = .015$ inch, test to see if marines have right arms that are longer than left arms. Use $\alpha = .05$.

11-35 10 students took a statistics test, and then they attended Dr. Fleece's Quickie Course in Imperative Statistics. After having completed the course, they took a test equivalent to the first one. Here are their grades.

student	before course	after course
Re	15	18
Jeanne	10	14
Barbe	13	15
Tommy	19	18
Jo	20	18
Anne	14	16
Mary Frances	3	17
John	18	17
Paul	16	14
Tim	16	15

Does the Dr.'s course significantly improve grades? Use $\alpha = .05$.

11-36 A sample of fully grown cats were fed a diet of Fatkat brand food. The results of the diet are shown in the following table.

cat	weight before	weight after
Tabby	12	7
Maurice	7	7
Samantha	5	7
Snow White	7	6
Toby	8	9
Licorice	10	9
Felix	3	3
Krazy	4	5
Chris	9	9
Apples	6	7
Pucini	8	8
Tiger	7	8

Test at the .05 significance level to see if Fatkat affects the cats' weights.

CLASS SURVEY

Consider this class as a random sample of your school. Are your class data strong enough to show that the average height of the males in this school is greater than the average height of the females in this school?

FIELD PROJECTS

Do a one-sample or a two-sample field project similar to those in Chapter 10, but using small samples, or do a project using a paired-differences design.

SAMPLE TEST FOR CHAPTERS 8, 9, 10, AND 11

1. A statistical hypothesis is a statement about a (parameter, statistic).
2. What is a Type I error? What does it have to do with "significance level"?
3. Why don't we conduct hypothesis tests using $\alpha = 0$?
4. We are testing to see if two populations have the same mean. We wish to use small samples and to be able to use a standard t test. What two assumptions must be reasonable about the two populations?

For Problems 5 to 9:
(a) *State a null and an alternative hypothesis.*
(b) *State a decision rule.*
(c) *Give a conclusion in good English mentioning the specific variables in the problems.*

5. A carnival spinning wheel is one-quarter red. This means that the wheel should stop on red about one-fourth of the time. A student wonders if the wheel is honest. What would your decision be if it came up red 12 out of 60 trials? $\alpha = .05$. Is this a one- or a two-tail test?

6. A study of a random sample of 100 parking receipts at a large airport indicated that the average time a car stayed in the "short-term" lot was 2.6 hours, with $s = 45$ minutes. Does this evidence imply, at $\alpha = .05$, that the true mean parking time is more than 2 hours and 15 minutes? *Hint:* Be careful of units. You might want to change all values to minutes, or all to hours.

7. We wish to compare two new models of solar-powered cooking units. The test is to see how long each kind takes to boil one quart of water (which is at 10°C to begin with). Five trials are done on each type of unit. The results are as shown.

	time to boil water, minutes
Sunny Boil	4.7, 5.1, 5.5, 5.7, 4.0
Sol-Heato	4.2, 4.2, 4.7, 4.5, 4.4

Is this conclusive evidence that one is superior to the other? Use $\alpha = .05$.

8. In a study for her PhD thesis, Dr. Payne subjected 100 volunteers to the following experiment. The 100 people were split randomly into two groups of 50. In group 1 they were shown a film about heroic soldiers who endured suffering and emerged victorious. In group 2 they were shown a similar film, but in it the soldiers finally lost the battle. Then each volunteer was asked to endure a very uncomfortable noise as long as he or she could. Do these results indicate that the films had any effect on the volunteers' behavior? Would you use a one-tail or a two-tail test? $\alpha = .05$.

group	minutes of noise endured	
	m	s
1	4.4	1
2	3.6	1

Hypothesis testing with sample means: small samples

9. Ten sets of 8-year-old identical male twins were used in a study of the effect of teaching techniques. One twin in each pair was taught some vocabulary words by computer tutorial without teacher assistance. The other was taught by the teacher without machine aids. At the end they were all given the same test. The results are shown below.

names	scores with computer	scores with teacher
Abe and Babe	116	123
Bob and Rob	112	170
Clark and Mark	100	140
Don and Ron	186	108
Ed and Fred	173	163
Frank and Hank	198	153
Garry and Larry	178	119
Harry and Barry	140	140
Jake and Bake	173	171
Kenneth and Percival	159	181

(i) Why were twins used?
(ii) What assumptions were made about their vocabulary skills at the start of the experiment?
(iii) At $\alpha = .05$, what does the study indicate?
Do the analysis two ways: as a paired t test and as an unpaired t test.

10. Your Aunt Tilly is not mathematically inclined. Upon hearing that you were studying statistics in college, she wrote to you asking just what is a hypothesis test? Answer her as clearly as you can.

Confidence Intervals

12 CONFIDENCE INTERVALS FOR PROPORTIONS IN A BINOMIAL POPULATION

An advertising agency wants to know the percentage of families in Nassau County that own a color TV. It would be very difficult and time-consuming to find the exact answer, because that would mean checking every family in the county. So the agency selects a random sample of 500 families in the county instead. Suppose they find that 340 of the families own a color TV. This is 340/500, or 68 percent. They would conclude that *about* 68 percent of the families in the county own a color TV. The single best estimate for the true answer is 68 percent. They write $\hat{p} = .68$.

In most statistical applications it is not enough to say that the correct value of p is "about" 68 percent. After all, that is a rather vague statement. Just what do we mean by "about" 68 percent? Do we mean that we believe the correct value is some value between 58 percent and 78 percent? Or do we mean that we believe the correct value is between 67 percent and 69 percent? When we give a range of values that we think includes the true value of some population parameter, this range of values is called an **interval estimate**. Such an interval is usually assigned a probability, and then it is called a **confidence interval**.

The higher the probability assigned, the more confident we are that the interval does, in fact, include the true value.

Let us develop this idea further in the following example.

EXAMPLE 12-1 Using a random sample of 500 families, compute an interval estimate for the percentage of families in Nassau County that own a color TV. We want this

Confidence intervals

interval to have a probability of .95 of containing the true percentage. In short, find the 95 percent confidence interval for the percentage of families that own a color TV.

SOLUTION Imagine that we *repeatedly* pick 500 families at random from Nassau County and record the number that own a color TV. We would get a table like Table 12-1.

Table 12-1

sample number	n	number with color TV	$\hat{p}$ percentage with color TV
1	500	340	68
2	500	345	69
3	500	304	60.8
.	.	.	.
.	.	.	.
300	500	375	74
.	.	.	.
1000	500	320	64
.	.	.	.

We let p equal the true percentage of families that own a color TV. If $np > 5$, and $nq > 5$, then it follows from the central limit theorem that, in the long run, this distribution of percentages $\hat{p}$ would be approximately normal with mean equal to p and standard deviation equal to $\sqrt{pq/n}$, where $n = 500$ families and p is the true percentage of families who own a color TV. Note that because the distribution is normal (Figure 12-1), the sample values $\hat{p}$ will tend to cluster near to the true value p.

Figure 12-1

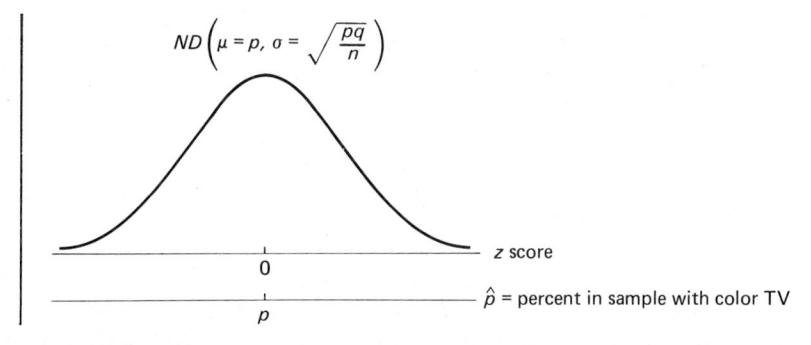

$$ND\left(\mu = p, \sigma = \sqrt{\frac{pq}{n}}\right)$$

z score
0

$\hat{p}$ = percent in sample with color TV
p

In reality we do the experiment only once. Suppose that when we check the 500 families, we find that 68 percent have color TVs. On the basis of this we want to compute a confidence interval for p (which we assume is near 68 percent). We have picked one of the many possible random samples that could have been selected from the county. We are faced with two possibilities: either we have picked a rare sample with a value of $\hat{p}$ far from p, or we have picked a more common sample with a value of $\hat{p}$ near p. Statisticians will always assume that one of the more common ones has been chosen. For instance, if we are computing the 95 percent confidence interval, we will assume that we have picked a sample with a value of $\hat{p}$ which occurs in the most common 95 percent of the cases. (If we are computing the 99 percent confidence interval, we will assume that we have one of the most common 99 percent of the cases.)

Let us follow through the reasoning which leads to the 95 percent confidence interval. You can find any other confidence interval by similar reasoning. Recall that our sample yielded $\hat{p} = .68$. Therefore we assume that $\hat{p} = .68$ is within the middle 95 percent of the normal distribution based on the true value of p (Figure 12-2).

Figure 12-2

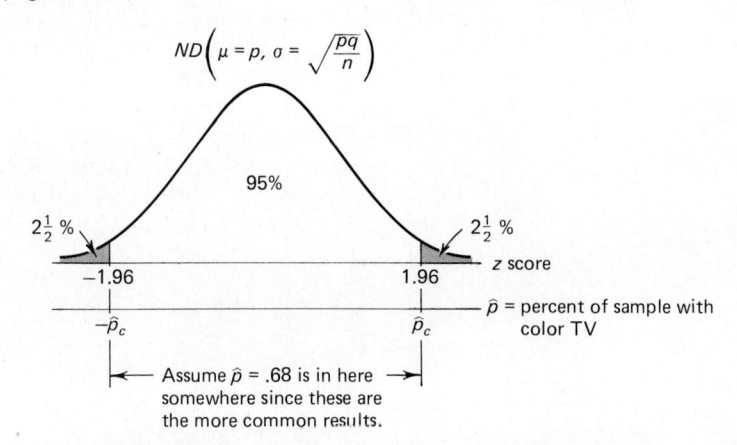

$$ND\left(\mu = p, \sigma = \sqrt{\frac{pq}{n}}\right)$$

95%

$2\frac{1}{2}\%$ $2\frac{1}{2}\%$

−1.96   1.96   z score

$-\hat{p}_c$ $\hat{p}_c$ $\hat{p}$ = percent of sample with color TV

Assume $\hat{p} = .68$ is in here somewhere since these are the more common results.

If we assume that $\hat{p} = .68$ is somewhere in this interval, then $\hat{p} = .68$ is within 1.96 standard deviations of the true value of p. Symbolically this is written

$$p - 1.96\sigma \leq .68 \leq p + 1.96\sigma$$

By algebra, this is equivalent to

$$.68 - 1.96\sigma \leq p \leq .68 + 1.96\sigma$$

Therefore, to get our 95 percent confidence interval for p we just add 1.96σ to our sample value and subtract 1.96σ from our sample value of $\hat{p}$. We can summarize this in a formula:

95% confidence interval for p: p lies between $\hat{p} - 1.96\sigma$ and $\hat{p} + 1.96\sigma$

In a real application, of course, we do not know the exact value of σ because it depends on p. So we use

$$\hat{\sigma} = \sqrt{\frac{\hat{p}\hat{q}}{n}}$$

Therefore the formula for the 95 percent confidence interval that we actually use is

95% confidence interval for p: p lies between $\hat{p} - 1.96\hat{\sigma}$ and $\hat{p} + 1.96\hat{\sigma}$

Let us apply this formula to our data. We had $n = 500$, $\hat{p} = .68$. Therefore $\hat{q} = 1 - \hat{p} = .32$. Since $n\hat{p} = 340 > 5$ and $n\hat{q} = 160 > 5$, the distribution of $\hat{p}$ is approximately normal.

Confidence intervals

$$\hat{\sigma} = \sqrt{\frac{\hat{p}\hat{q}}{n}}$$

$$= \sqrt{\frac{(.68)(.32)}{500}} = \sqrt{\frac{.2176}{500}} = \sqrt{.0004} = .02$$

We find $1.96\hat{\sigma} = 1.96(.02) = .04$. Therefore we estimate that p is between $.68 - .04$ and $.68 + .04$, or p is between .64 and .72. (Some authors write this as $.64 \leq p \leq .72$.)

Conclusion. On the basis of our sample data, the 95 percent confidence interval for the percentage of families that own color TVs is 64 percent to 72 percent. For this problem we have answered our opening question, "What do we mean by 'about 68 percent'?" In this problem, "about 68 percent" means "between 64 percent and 72 percent."

Further comments. The interval we arrived at is a 95 percent confidence interval because the estimation *procedure* we used to get it has probability equal to .95 or producing an interval which contains the true percentage. (See Table 12-2.) In the procedure outlined in Table 12-2, we find that 95 percent of the confidence intervals we calculate do "capture" the true parameter.

Table 12-2

sample number	n	number with color TVs	percentage with color TVs	95% confidence interval
1	500	340	68	.64 to .72
2	500	345	69	.65 to .73
3	500	304	60.8	.57 to .65
.	.	.	.	.
.	.	.	.	.
.	.	.	.	.
300	500	375	74	.70 to .78
.	.	.	.	.
.	.	.	.	.
1000	500	320	64	.60 to .68
.	.	.	.	.
.	.	.	.	.

The formulas for confidence intervals for p are:

95% confidence interval: p lies between $\hat{p} - 1.96\hat{\sigma}$ and $\hat{p} + 1.96\hat{\sigma}$

99% confidence interval: p lies between $\hat{p} - 2.58\hat{\sigma}$ and $\hat{p} + 2.58\hat{\sigma}$

In general: p lies between $\hat{p} - z_c\hat{\sigma}$ and $\hat{p} + z_c\hat{\sigma}$

where z_c corresponds to the particular confidence interval we seek. The value of z is the same value that you have been using in *two*-tail hypothesis tests.

DETERMINING THE SAMPLE SIZE NEEDED FOR A DESIRED CONFIDENCE INTERVAL

You are estimating the percentage of left-handed students at State University. What size sample is needed to ensure that the limits of a 95 percent confidence interval are no more than, say, 3 percent away from the correct percentage? That is, the quantity 1.96σ, which we add to and subtract from our sample value, should be no more than 3 percent. Algebraically we write $1.96\sigma \le .03$. Now since

$$\sigma = \sqrt{\frac{pq}{n}}$$

we have

$$1.96\sqrt{\frac{pq}{n}} \le .03$$

Solving for n, we get

$$n \ge \left(\frac{1.96}{.03}\right)^2 pq$$

We don't know p and q, but it is true that no matter what they are, their product pq cannot be bigger than .25. (Try it! Remember p plus q must equal 1.) So if we replace pq by .25, we get a value of n that meets our requirements:

$$n \ge \left(\frac{1.96}{.03}\right)^2 (.25) = 1067.1$$

Any sample bigger than this, that is, $n \ge 1068$, will give us our required accuracy.

In general, the formula for n needed to achieve a desired accuracy is

$$n \ge \left(\frac{z_c}{a}\right)^2 (.25)$$

where a is the desired accuracy.

EXAMPLE 12-2 You want to know the percentage of students whose original parents are divorced. You want to be within 5 percent of the correct value on a 95 percent confidence interval. What size sample do you need?

$$n = \left(\frac{z_c}{a}\right)^2 (.25)$$

$$= \left(\frac{1.96}{.05}\right)^2 (.25) = 384.16$$

■ Any random sample of size 385 or more is sufficient.

EXERCISES

12-1 When speaking about a 95 percent confidence interval, why is it not correct to say that $\alpha = .05$?

12-2 Explain the 2.58 in the formula for the 99 percent confidence interval for p.

12-3 Find a formula for the 90 percent confidence interval for p.

Confidence intervals

12-4 In a medical screening test for a certain disease some proportion of the patients will be *false positives.* That is, they will show a reaction to the test, but on further examination will be found not to have the disease. When doctors design a screening test, they need an estimate of how often false positives will occur in continued use of the test. They must perform a pilot study to get such an estimate. In one such pilot study 10 out of 100 positives were false positives. Give a 95 percent confidence interval for the false positive rate of the test.

12-5 You read in a newspaper poll that 53 percent of college students have used illegal drugs. Upon checking you learn that the sample size used in the research was $n = 100$. Find the 99 percent confidence interval for this percentage.

12-6 A newspaper reports that of 1064 soldiers in the 173d Airborne Brigade, 16 percent said they used marijuana "about every day" or "more often." Find the 95 percent confidence interval for this estimate.

12-7 Private Walter Parte, one of General Custer's privates, took a random sample of 50 men in his regiment. Of them, 30 felt that they had the enemy outnumbered. Find the 95 percent confidence interval for the proportion of men who felt this way.

12-8 A random sample of 90 prisoners in a large state prison revealed that 20 percent were college graduates. Find the 99 percent confidence interval in the population sampled for the percentage of prisoners who are college graduates.

12-9 Professor Kwizee's tests have contained 100 true-false questions so far this term; 80 of them have been true. Presuming that he will continue in this vein, find the 90 percent confidence interval for the proportion of "trues" on the next test.

12-10 If the early returns on election night are considered as a random sample, and Mayor Fogbottom is leading her opponent with 52 percent of the vote, based on the first 400 votes counted, should she be 99 percent confident of victory?

12-11 At a certain beach in New Jersey there is a local rule which says that only town residents may use the beach. One Sunday, 80 people (at random) on the beach are checked. It turns out that 68 are residents.
(a) What is the 95 percent confidence interval for the percentage of nonresidents using the beach?
(b) Based on your answer to part (a), if there are 700 people on the beach, about how many are nonresidents?

12-12 Sanford R. Brochure, the noted mail-advertising magnate, wants to verify the percentage of the population of Upperdownunder that discards the mail he sends them without even opening it. How many residents should be sampled if he wants to be 95 percent sure that he is within about 2 percent of the correct value of p?

12-13 Not every binomial problem is as simple as a heads-tails or yes-no problem. An 1800-step computer program simulates the joint use of sonar by a navy destroyer and a helicopter to locate an enemy submarine. At the end of each run of this program either the submarine has been spotted or it has escaped.
(a) When the program was run 100 times, the submarine was located 32 times. Estimate $p = P$(a submarine is located). Find the 95 percent confidence interval for your estimation.

(b) Repeat part (a) for the data $n = 500$ and the submarine located 139 times.
(c) Repeat part (a) for the data $n = 1000$ and the submarine located 297 times.
(d) We know that computer time costs money. On the basis of your answers to parts (b) and (c) alone, was it worthwhile running the program 1000 times, or would 500 times have been just as useful?
(e) Show that for $n = 38,000(!)$ we can estimate p to within 1 percent.

12-14 Milly Meter, a theater manager, wants to know what percentage of the elementary school children in West Islip have seen *Snow White*.
(a) What size sample does she need to be 95 percent sure that she is within 2 percent of the correct value?
(b) What size sample does she need to be 95 percent sure that she is within 1 percent of the correct value?

12-15 Coolo Keeno watches a lot of TV. She wishes to estimate the percentage of scenes that last less than 5 seconds. How many scenes must she time if she wishes to be 99 percent sure that she is within 5 percent of the correct value?

CONFIDENCE INTERVALS FOR MEANS BASED ON LARGE SAMPLES

To estimate the mean of a population we often use the mean of a random sample. For example, the average time it takes for an anesthetic to take effect could be estimated by trying it on a random sample of 100 patients and finding the mean for the sample.

If we found the mean of the sample to be, say, $m = 4.6$ minutes, we could use this as our best guess at the mean of the population. We would expect, if we put the anesthetic into general use, that the mean time to take effect would be *about* 4.6 minutes. To give meaning to the word "about," we can find the 95 percent confidence interval for the mean.

To do this when the sample size n is greater than 30, we use the central limit theorem (Chapter 10). We reason that the sample of patients we studied is an ordinary sample, one of the usual 95 percent of the samples with m near the true mean, and not one of the very rare 5 percent of the samples. Thus, we can be reasonably sure that the true mean is not more than 1.96 standard deviations away from our sample mean. This is illustrated in Figure 12-3.

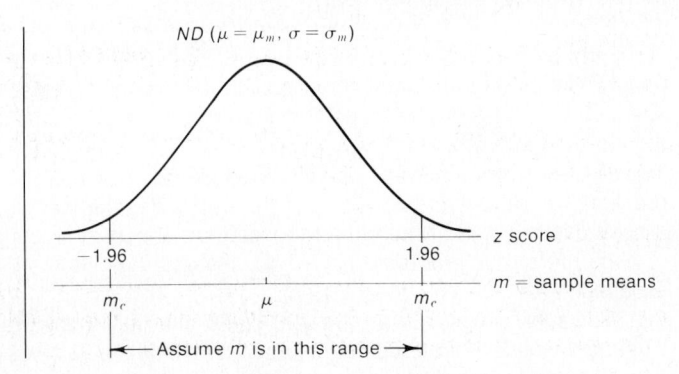

Figure 12-3

ND ($\mu = \mu_m$, $\sigma = \sigma_m$)

−1.96   1.96   z score

m_c μ m_c m = sample means

← Assume m is in this range →

Confidence intervals

Following the reasoning that we used in the binomial situation at the beginning of this chapter, we are led to similar formulas for the confidence intervals of population means.

> The formulas for confidence intervals for μ are:
> 95% confidence interval: μ lies between $m - 1.96 s_m$ and $m + 1.96 s_m$
> 99% confidence interval: μ lies between $m - 2.58 s_m$ and $m + 2.58 s_m$
> In general: μ lies between $m - z_c s_m$ and $m + z_c s_m$
> where z_c is the appropriate critical z score.

EXAMPLE 12-3 Suppose that our sample of 100 patients led us to compute $m = 4.6$ minutes with $s = 1.1$ minutes. What is the 95 percent confidence interval for the average time it takes for the anesthetic to take effect?

SOLUTION

$$s_m = \frac{s}{\sqrt{n}}$$

$$= \frac{1.1}{\sqrt{100}} = \frac{1.1}{10} = .11$$

$n = 100 > 30$

Hence we find $1.96 s_m = 1.96(.11) = .22$. Therefore, our confidence interval for μ is between $4.6 - .22$ and $4.6 + .22$, or μ is between 4.4 and 4.8.

This means that we are 95 percent sure that the average for the entire population of potential patients is between 4.4 and 4.8 minutes. ■

EXAMPLE 12-4 Find the 99 percent confidence interval for the average age of workers in the A-Maize toothpick factory if a random sample of only 100 workers there had a mean age of 35.2 with $s = 10.3$.

SOLUTION

$$s_m = \frac{s}{\sqrt{n}}$$

$$= \frac{10.3}{\sqrt{100}} = \frac{10.3}{10} = 1.03$$

$n = 100 > 30$

Therefore, we find $2.58 s_m = 2.58(1.03) = 2.7$, and hence, the confidence interval for μ lies between $35.2 - 2.7$ and $35.2 + 2.7$, or μ lies between 32.5 and 37.9.

This means that we are 99 percent sure that the average age for all the A-Maize workers is between 32.5 and 37.9 years. ■

EXERCISES

12-16 A rule of thumb used by some people to obtain a quick confidence interval is to simply go two standard deviations in each direction from the mean. What percent confidence interval does this rule yield?

Understanding statistics

12-17 Find the 95 percent confidence interval for the mean weight of a Venusian if a random sample of 36 Venusians weighed 16, 22, 31, 28, 15, 20, 20, 21, 22, 35, 28, 27, 25, 24, 20, 18, 19, 31, 17, 18, 20, 15, 25, 24, 27, 18, 20, 31, 29, 23, 20, 20, 19, 31, 30, 20.

12-18 Find the 99 percent confidence interval for the mean number of persons riding in a subway car between midnight and 6 A.M. if a random sample of 40 subway cars had the following number of persons: 0, 0, 1, 3, 7, 13, 15, 20, 20, 20, 23, 25, 29, 30, 35, 35, 36, 36, 36, 36, 37, 40, 41, 41, 41, 43, 44, 47, 50, 56, 59, 60, 60, 61, 63, 69, 70, 71, 89, 103.

12-19 Find the 95 percent confidence interval for the mean salary of teachers in Constick County if a random sample of 100 teachers had a mean salary of $14,000 with $s = \$1000$.

12-20 Find the 99 percent confidence interval for the mean number of hours of TV watched by children per day if a random sample of 49 children had a mean of 4.6 hours with $s = 2.9$ hours.

12-21 Find the 95 percent confidence interval for the mean number of pieces of junk mail received per week if a random sample of 1000 homes had a mean of 23.1 pieces with an estimated standard deviation of 4.1 pieces.

12-22 Find the 98 percent confidence interval for the mean number of minutes of commercials per hour if a random sample of 48 hours of TV yielded a mean of 15.2 minutes per hour and $s = 1.5$ minutes per hour.

12-23 Find the 95 percent confidence interval for the average time spent waiting for a prescription to be filled at Rex's Rx Center if a random sample of 64 customers showed an average waiting time of 12.3 minutes with a standard deviation of 5.0 minutes.

12-24 A sample of 400 students turned in statistics projects. The average sample size was 110.3 and the standard deviation was 12.2. Find the 99 percent confidence interval for the average sample size of student projects.

12-25 The commissioner of the little league in Spearfish, South Dakota, is planning the annual picnic. He knows from past years that the average number of hot dogs consumed is 6.5 per person, with a 95 percent confidence interval from 6.1 to 6.9 hot dogs. He expects all 507 little leaguers to show up for the picnic.
(a) Using the average 6.5, how many hot dogs should he order?
(b) If he wants to be sure that he does not run out of hot dogs as he did last year (before the riot), how many should he order?
(c) If he orders as in part (b), using the high end of the interval, and it turns out that the mean number of hot dogs this year is near the low end of the interval, how many hot dogs will be left over?
(d) If he does get stuck as in part (c), and he has to eat 2 hot dogs a day until the leftovers are gone, how long will it take him to finish the hot dogs?

CONFIDENCE INTERVALS FOR MEANS BASED ON SMALL SAMPLES

In this section dealing with small samples we will follow the same procedures used above to find confidence intervals, except that we will use *critical t scores* instead of critical z scores. Recall that when using t scores we must have approximately normal distributions for our original data.

Thus our formula for a confidence interval for a mean becomes:

μ is between $m - t_c s_m$ and $m + t_c s_m$

EXAMPLE 12-5 A lawyer must drive from a northern suburb to Chicago's Loop every day to get to work. In order to decide whether he should take the train instead, he computes his gas mileage each day for 15 days. He finds that he has a mean mileage of 12.2 miles per gallon, with an estimated standard deviation of 2.1 miles per gallon. Find the 95 percent confidence interval for his mileage, assuming that gas mileage is normally distributed.

SOLUTION $s = 2.1$ and $n = 15$. Therefore,

$$s_m = \frac{s}{\sqrt{n}}$$

$$= \frac{2.1}{\sqrt{15}} = .54$$

Degrees of freedom $= n - 1 = 14$

The critical values of t for $\alpha = .05$ (corresponding to the 95 percent confidence interval) with 14 degrees of freedom are ± 2.14. Therefore we find $t_c s_m = 2.14(.54) = 1.2$. Thus, the confidence interval for μ lies between $12.2 - 1.2$ and $12.2 + 1.2$, or μ lies between 11.0 and 13.4.

Thus he is 95 percent confident that the true gas mileage lies between 11.0 and 13.4 miles per gallon. He can use this to decide whether it is cheaper to drive or to take the train. ■

EXERCISES

In these exercises assume that the populations being sampled are approximately normal.

12-26 If you mistakenly used a z score instead of a t score for a small-sample confidence interval, would the resulting interval be too wide or too narrow?

12-27 In an overcrowded computer lab students must wait to get at a computer terminal. John De Newman recorded the times he waited for his last 10 visits as: 15, 14, 12, 15, 15, 16, 14, 15, 16, 14 minutes. Estimate his average waiting time and find the 99 percent confidence interval for your estimate.

12-28 A consumer magazine wants to estimate the life of Extra-Strong light bulbs. With 25 randomly purchased bulbs they find the mean life to be 998 hours, with a standard deviation s of 30 hours. Find the 95 percent confidence interval for the mean life of the bulbs.

12-29 A doctor desires to estimate the mean age of women having abortions in New York City municipal hospitals. She finds that the mean age in a random sample of 19 patients is 29.7 years, with s equal to 3.2 years. Find the 99 percent confidence interval for the mean age of women having abortions in these hospitals.

12-30 The Gypsy Taxi Cab Company of Brooklyn, New York, desires to know

Understanding statistics

the mean life of tires on its cabs. A random sample of 28 tires from the cabs had a mean life of 17,821 miles, with a standard deviation of 1206 miles. Find the 95 percent confidence interval for the life of the tires.

12-31 An efficiency expert for a chalkboard eraser company gathers a random sample of 16 workers' hourly production. He finds that the mean number of erasers produced per worker per hour is 123.1, with s equal to 8.0. Find the 90 percent confidence interval for the mean number of erasers produced by a worker in 1 hour.

12-32 Peggy Babbcock of Fredericksburg, Virginia, won the rapid speaking contest three times, rapidly using the phrase "red leather, yellow leather." Her rate of speaking was 7.3, 7.1, and 7.2 syllables per second. Find the 95 percent confidence interval for her average rate of speaking at contests. (If you can say her name three times rapidly, you will get second prize.)

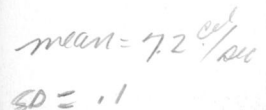

TWO-SAMPLE CONFIDENCE INTERVALS FOR DIFFERENCES

Just as we have estimated the value of a binomial probability p or a population mean μ by taking a sample from one population, so too we can estimate the difference between two probabilities $p_1 - p_2$, or the difference between two means $\mu_1 - \mu_2$, by taking separate and independent samples from each of *two* populations.

DIFFERENCES BETWEEN TWO PROPORTIONS

EXAMPLE 12-6 Suppose we are interested in the difference between the proportion of blondes at Men's Institute of Tennis who wear contact lenses and the proportion of redheads who wear contact lenses. If $p_1 = P$(a blonde at MIT wears contact lenses) and $p_2 = P$(a redhead at MIT wears contact lenses), then we wish to estimate the value of the parameter $p_1 - p_2$.

We could gather two random samples of students at MIT, one a sample of blondes, and the other a sample of redheads. If 90 of 300 blondes wore contact lenses while 24 of 200 redheads wore contact lenses, we would estimate p_1 and p_2 by

$$\hat{p}_1 = \frac{90}{300} = .30 \quad \text{and} \quad \hat{p}_2 = \frac{24}{200} = .12$$

Then our estimate of $p_1 - p_2$ would be $\hat{p}_1 - \hat{p}_2 = .30 - .12 = +.18$. Therefore, about 18 percent more blondes than redheads wear contact lenses. To calculate the 99 percent confidence interval for this estimate we note that $n_1\hat{p}_1$, $n_1\hat{q}_1$, $n_2\hat{p}_2$, and $n_2\hat{q}_2$ are equal to the four outcomes 90, 210, 24, and 176. Since these outcomes are all greater than 5, a distribution of such estimates would be approximately normal, and the middle 99 percent will be between $z = -2.58$ and $z = 2.58$. If we let $d\hat{p}$ symbolize $\hat{p}_1 - \hat{p}_2$, then

$$\mu_{d\hat{p}} = p_1 - p_2 \quad \text{and} \quad \hat{\sigma}_{d\hat{p}} = \sqrt{\frac{\hat{p}_1\hat{q}_1}{n_1} + \frac{\hat{p}_2\hat{q}_2}{n_2}}$$

Note: The question is: how different are p_1 and p_2? We are not assuming that p_1 equals p_2 and hence we do *not* pool our sample results as we did in Chapter 9.

Confidence intervals

Our confidence interval is given by:

$p_1 - p_2$ is between $d\hat{p} - z_c\hat{\sigma}$ and $d\hat{p} + z_c\hat{\sigma}$.

Computing $\hat{\sigma}$ we have

$$\hat{\sigma} = \sqrt{\frac{.30(.70)}{300} + \frac{.12(.88)}{200}} = .035$$

We find $z_c\hat{\sigma} = 2.58(.035) = .09$. Thus, our 99 percent confidence interval for $p_1 - p_2$ lies between $.18 - .09$ and $.18 + .09$, or $p_1 - p_2$ lies between .09 and .27.

The difference is that anywhere from 9 to 27 percent more blondes at MIT wear contact lenses. We have conclusive evidence that a larger percentage of blondes wear contact lenses, but we cannot pin down the difference any more than to say it is somewhere between 9 and 27 percent. If we wish to have a narrower confidence interval with this size sample, then we will have to run more than a 1 percent risk of error. Correspondingly, if we wish to estimate the difference between the percentage of blondes and the percentage of redheads more closely and *not* increase the risk of error, we will have to get larger samples.

DIFFERENCES BETWEEN TWO MEANS

The confidence interval for $\mu_1 - \mu_2$ is given by:

$\mu_1 - \mu_2$ is between $dm - z_c s_{dm}$ and $dm + z_c s_{dm}$ or

$\mu_1 - \mu_2$ is between $dm - t_c s_{dm}$ and $dm + t_c s_{dm}$

depending upon whether or not both n_1 and n_2 are larger than 30.

EXAMPLE 12-7 *Large samples* Find the 95 percent confidence interval for the difference between the average size of a window in a one-family house built over 10 years ago and the average window size in a newer house if two random samples gave the following data:

400 windows in older houses,
$m_1 = 15.6$ square feet, $s_1 = 20.2$ square feet

500 windows in newer houses,
$m_2 = 19.0$ square feet, $s_2 = 24.8$ square feet

SOLUTION Since $n_1 = 400$ and $n_2 = 500$ are both larger than 30, we have an approximately normal distribution of differences and can use $z_c = \pm 1.96$, and

$$s_{dm} = \sqrt{\frac{s_1^2}{n_1} + \frac{s_2^2}{n_2}}$$

$$= \sqrt{\frac{(20.2)^2}{400} + \frac{(24.8)^2}{500}} = 1.50$$

We find $z_c s_{dm} = 1.96(1.50) = 2.9$. Hence, our confidence interval for $\mu_1 - \mu_2$ lies between $-3.4 - 2.9$ and $-3.4 + 2.9$, or $\mu_1 - \mu_2$ lies between -6.3 and -0.5.

We are 95 percent sure that the average size of windows in newer houses is at least 0.5 square feet, but at most 6.3 square feet, larger than the average size of windows in houses over 10 years old. ∎

EXAMPLE 12-8

Small samples Corresponding to our work in Chapter 11 on small-sample hypothesis tests, we will present an approach for the case in which the assumptions that both populations are normal and that $\sigma_1 = \sigma_2$ are reasonable. This allows us to use the t distribution in our analysis. Under these assumptions we use the pooled version of s_{dm}. You should see a more extensive text for an approach if this assumption is not reasonable.

(a) In a pilot study of the time it takes to fly from earth to Saturn, a sample of 10 human pilots averaged 8.9 velans with $s = 1.1$. Forty Saturnian pilots averaged 7.1 velans with $s = 1.2$. Estimate the difference between the average times of these two populations.

(b) Find a 95 percent confidence interval for your estimate.

SOLUTION

(a) $m_1 - m_2 = 8.9 - 7.1 = 1.8$. Our best single estimate is that the average time of Earth pilots is 1.8 velans more than the average time of the Saturnian pilots.

(b) $s_p^2 = \dfrac{(n_1 - 1)s_1^2 + (n_2 - 1)s_2^2}{n_1 + n_2 - 2}$

$= \dfrac{9(1.1)^2 + 39(1.2)^2}{48} = \dfrac{67.05}{48} = 1.4$

$s_{dm} = \sqrt{s_p^2 \left(\dfrac{1}{n_1} + \dfrac{1}{n_2}\right)} = \sqrt{1.4 \left(\dfrac{1}{10} + \dfrac{1}{40}\right)} = .42$

Since $n_1 = 10$ and is smaller than 30, we have a t distribution with $n_1 + n_2 - 2 = 10 + 40 - 2 = 48$ degrees of freedom. Looking up the critical t scores for the middle 95 percent of the outcomes in Table C-5, we use 50 degrees of freedom, the entry nearest to 48, and find $t_c = \pm 2.01$. We find $t_c s_{dm} = 2.01(.42) = .84$. Thus, our confidence interval $\mu_1 - \mu_2$ lies between $1.8 - .84$ and $1.8 + .84$, or $\mu_1 - \mu_2$ lies between 1.0 and 2.6.

We are 95 percent sure that human pilots average at least 1.0 velans more than Saturnian pilots. ∎

STUDY AIDS

VOCABULARY

1. Interval estimate
2. Confidence interval

Confidence intervals

FORMULAS

Distribution of sample proportions

1. $\mu = p$ 2. $\sigma = \sqrt{\dfrac{pq}{n}}$ 3. $\hat{\sigma} = \sqrt{\dfrac{\hat{p}\hat{q}}{n}}$

4. p lies between $\hat{p} - z_c\hat{\sigma}$ and $\hat{p} + z_c\hat{\sigma}$

Distribution of sample means

5. $s_m = \dfrac{s}{\sqrt{n}}$

6. μ lies between $m - z_c s_m$ and $m + z_c s_m$ (large samples)
7. μ lies between $m - t_c s_m$ and $m + t_c s_m$ (small samples)

Differences between two proportions

8. $\hat{\sigma} = \sqrt{\dfrac{\hat{p}_1\hat{q}_1}{n_1} + \dfrac{\hat{p}_2\hat{q}_2}{n_2}}$

9. $p_1 - p_2$ lies between $d\hat{p} - z_c\hat{\sigma}$ and $d\hat{p} + z_c\hat{\sigma}$

Differences between two means

10. $s_{dm} = \sqrt{\dfrac{s_1^2}{n_1} + \dfrac{s_2^2}{n_2}}$ (large samples)

11. $s_p^2 = \dfrac{(n_1 - 1)s_1^2 + (n_2 - 1)s_2^2}{n_1 + n_2 - 2}$

12. $s_{dm} = \sqrt{s_p^2\left(\dfrac{1}{n_1} + \dfrac{1}{n_2}\right)}$ (small samples)

13. $\mu_1 - \mu_2$ lies between $dm - z_c s_{dm}$ and $dm + z_c s_{dm}$ (large samples)
14. $\mu_1 - \mu_2$ lies between $dm - t_c s_{dm}$ and $dm + t_c s_{dm}$ (small samples)

EXERCISES

12-33 Suppose that data from two samples are used to compute a 95 percent confidence interval for the difference between the means of the two populations. These same data are also used to perform a hypothesis test at the analogous .05 significance level. If the confidence interval goes from a negative value to a positive value, would you generally expect to reject the null hypothesis or not to reject it?

12-34 Dr. Showvan spent years gathering the IQ data below.

	male	female
n	10,000	10,000
m	100.3	99.9
s	10.8	11.1

He then performed a test on the difference between the average IQ of American males and the average IQ of American females. He calculated

$$dm_c = 0 \pm 1.96 \sqrt{\frac{(10.8)^2}{10,000} + \frac{(11.1)^2}{10,000}} = \pm .30$$

$$dm = 0.4$$

Hence he concluded that the average IQ of American males is higher than the average IQ of American females with $\alpha = .05$. He claims that his results are "statistically significant."

(a) Show that the 99 percent confidence interval for this difference is $.0 < dm < .8$.

(b) Comment on the importance of a difference of at most 8/10 of a point on an IQ test.

(c) Do "statistical significance" and "important" have the same connotation?

12-35 A teacher desires to estimate the difference in reading levels between children from two-parent homes and children from one-parent homes in his community. Using two random samples of fourth-grade youngsters, he finds that 19 children from two-parent homes had a mean reading level m_1 of 5.1 with standard deviation $s_1 = 1.4$, and that 13 children from one-parent homes had a mean reading level m_2 of 3.8 with standard deviation $s_2 = 2.1$. Find a 99 percent confidence interval for the difference in the reading levels.

12-36 The Gypsy Taxi Cab Company of Brooklyn, N.Y., is still checking tires. They now want to know if cab drivers under 25 are harder on tires than older drivers. Of their 76 cabs 32 are driven exclusively by the younger drivers and the remaining 44 by the older drivers. The younger drivers average 17,482 miles for a set of tires with a standard deviation of 1320 miles, while the older drivers average 17,728 miles with a standard deviation of 981 miles. Find a 99 percent confidence interval for the estimate of the true difference in mileage.

12-37 *Gotcha,* the consumer magazine, is testing the lives of two kinds of flashlight batteries. BRITELITE claims to give more life in normal use but is more costly than ordinary batteries. They randomly purchase 50 BRITELITE batteries and 50 ordinary batteries. The BRITELITE batteries are found to have a mean life of 17.5 months of normal use with a standard deviation of 1.1 months. The ordinary batteries have a mean life of 14.7 months with a standard deviation of 1.3 months. Find a 95 percent confidence interval for the time difference in the lives of the batteries.

12-38 Albert Sechsauer, an efficiency expert, claims that an additional afternoon break in a factory will result in more production on an assembly line. In the prior month with 21 working days, the line produced a mean of 72.3 items per day with a standard deviation of 3.4 items. In the 22 working days of the first month with the break, the line produced a mean of 70.6 items with a standard deviation of 2.1 items.

(a) Find a 99 percent confidence interval for the estimate of the true difference between the two production averages.

(b) Without the break, in 21 days the line had a mean of 3.1 defective items per day with a standard deviation of .43 item. In the 22 days after the break the line had a mean of 2.4 defective items with a standard deviation of .53 item. Find a 99 percent confidence interval for the estimate of the difference between the number of defects.

12-39 Find a 99 percent confidence interval for the difference between the percentage of Jewish students who joined Hillel and the percentage of Catholic students who joined Newman at Luther University, if a random sample of 500

Confidence intervals

persons showed that 20 of 80 Jewish students belonged to Hillel, while 24 of 120 Catholic students belonged to Newman.

12-40 A sample of 500 high school students and 500 adults in Reading, Pennsylvania, showed that 350 students and 250 adults owned library cards. Estimate the difference between the percentage of each group that owns a library card and find a 95 percent confidence interval for this difference.

CLASS SURVEY

Find a confidence interval for the percentage of students in your school that are left-handed.

FIELD PROJECTS

Estimate the mean of a population, the difference between the means of two populations, the proportions of a binomial population, or the difference between two proportions. Find a confidence interval for your estimate. Some examples which other students have tried on campus include estimates of:

1. Average age of faculty
2. Difference between average age of male and female faculty
3. Average amount of coin change carried by students
4. Percentage of students who use public transportation to get to campus

Chi-Square Tests

13

In Chapter 9 we saw how we could compare two proportions. For example, we could compare men and women in some community on their attitudes toward building a nuclear power plant nearby. If we should find only a small difference between the percentage of men and the percentage of women who are in favor of building the plant, we would say that the difference is not statistically significant. Another way to express the finding is to say that men and women respond similarly to the question, or that a person's sex *has nothing to do* with his or her attitude on building the plant. In the technical language of statistics, we can say that sex and opinions on building the nuclear plant are *independent*. In this chapter we shall expand on this idea of statistical independence.

EXAMPLE 13-1 Suppose we took a random sample of 200 people in this community and found that of 50 men and 150 women sampled, 60 favored building the nuclear power plant, 100 were against it, and 40 had no opinion. We can summarize these results in Table 13-1, called a **contingency table.**

Table 13-1

	men	women	row totals
in favor of building plant			60
against building plant			100
no opinion			40
column totals	50	150	200 = sample size

Chi-square tests

If sex and opinions on building the nuclear power plant are independent, the *same* percentage of men and women should be in favor of it (that is, men are no more likely than women to be for building it). Notice that 60 of the 200 people interviewed, or 30 percent, were in favor of building the plant. Therefore, if sex and opinions on nuclear power plants are independent, we expect about 30 percent of the males *and* 30 percent of the females to be in favor. Similarly, we expect 100/200, or 50 percent, of the males and 50 percent of the females to be against. Also, we expect 40/200, or 20 percent, of males and 20 percent of the females to have no opinion. Thus, our **expected results** under the null hypothesis of independence should be as listed in Table 13-2.

Table 13-2

	men	women	row totals
in favor	$\frac{60}{200}$ of 50	$\frac{60}{200}$ of 150	60
against	$\frac{100}{200}$ of 50	$\frac{100}{200}$ of 150	100
no opinion	$\frac{40}{200}$ of 50	$\frac{40}{200}$ of 150	40
column totals	50	150	200 = sample size

Numerically, we now have Table 13-3.

Table 13-3 Expected Results

	men	women	row totals
in favor	[1] 15	[2] 45	60
against	[3] 25	[4] 75	100
no opinion	[5] 10	[6] 30	40
column totals	50	150	200 = sample size

For convenience, we have numbered all **cells** or boxes of the contingency table from left to right.

A simpler way to obtain these expected results would be to multiply the column total times the row total for each cell in the table and divide by the sample size. For example, in cell 1 (men in favor) we have column total = 50, row total = 60, and sample size = 200; therefore, the expected result is

$$\frac{50(60)}{200} = 15$$

In cell 2 (women in favor) we have

$$\frac{150(60)}{200} = 45$$

In cell 3 (men against) we have

$$\frac{50(100)}{200} = 25$$

Understanding statistics

In cell 4 (women against) we have

$$\frac{150(100)}{200} = 75$$

In cell 5 (men with no opinion) we have

$$\frac{50(40)}{200} = 10$$

In cell 6 (women with no opinion) we have

$$\frac{150(40)}{200} = 30$$

In general, to find the expected result for any cell of the table we use the following formula:

$$\text{Expected result} = \frac{\text{column total} \cdot \text{row total}}{\text{sample size}}$$

In Table 13-4, we give the **observed results** obtained from our random sample of 200 people.

Table 13-4 Observed Results

	men	women	row totals
in favor	1. 17	2. 43	60
against	3. 22	4. 78	100
no opinion	5. 11	6. 29	40
column totals	50	150	200 = sample size

Let us now examine the difference between the observed results and the expected results (Table 13-5).

Table 13-5

cell	category	observed O	expected E	difference $O - E$
1	men in favor	17	15	2
2	women in favor	43	45	−2
3	men against	22	25	−3
4	women against	78	75	3
5	men no opinion	11	10	1
6	women no opinion	29	30	−1

Are these differences 2, −2, −3, 3, 1, and −1 large or small? If they are large, we will claim that opinions on building nuclear power plants are related to sex. If they are small, we will be unable to make this claim. We are looking for one number which will indicate whether these differences are large. Note two problems. The first problem is that the sum of these differences is zero, and this sum will always be zero. For this reason we cannot use the mean of these differences. As you recall, we encountered this problem before when we studied the standard deviation. As we did at that time, we will now work with the squares of the differences. They are shown in Table 13-6.

Chi-square tests

Table 13-6

cell	observed O	expected E	difference $O - E$	difference squared $(O - E)^2$
1	17	15	2	4
2	43	45	-2	4
3	22	25	-3	9
4	78	75	3	9
5	11	10	1	1
6	29	30	-1	1

The second problem is this. The squared difference of 4 in cell 1 is the same as the squared difference in cell 2. However, in cell 1 the expected result was 15, while in cell 2 the expected result was 45. A squared difference of 4 is more important when you expected 15 than when you expected 45; 4/15 is 27 percent, while 4/45 is only 9 percent. To take relative size into account, we divide each squared difference by the expected result for its cell. The sum of these numbers is the statistic X^2, which is our sample estimate of a parameter called **chi square**, χ^2. As usual, we use the English X^2 for our statistic and the Greek χ^2 for our parameter. In symbols, the formula for this statistic is

$$X^2 = \Sigma \frac{(O - E)^2}{E}$$

We now compute X^2 for the above problem. Our results are shown in Table 13-7.

Table 13-7

cell	O	E	$O - E$	$(O - E)^2$	ratio of differences squared to expected result $(O - E)^2/E$
1	17	15	2	4	4/15 = .27
2	43	45	-2	4	4/45 = .09
3	22	25	-3	9	9/25 = .36
4	78	75	3	9	9/75 = .12
5	11	10	1	1	1/10 = .10
6	29	30	-1	1	1/30 = .03
			sum = 0		sum = .97

Therefore, $X^2 = \Sigma \frac{(O - E)^2}{E} = .97$

If the expected results turn out to be equal to or very close to the observed results, then the differences would be zero or near zero, and so the value of X^2 would be near zero. This result should occur if there is no relationship between the variables.

On the other hand, if X^2 is far away from zero, there is a high probability that the variables are not independent, and there is some statistical relationship between them.

In these chi-square tests of independence our null hypothesis will be that the **variables are independent**; that is,

H_0: opinions on building nuclear power plants and one's sex are independent

Our motivated or alternative hypothesis is that the variables are dependent; that is,

H_a: opinions on building nuclear power plants and one's sex are not independent

These tests of independence are always one-tailed, for we wish to see if our X^2 statistic is significantly greater than 0. In this problem we must now decide whether .97, the value of X^2, is significantly larger than zero. As we did in previous chapters with the normal and t distributions, we compare a number based on sample data with a critical value. Table C-7 lists critical values of the statistic X^2 for different significance levels and degrees of freedom for one-tail tests. The theoretical distribution from which these values are taken is called the **chi-square distribution.** (This is similar to an earlier situation where we called the distribution of the statistic $\hat{p}$ a binomial distribution.) Table C-7 is called a chi-square table, and problems such as this power plant one are referred to as chi-square problems. We can assume that it is reasonable for us to use these critical values if our sample size is large enough. This will usually be the case if each expected value is greater than 5. Occasionally it may be necessary to combine categories to achieve expected values which are all greater than 5.

For a 3 × 2 table there are 2 degrees of freedom. (The way to determine the number of degrees of freedom will be discussed in the next section.) Let us perform a chi-square test at the .05 significance level. Looking at Table C-7, we find the critical value for 2 degrees of freedom and $\alpha = .05$ to be $X_c^2 = 5.99$.

Since the computed value, .97, is smaller than X_c^2, the differences we found between the observed and expected results are not large enough to reject the hypothesis of independence at the .05 significance level. Thus, we have been unable to show a statistical relationship between opinions on building nuclear power plants and sex.

DEGREES OF FREEDOM IN A CONTINGENCY TABLE

In the above example, we said that a 3 × 2 table has 2 degrees of freedom. Let us illustrate this concept. In the previous discussion you were told that of the 200 people, 50 were men and 150 were women; it was also stated that 60 of them favor building a nuclear power plant, 100 oppose it, and the remaining 40 have no opinion. Now suppose we have only one cell of Table 13-4, the observed results. For example, cell 1 indicates that 17 men favor building a nuclear power plant. From this information you could figure out what the entry would have to be in cell 2 in order to have the correct total of 60 (Table 13-8).

Table 13-8

	men	women	row totals
in favor	1 17	2	60
against	3	4	100
no opinion	5	6	40
column totals	50	150	200 = sample size

Clearly, you can see that the entry in cell 2 must be 60 − 17 = 43. However, cells 3 to 6 are not yet determined. If we had one of these cells given, such

Chi-square tests

as 22 people in cell 3, then all the remaining cells could be determined (Table 13-9).

Table 13-9

	men	women	row totals
in favor	1 17	2	60
against	3 22	4	100
no opinion	5	6	40
column totals	50	150	200 = sample size

Now we solve for the remaining cells as follows:

Cell 2:   60 − 17 = 43
Cell 4:   100 − 22 = 78
Cell 5:   50 − (17 + 22) = 50 − 39 = 11
Cell 6:   40 − 11 = 29

Thus you would have obtained the results shown in Table 13-10.

Table 13-10

	men	women	row totals
in favor	1 17	2 43	60
against	3 22	4 78	100
no opinion	5 11	6 29	40
column totals	50	150	200 = sample size

We say that a 3 × 2 table has 2 degrees of freedom because with 2 cells given, all the other cells can be determined by knowing the totals.

Suppose we had a 4 × 3 table with the totals given, as shown in Table 13-11.

Table 13-11

			row totals
1	2	3	200
4	5	6	200
7	8	9	100
10	11	12	500
column totals 400	200	400	1000 = sample size

How many cells would you need to know before you could fill them all in? You can see that if we take 6 numbers, leaving the last row and the last column blank, we can determine the remaining numbers (Table 13-12).

Table 13-12

			row totals
1 50	2 60	3	200
4 70	5 50	6	200
7 30	8 30	9	100
10	11	12	500
column totals 400	200	400	1000 = sample size

Understanding statistics

Thus we get

Cell 3:	200 − (50 + 60) =	90
Cell 6:	200 − (70 + 50) =	80
Cell 9:	100 − (30 + 30) =	40
Cell 10:	400 − (50 + 70 + 30) =	250
Cell 11:	200 − (60 + 50 + 30) =	60
Cell 12:	500 − (250 + 60) =	190

so that the complete table is as shown in Table 13-13.

Table 13-13

			row totals
¹ 50	² 60	³ 90	200
⁴ 70	⁵ 50	⁶ 80	200
⁷ 30	⁸ 30	⁹ 40	100
¹⁰ 250	¹¹ 60	¹² 190	500
column totals 400	200	400	1000 = sample size

Therefore, a 4 × 3 table has 6 degrees of freedom. In general, if a table has R rows and C columns, then it has $(R − 1)(C − 1)$ degrees of freedom. For example,

table size	$(R − 1)(C − 1)$	degrees of freedom
5 × 2	$(5 − 1)(2 − 1) = 4 \cdot 1$	4
4 × 3	$(4 − 1)(3 − 1) = 3 \cdot 2$	6
6 × 7	$(6 − 1)(7 − 1) = 5 \cdot 6$	30

We write

Degrees of freedom = $(R − 1)(C − 1)$

A SHORTER WAY TO COMPUTE X^2

The previous discussion about computing X^2 led to the formula

$$X^2 = \Sigma \frac{(O − E)^2}{E}$$

As we have seen before, mathematicians often find an equivalent formula which is easier to use. A more convenient formula for X^2 is

$$X^2 = \Sigma \left(\frac{O^2}{E} \right) − \Sigma O$$

We illustrate this by redoing the calculation of X^2 for Example 13-1. Our data are shown in Table 13-14. Note that ΣO is simply the sample size n.

Chi-square tests

Table 13-14

cell	O	E	O²	O²/E
1	17	15	289	19.27
2	43	45	1,849	41.09
3	22	25	484	19.36
4	78	75	6,084	81.12
5	11	10	121	12.10
6	29	30	841	28.03
	$\Sigma O = 200$			$\Sigma\left(\dfrac{O^2}{E}\right) = 200.97$

Therefore,

$$X^2 = \Sigma\left(\dfrac{O^2}{E}\right) - \Sigma O$$
$$= 200.97 - 200 = .97$$

Note that this is the same value as that computed with the previous formula.

EXAMPLE 13-2 A sampling on a university campus about the students' various places of residence led to the following results (Table 13-15).

Table 13-15

Observed Results

major field	live with parents	live on campus	other arrangements	row totals
business	30	6	14	50
education	7	17	6	30
fine arts	9	23	43	75
social science	6	17	22	45
column totals	52	63	85	200 = sample size

To determine if there is a relationship between major field and place of residence, test the null hypothesis that residence is independent of major field. Use the .01 significance level.

SOLUTION

H_a: residence and major field are dependent
H_0: residence and major field are independent
Degrees of freedom = $(4-1)(3-1) = 6$

Since there are 6 degrees of freedom, we must use our formula for at least 6 of the expected results E, and then the rest can be found by subtraction.

$$\text{Expected result} = \dfrac{\text{column total} \cdot \text{row total}}{\text{sample size}}$$

Cell 1: $E = \dfrac{52(50)}{200} = 13$ Cell 4: $E = \dfrac{52(30)}{200} = 7.8$

Cell 2: $E = \dfrac{63(50)}{200} = 15.75$ Cell 5: $E = \dfrac{63(30)}{200} = 9.45$

Understanding statistics

Cell 7: $E = \dfrac{52(75)}{200} = 19.5$ Cell 8: $E = \dfrac{63(75)}{200} = 23.625$

NOTE: If you use a calculator, it is probably easier to continue to find the rest of the expected values by the formula.

Thus, the expected results can now be tabulated as shown in Table 13-16.

Table 13-16 Expected Results

major field	live with parents	live on campus	other arrangements	row totals
business	1 13	2 15.75	3	50
education	4 7.8	5 9.45	6	30
fine arts	7 19.5	8 23.625	9	75
social science	10	11	12	45
column totals	52	63	85	200 = sample size

We can find the rest of the table by subtraction:

Cell 3: $E = 50 - (13 + 15.75) = 21.25$
Cell 6: $E = 30 - (7.8 + 9.45) = 12.75$
Cell 9: $E = 75 - (19.5 + 23.625) = 31.875$
Cell 10: $E = 52 - (13 + 7.8 + 19.5) = 11.7$
Cell 11: $E = 63 - (15.75 + 9.45 + 23.625) = 14.175$
Cell 12: $E = 45 - (11.7 + 14.175) = 19.125$

You can check the above by adding the cells in the last column under "other arrangements" to see if you get the column total of 85. This column acts as a check since we did not use it in our calculations. See Table 13-17.

Table 13-17

cell	O	E	O^2	O^2/E
1	30	13	900	69.23
2	6	15.75	36	2.29
3	14	21.25	196	9.22
4	7	7.8	49	6.28
5	17	9.45	289	30.58
6	6	12.75	36	2.82
7	9	19.5	81	4.15
8	23	23.625	529	22.39
9	43	31.875	1849	58.01
10	6	11.7	36	3.08
11	17	14.175	289	20.39
12	22	19.125	484	25.31

$E > 5$ $\Sigma\left(\dfrac{O^2}{E}\right) = 253.75$

Notice that all the expected values are greater than 5. Thus, the sample is large enough for us to use the critical values in Table C-7.

Chi-square tests

Since the sample size is 200, $\Sigma O = 200$. Therefore,

$$X^2 = \Sigma \left(\frac{O^2}{E}\right) - \Sigma O$$
$$= 253.75 - 200 = 53.75$$

Since we have 6 degrees of freedom and are using the .01 significance level, our critical value X_c^2 from Table C-7 is 16.81. Since our value of X^2 from the experiment, 53.75, is larger than 16.81, we reject the null hypothesis of independence and claim a relationship between major field and place of residence.

Conclusion Looking at the last column in Table 13-17 you can see that two entries are rather large and therefore contribute heavily to our finding. The value 69.23 for cell 1 indicates that an exceptionally high proportion of business students live with their parents. The value 58.01 for cell 9 indicates that an exceptionally high proportion of fine arts students live in "other arrangements." We see that major field and living arrangements are not independent. These two categories shed light on the nature of the dependence.

2 × 2 CONTINGENCY TABLES

A 2 × 2 chi-square analysis may also be done as a two-sample binomial hypothesis test of the equality of two proportions (see Chapter 9). In fact, there is an exact algebraic equivalence between the two approaches. The value of X^2 in the 2 × 2 test is equal to the square of the value of z you get in the dp test from the formula

$$z = \frac{\hat{p}_1 - \hat{p}_2}{\sigma_{d\hat{p}}}$$

You can notice that the critical values for X^2 for 1 degree of freedom are in fact the squares of the critical values of z. For example, in a two-tail test with $\alpha = .05$, $z_c = 1.96$, while X_c^2 for the 2 × 2 test at $\alpha = .05$ is 3.84, and $1.96^2 = 3.84$.

The 2 × 2 chi-square test is one of the most commonly used tests in applied statistics. This is due to the fact that in so many experiments the most basic questions involve splitting variables into just two categories, such as "success, failure," "male, female," "old, young." Here is a typical example.

EXAMPLE 13-3 Two medicines are being compared regarding a particular side effect; 60 similar patients are split randomly into two groups, one on each drug. The results are presented in Table 13-18.

Table 13-18

Observed Values

	side effects	no side effects	row totals
drug A	10	20	30
drug B	5	25	30
column totals	15	45	60

Understanding statistics

As a two-sample binomial test we would test the hypothesis that the probability of side effects with drug A equals that with drug B. As a chi-square test we test the equivalent hypothesis that drug type and side effects are independent, that is, which drug a patient takes makes no difference as far as the likelihood of side effects is concerned.

SOLUTION H_0: Drug types and side effects are independent
H_a: Drug types and side effects are dependent

The expected values are presented in Table 13-19.

Table 13-19 | Expected Values

	side effects	no side effects	row totals
drug A	7.5	22.5	30
drug B	7.5	22.5	30
column totals	15	45	60

Calculation of X^2

cell	O	E
1	10	7.5
2	20	22.5
3	5	7.5
4	25	22.5

$E > 5$

$$X^2 = \Sigma \frac{O^2}{E} - \Sigma O$$

$$= \frac{10^2}{7.5} + \frac{20^2}{22.5} + \frac{5^2}{7.5} + \frac{25^2}{22.5} - 60$$

$$= 13.33 + 17.78 + 3.33 + 27.78 - 60$$

$$= 62.22 - 60 = 2.22$$

With $\alpha = .05$ and 1 degree of freedom, $X_c^2 = 3.84$.

Decision Rule Reject the null hypothesis if $X^2 > 3.84$.

Outcome $X^2 = 2.22$

Conclusion We fail to reject the null hypothesis. We do not have enough evidence to show that drug types and likelihood of side effects are dependent; they may well be independent.

SIMPLIFIED CALCULATION IN THE 2 × 2 CASE

It can be shown by algebra that the value of X^2 in the 2 × 2 case can be calculated directly from the observed values without first figuring out the expected values. Here is a sample 2 × 2 table of observed values and the simplified formula.

Chi-square tests

a	b	g
c	d	h
e	f	n

The formula for X^2 is:

$$X^2 = \frac{(ad - bc)^2 n}{efgh}$$

We illustrate this with the data just used (Table 13-20).

Table 13-20

	side effects	no side effects	
drug A	10	20	30
drug B	5	25	30
	15	45	60

$$X^2 = \frac{[10(25) - 5(20)]^2 60}{15(45)(30)(30)} = \frac{150^2(60)}{607{,}500} = 2.22$$

This is the same value we obtained above.

CORRECTION FACTOR

Some statisticians argue for using what is called the Yates' correction factor in the 2 × 2 chi-square problem. The use of the correction factor makes the value of X^2 smaller. The purpose is to ensure that the probability of a Type I error is not more than what it is stated to be. If you wish to use the correction factor in order to be on the conservative side, the simplified formula becomes

$$X^2 = \frac{\left(|ad - bc| - \frac{n}{2}\right)^2 n}{efgh}$$

where $|ad - bc|$ is the absolute value of $ad - bc$.

CONTINGENCY TABLES WITH ONLY ONE ROW

All the examples so far have had tables with at least two rows. We can do chi-square tests on data in contingency tables with only <u>one row (or, equivalently, only one column)</u>. The expected values in the experiment will come from the null hypothesis, and the formula for the degrees of freedom will be simpler.

Our examples will be the simplest of tests, called **goodness of fit** tests. Such tests are used to determine whether sample observations fall into categories in the way they "should" according to some ideal model. When they come out as expected, we say that the data fit the model. The chi-square statistic helps us to decide whether the fit of the data to the model is good.

EXAMPLE 13-4 A carnival wheel of fortune for a July 4 fair is divided into 5 equal areas colored red, blue, red, white, and blue. The wheel is spun 50 times and the results are 25 red, 18 blue, and 7 white. Should you decide that the wheel is biased at the .05 significance level?

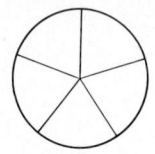

SOLUTION H_0: wheel is fair
H_a: wheel is biased

We always obtain the expected values from the null hypothesis. However, the *method* used when we had more than one row and more than one column doesn't apply when we have only one row or only one column. In this particular problem, because the null hypothesis says the wheel is fair we expect red = 2/5 (50) = 20, blue = 2/5 (50) = 20, and white = 1/5(50) = 10.

Expected Results

red	blue	white
20	20	10

Observed Results

red	blue	white
25	18	7

We can now combine these results into a more complete table (Table 13-21) and find X^2.

Table 13-21

cell	O	E	O^2	O^2/E
1	25	20	625	31.25
2	18	20	324	16.2
3	7	10	49	4.9

$\Sigma O = 50 \quad E > 5 \quad \Sigma O^2/E = 52.35$
$\Sigma O = 50$
$X^2 = 2.35$

DEGREES OF FREEDOM FOR A CONTINGENCY TABLE WITH ONE ROW

If we wish to fill in the 3 numbers in a 3 × 1 contingency table so that their total is 50, it should be clear that we can arbitrarily pick any 2 of them. Therefore, in our wheel of fortune problem we have 2 degrees of freedom.

Note. For the simplest problems where the data are displayed in a $C \times 1$

Chi-square tests

contingency table, the data should be analyzed using a chi-square test with $C - 1$ degrees of freedom. We have deliberately included only this type of problem in the text. More complex problems, where certain parameters have to be estimated from the data before the chi-square statistic is calculated, have fewer degrees of freedom. See a more advanced text for a discussion of this topic.

For the wheel of fortune problem above we have:

Degrees of freedom $= 3 - 1 = 2$

With $\alpha = .05$,

$X_c^2 = 5.99$

Decision rule Reject the null hypothesis if the outcome X^2 is greater than 5.99.

Outcome $X^2 = 2.35$

Conclusion I fail to reject the null hypothesis. The wheel may be honest.

GUIDELINES FOR A CHI-SQUARE TEST

1. Each observation in our sample falls into one and only one cell of the table.
2. The sample size is large enough so that the expected value of every cell of the table is larger than 5.
3. We compute X^2.
4. The experimental result X^2 as calculated above is compared to the critical value X_c^2 from Table C-7. Which value X_c^2 we choose depends on the significance level and the degrees of freedom.
5. If the experimental result X^2 is larger than our critical value X_c^2, then we reject the null hypothesis.

STUDY AIDS

VOCABULARY

1. Contingency table
2. Expected result
3. Cell
4. Observed result
5. Chi square
6. Independence of variables
7. Chi-square distribution
8. Goodness of fit
9. Statistical relationship between variables

SYMBOLS

1. O 2. E 3. χ^2, X^2 4. R 5. C

FORMULAS

1. Expected result $= \dfrac{\text{column total} \cdot \text{row total}}{\text{sample size}}$

2. $X^2 = \Sigma \left(\dfrac{O^2}{E} \right) - \Sigma O$ or $X^2 = \Sigma \dfrac{(O - E)^2}{E}$

3. $X^2 = \dfrac{(ad-bc)^2 n}{efgh}$

4. Degrees of freedom = $(R-1)(C-1)$, if $R \neq 1$
5. Degrees of freedom = $C-1$, if $R = 1$

EXERCISES

13-1 (a) A friend of yours is planning a chi-square field project on people's attitudes toward the construction of a new state highway. She shows you this contingency table which she intends to use. What mistake has she made?

	for	against	undecided	row totals
under age 30				150
over age 30				150
column totals	100	100	100	$n = 300$

(b) After she corrected her error (on your advice) her results were as follows.

	for	against	undecided	row totals
under age 30	100	40	10	150
over age 30	50	60	40	150
column totals	150	100	50	$n = 300$

With these data she computed $X^2 = 38.67$. State her null hypothesis and her conclusions. Use $\alpha = .05$.

(c) Another friend tried a similar project. His data were:

	for	against	undecided	row totals
under age 30	15	13	2	30
over age 30	9	9	2	20
column totals	24	22	4	$n = 50$

Show that not all Es are greater than 5. What can your friend do in order to complete the chi-square analysis?

13-2 For each of the following contingency tables find the degrees of freedom and the appropriate critical value of X^2 at the indicated significance level.
(a) 8×2, $\alpha = .05$
(b) 7×1, $\alpha = .01$
(c) 3×4, $\alpha = .05$
(d) 5×5, $\alpha = .01$

13-3 Students in Uptudate College's English Comp courses may use a word processor for their assignments if they wish. The following contingency table reports their grades last semester.

grades	always used word processor	sometimes used word processor	never used word processor
A or B	38	20	2
C or D	20	18	22
F	3	21	36

(a) State the null and alternative hypotheses.
(b) Find the degrees of freedom.
(c) Find the critical value of X^2, using $\alpha = .05$.
(d) Find the expected results.
(e) Find X^2.
(f) State a conclusion.
(g) Does this show that using a word processor improves grades?

13-4 The Maid of White Wheat Tavern in Agoura, California, is offering a special on a half dozen of their rum drinks. Damon, the barkeep, wants to see if there is any connection between a patron's dress and the drink ordered. He collected the following data last week, and gave it to his friend Pythias, who is a statistics student, to analyze. At the .05 significance level, what should Pythias conclude?

drink ordered	T-shirts, shorts, jeans, etc.	sports shirts, skirts, dress pants, etc.
strawberry daiquiri	10	21
piña colada	10	12
strawberry colada	9	15
Mai-Tai	14	11
Chi-Chi	7	20
Long Island iced tea	25	6

13-5 Eddy Torre notices that various authors prefer different symbols for footnotes and for cross referencing. Wondering if there is any connection between the type of manuscript and the symbols, he gathers the following data for the three most popular[†] symbols used: the asterisk, the octothorpe, and the dagger.

type of manuscript	*	#	†
historical	74	42	51
physical sciences	73	29	93
social sciences	102	68	78
computing	41	111	28

Perform a chi-square hypothesis test on these data allowing for a 5 percent probability of a Type I error.

13-6 A sample of Franciscans were questioned as to their smoking habits. Based upon the data below, are sex and smoking independent in this community? Use $\alpha = .05$.

priests and Brothers, smokers	100
priests and Brothers, nonsmokers	250
nuns, smokers	50
nuns, nonsmokers	100

† Excluding numerals.

13-7 Because of a great number of applicants, the director of admissions at a private university must establish a new standard for admission. It was suggested that he turn down anyone whose college board scores averaged less than 600, but he believes that there is no significant difference between previously admitted students who averaged over 600 and those who averaged under 600 in their potential to graduate. To determine who was right, he took a random sample of records of students who entered the university 5 years ago.

	graduated	withdrew voluntarily	flunked out	row totals
scored 600 or more	48	7	5	60
scored under 600	76	13	11	100
column totals	124	20	16	160 = sample size

Do the data present enough evidence to establish that 600 is a good cutoff point for admission? Use $\alpha = .05$.

13-8 In a congressional district in Los Angeles a random sample of voters was asked which of the three candidates they voted for. The results were as follows.

candidates	white	black	Mexican-American	row totals
C. Chavez	2	4	19	25
R. Brown	3	28	4	35
R. Milhaus	25	13	2	40
column totals	30	45	25	100 = sample size

Do the data present sufficient evidence to claim that voting habits depend on ethnic background? Test at the .01 significance level.

13-9 A mathematics teacher, Professor Cirkel, wanted to know if there is a relationship between grades in Calculus 1 and the ability to pass Calculus 2. A random sample of students who completed Calculus 2 was taken with the following results.

grade in Calculus 1	passed Calculus 2	failed Calculus 2	row totals
A	14	1	15
B	18	4	22
C	26	12	38
D	5	15	20
column totals	63	32	95 = sample size

What would you claim about the relationship of grades in Calculus 1 and passing Calculus 2? Use $\alpha = .05$.

13-10 A psychologist was studying patterns of selfishness in various family groups. As part of her study she gave a test designed to measure selfishness to a random sample of 600 women who said that they were not going to have any more children. Here are her results.

Women's Score on Selfishness Exam

number of children	low	medium	high	row totals
0	30	40	50	120
1	40	50	60	150
2	60	50	40	150
3	50	30	20	100
more than 3	40	30	10	80
column totals				600

At $\alpha = .05$ does this indicate a relationship between score on selfishness exam and number of children?

13-11 In horse races at Upsand Downs racetrack, is the order of winning independent of the pole positions at the start of the race? Use $\alpha = .01$. The results of 100 horse races were as follows.

pole position	win	place	show
1	30	19	11
2	16	12	16
3	20	18	16
4	8	17	11
5	9	9	16
6	6	12	9
7	6	6	16
8	5	7	5

13-12 Jacob J. Kubb, M.A., brought 32 boys on a nature study hike. Of them, 16 had older brothers or sisters and 16 did not. He noted that of the first 16 to ask questions, 10 had older brothers or sisters. Should Master Kubb accept this as evidence at the .05 significance level that there is a connection between having older siblings and asking questions?

13-13 Ms. Edwards and Mr. Kelvin teach kindergarten. They each have 20 students. Their approaches to zippers, overshoes, etc., at dismissal time are very different. They observe that by the time half the children are able to dress themselves, 11 are from Ms. Edwards' class. Does this evidence indicate any difference between the two approaches at $\alpha = .05$?

13-14 Tom performed the following field project. He walked directly toward a person who was approaching him in the middle of the quadrangle on campus. He recorded the sex of each person and whether they turned to the right or to the left. (He disregarded two collisions, one fight, and one person whose sex he could not determine right away.) His data were as follows.

	turned to left	turned to right
male	57	43
female	43	57

(a) Using $\alpha = .05$, finish Tom's project without using the correction factor for 2×2 contingency tables.
(b) Tim, too lazy to gather his own data, copied Tom's information. To make his project appear different, Tim used the correction factor. Finish Tim's project.
(c) Chet, aware of what Tom and Tim were doing, solved the problem as a binomial two-sample problem, taking as his null hypothesis that the same percentage of males and females turned left. Finish Chet's project.
(d) Compare the results in the Tom-Tim-Chet problem.

13-15 In exercise 9-21 we posed two questions for the data

	patient is visited regularly	patient is not visited regularly
patient has grandchildren	20	40
patient does not have grandchildren	10	30

Now we ask a third question: are having grandchildren and being visited regularly independent? With α still .01, as it was in Chapter 9, we find $X^2 = .79$ and X_c^2 for 1 degree of freedom is 6.635. Since .79 is less than 6.635, we fail to reject.
(a) In Exercise 9-21(a) we found $\mu = 0$, $\hat{\sigma} = .107$, and $d\hat{p} = .0952$. Show that the z score for $d\hat{p} = .0952$ is .89.
(b) In Exercise 9-21(b) we found $\mu = 0$, $\hat{\sigma} = .0935$, and $d\hat{p} = .0833$. Find the z score for $d\hat{p} = .0833$.
(c) We found $X^2 = .79$ above. Find $X = \sqrt{.79}$.
(d) $X_c^2 = 6.635$. Find X_c.
(e) Compare your answers in parts (c) and (d) with z and z_c in parts (a) and (b).
(f) What can you conclude about the three questions that we asked for these data?

13-16 In a certain game a coin is tossed twice. Mary is tossing the coin, and John is getting suspicious about the results. To test for fairness, he decides to do a 3×1 chi-square test. He incorrectly uses the following expected values for the next 120 games.

no heads	1 head	2 heads
40	40	40

$n = 120$

What are the correct expected values?

13-17 Dr. Noit Ahl theorized that 20 percent of students drink vodka, 10 percent drink gin, 40 percent drink beer, and the rest abstain from alcoholic beverages. At a recent party you observed the following data.

vodka	gin	beer	abstain
20	10	40	10

Should you reject his theory, using $\alpha = .05$?

13-18 In Example 5-2 we discussed Marty's spinning a spinner 4 times. Marty repeated this game 810 times and observed the following data.

	number of wins in every 4 spins				
	4	3	2	1	0
number of times it happened	10	100	200	300	200
number of times it was expected					

(a) Complete the chart above.
(b) Using $\alpha = .01$, should he consider the spinner biased?

13-19 Is the distribution of boys and girls in families with 2 children a random binomial distribution with equal numbers of boys and girls? A random sample of 1000 families showed 100 with 2 boys, 600 with 1 boy and 1 girl, and 300 with 2 girls. Use $\alpha = .05$.

13-20 Using $\alpha = .01$, are two monopoly dice fair if the last 360 tosses produced the following results.

	total on each toss										
	2	3	4	5	6	7	8	9	10	11	12
number of times total occurred	4	16	24	46	55	70	57	43	25	15	5

13-21 If a particular species of bug is distributed *randomly* in a wooded area, then probability theory can be used to predict how many bugs should be found under rocks picked up at random in the area. For 100 rocks the predictions are as follows.

	number of bugs under rock					
	0	1	2	3	4	5
expected frequency	38	35	17	6	3	1

An ecology student went to such a wooded area to count bugs under 100 rocks and found the following.

	number of bugs under rock					
	0	1	2	3	4	5
actual count	33	3	15	33	15	1

Interpret these results by a chi-square test.

13-22 In 1975 the Bicentennial Committee surveyed thousands of federal employees. Without being identified as such, a portion of the Declaration of Independence was given to the government employees to read, and they were then asked the following questions. (1) Would you sign such a declaration? (2) Do you recognize this quote? The data collected are as follows.

	employees of the Pentagon	employees of the Congress	other
would sign	41	94	30
would not sign	159	106	55

	recognized quote	failed to recognize
would sign	121	615
would not sign	130	1434

Analyze these two sets of data by chi-square tests at $\alpha = .01$ and interpret your results.

13-23 Sue asked a random sample of students and former students of Lax University if they would pose in the nude for the centerfold of a magazine. Of 100 current students 30 said that they would, but of 200 former students only 40 would pose nude.

(a) Find the 95 percent confidence interval for the difference between the proportion of students who would pose in the nude and the proportion of former students who would.

(b) Do a two-sample binomial hypothesis test on these data with $\alpha = .05$.

(c) Do a chi-square hypothesis test on these data with $\alpha = .05$.

13-24 (a) Little Joe created a random number table by tossing two dice. He counted the number of times each number appeared. He let a toss of a 10 correspond to 0, a toss of an 11 correspond to a 1, and he disregarded all 12s. His dice tosses were

4, 6, 7, 11, 8, 9, 10, 5, 8, 7, 7, 4, 6, 9, 4, 12, 7, . . .

giving him the random digits

4, 6, 7, 1, 8, 9, 0, 5, 8, 7, 7, 4, 6, 9, 4, 7, . . .

He continued this until he had a total of 100 outcomes. Counting the number of times each digit appeared, he found the following.

						digit				
	0	1	2	3	4	5	6	7	8	9
number of appearances	10	8	4	7	9	11	13	16	12	10

Test the hypothesis that all the digits are equally likely to appear. Use $\alpha = .05$.

(b) In part (a), probability theory of randomly tossed dice leads us to the following expected values.

						digit				
	0	1	2	3	4	5	6	7	8	9
number of appearances	8.6	5.7	2.6	5.7	8.6	11.4	14.3	17.1	14.3	11.4

Test this theory at the .01 significance level.

13-25 (a) A newspaper report listed the total number of homicides each month in New York over a 10-year period. Use these data to test the hypothesis that homicides in all months are equally likely. Use $\alpha = .01$.

month	number of homicides	month	number of homicides
January	834	July	1024
February	744	August	985
March	789	September	875
April	829	October	973
May	867	November	869
June	823	December	1042

(b) Test the hypothesis that for a whole year the percentage of murders that occur in December is greater than the percentage occurring in July.
(c) True or false? For *this* period of time the average number of murders per *day* in February is more than the average number per *day* in March. (Assume 3 leap years in this 10-year period.)

13-26 A group of 50 people each toss a penny 7 times.
(a) Assuming the coin is fair and using Pascal's triangle, figure out what *percentage* of the people should get 0 heads, 1 head, 2 head, ..., 7 heads.
(b) About *how many* people should be in each of these categories?
(c) The observed results in one group are listed below. Is this evidence, at the .05 significance level, that the coin is not fair? (You may have to combine some categories if all the expected values are not more than 5.)

number of heads	number of people
0	0
1	0
2	3
3	7
4	12
5	13
6	11
7	4
	$n = 50$

13-27 In Chapter 11 we often made the simplifying assumption that the population was normal. This exercise demonstrates how we can use chi-square to do a goodness of fit test for normalcy. A quality control engineer for an engineering firm, which makes rocket engines, tests a sample of 260 relay switches taken from a production line. Supposedly they operate at a mean speed of 50 microseconds with a standard deviation of 10 microseconds.
(a) If the population is normal with a mean of 50 and a standard deviation of 10, and $n = 260$, how many values would you expect to fall into each of the following categories?
 (1) Less than 30 ($z < -2$).
 (2) Between 30 and 40 ($-2 < z < -1$).
 (3) Between 40 and 50 ($-1 < z < 0$).
 (4) Between 50 and 60 ($0 < z < 1$).

(5) Between 60 and 70 ($1 < z < 2$).
(6) More than 70 ($2 < z$).

(b) The data for the sample of 260 relay speeds were as follows.

speed, microseconds	number of relay switches
less than 30	15
30 to 40	30
40 to 50	85
50 to 60	80
60 to 70	40
more than 70	10

Use the expected data found in part (a) to perform a 6 × 1 chi-square test with $\alpha = .05$. Is it reasonable to assume that these speeds are from a normal population?

13-28 At the annual meeting of the Association of Tavernkeepers of Bayonne, a group of 1000 bartenders were asked to pour out a shot of whiskey (1.5 ounces) without measuring. The results were grouped as follows.

	.4–.7	.7–1.0	1.0–1.3	ounces 1.3–1.6	1.6–1.9	1.9–2.2	2.2–2.5
number of bartenders	80	100	0	400	300	100	20

(a) The distribution of all bartenders' estimates of one shot of whiskey was claimed to be a normal distribution with $\mu = 1.6$ ounces and $\sigma = .3$ ounce. Assuming this claim to be true, how many of the 1000 bartenders would you expect in each cell of the contingency table shown above?

(b) Perform a chi-square goodness of fit test on this claim. Use $\alpha = .05$.

13-29 In a medical study of chronic pain 3 groups of patients were given different pills. In group 1, the pills were just sugar; the patients in group 2 got aspirin; the patients in group 3 got a new experimental medicine. Neither the patients nor the doctors knew which patients got which pills until the experiment was over. Such an experiment is called *double-blind*. Do a chi-square test at $\alpha = .05$, and interpret your results.

	got relief	did not get relief	row totals
sugar	70	30	100
aspirin	80	20	100
experimental medicine	85	15	100
column totals	235	65	300 = sample size

13-30 The Duodecimal Society of America tested a sample of 300 Americans about their beliefs concerning counting and measurement. The experiment went as follows. First, the people in the sample were given a short lecture clarifying three ideas.

Chi-square tests

1. How we naturally measure many things in twelves (inches, dozens, and grosses, months, hours, ounces in the Troy pound used by druggists, the new telephone dialing systems, etc.), yet we count with 10 symbols (0, 1, 2, 3, 4, 5, 6, 7, 8, 9).
2. How we could change our measurements to some artificial system based on 10, similar to one of the decimal metric systems used in various parts of the globe today.
3. How we could learn to count in dozens using twelve symbols (0, 1, 2, 3, 4, 5, 6, 7, 8, 9, *, #) and be able to retain our natural measurements.

After the lecture they were asked to pick one of these three responses as best expressing their beliefs:

A. The dozenal system of counting and measuring is easier, and hence the world will one day adopt it.
B. The dozenal system of counting and measuring is easier, but it will never be adopted by many people.
C. The dozenal system of counting and measuring is not any easier.

Their responses were gathered in the following contingency table.

profession	choice A	choice B	choice C	undecided
involves both measuring and arithmetic	45	35	10	10
involves arithmetic but not measurement	35	35	20	10
involves neither measurement nor arithmetic	30	30	30	10

Perform a hypothesis test on the data with $\alpha = .05$.

13-31 Repeat some of the exercises in Chapter 9 using the methods of this chapter.

CLASS SURVEY

1. It has been noted in many populations that sex and smoking are not independent. Test to see whether this is true on your campus.
2. Are hair coloring and eye coloring independent at your school? Do a chi-square hypothesis test using your sample data.

FIELD PROJECTS

Design and perform a chi-square test of your choosing. Some projects that students have done on campus include investigating the relationship between

1. Sex of student and student's major.
2. Class size and number of students who drop the course.
3. Students' grades and how close they sit to the teacher.
4. Faculty rank and amount of time faculty member spends with students.
5. Religion and membership in religious clubs on campus.

Understanding statistics

SAMPLE TEST FOR CHAPTERS 12 AND 13

1. Explain why a 100 percent confidence interval is useless.
2. A random sample of 200 people chosen from the large membership list of the American Therapist Association included 30 females.
(a) What is the 95 percent confidence interval for the percentage of females in the association?
(b) Under what conditions would you expect your answer in part (a) to estimate the percentage of females in the *profession*?
3. A 95 percent confidence interval is desired for the percentage of persons who perform perfectly. What size sample should be used if we want to be confident that our estimate is no more than 3 percent off the true value?
4. New York City officials monitored the 8 A.M. to 4 P.M. shift in the emergency medical services department June 16, 1980. This department receives emergency phone calls for ambulances. They reported that for 88 calls the average response time was 28.8 minutes.
(a) If we wish to compute a confidence interval based on these data, what statistical population would it refer to, that is, of what population do you think 28.8 is the sample mean? Give some conditions under which this would be reasonable. Why might it be unreasonable?
(b) If we assume that s is about 15 minutes, what is the 95 percent confidence interval?
5. We wish to estimate for a large industry the difference between the percentage of male and female sales managers who earn more than $25,000 per year. A random sample of 50 male and 50 female managers is chosen, and their salaries are recorded. It is found that 36 of the males earn more than $25,000 while 18 of the females do. Find the 95 percent confidence interval for the difference in percentages earning more than $25,000 per year. What are some of the reasons often given to explain why such an imbalance might exist?
6. The observed frequencies shown below are based on a random sample of father-son pairs in a large city. Interpret the results using a chi-square test. Use $\alpha = .05$.

son	father shorter than 5 feet 6 inches	father 5 feet 6 inches to 6 feet	father taller than 6 feet
shorter than 5 feet 6 inches	50	400	10
5 feet 6 inches to 6 feet	150	2000	200
taller than 6 feet	5	300	60

7. A random sample of students at Wayup High School was asked two questions. (1) Do you own more rock or more disco recorded music? (2) Are you for or against Proposition 31? Do the results indicate that answers to these two questions are statistically independent? Use $\alpha = .05$.

	Proposition 31 for	Proposition 31 against
more rock	20	60
more disco	30	90

Correlation and Prediction

14

CORRELATION COEFFICIENTS

It frequently happens that statisticians want to describe with a single number a relationship between two sets of scores. A number that measures a relationship between two sets of scores is called a **correlation coefficient.** There are several correlation coefficients for measuring various types of relationships between different kinds of measurements. In this text we will illustrate the basic concepts of correlation by discussing only the Pearson correlation coefficient, which is one of the more widely used correlation coefficients. The statistic is named for its inventor, Karl Pearson (1857–1936), one of the founders of modern statistics. It is denoted by r, and is used to measure what is called the *linear relationship* between two sets of measurements.

To explain how r works and what is meant by a linear relationship, we will look at a few overly simplified examples. It is unlikely that a real application of the correlation coefficient would be made with so few scores. Imagine that 6 students are given a battery of tests by a vocational guidance counselor with the results shown in Table 14-1.

Table 14-1

student	interest in retailing	interest in theater	math aptitude	language aptitude
Pat	51	30	525	550
Sue	55	60	515	535
Inez	58	90	510	535
Arnie	63	50	495	520
Gene	85	30	430	455
Bob	95	90	400	420

Understanding statistics

Figure 14-1

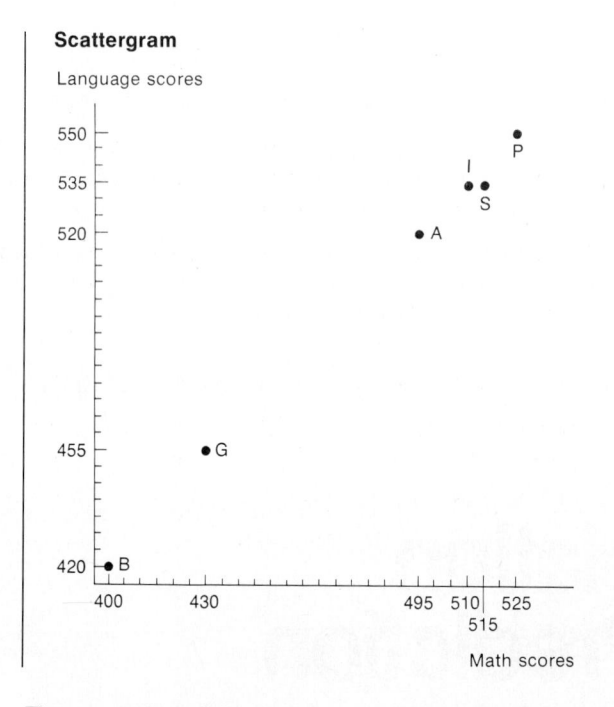

Scattergram

Language scores

```
550 ┤                                      • P
535 ┤                                    • I
                                          • S
520 ┤                              • A



455 ┤              • G


420 ┤• B
    └┬────────┬──────────────┬──┬──┬──
    400      430            495 510 525
                                515
```

Math scores

The counselor might want to see if there are any correlations among these sets of marks. For example, it looks as if the people who do well in math also do well in language skills. Let us draw a graph called a **scattergram** to investigate this relationship. To draw this scattergram, we first draw a vertical axis and a horizontal axis, one for the math scores and the other for the language scores. In the scattergram shown in Figure 14-1 we have put the math scores on the horizontal axis, but that is not important. We could have put them on the vertical axis. Notice how the math axis is labeled. The reported math scores ranged from 400 to 525, and so the math axis must be labeled in a way that all those scores can be recorded. We therefore chose to make a series of equally spaced marks labeled from 400 to 525 counting by 5 at a time. It is important that the marks be equally spaced. Similarly, the vertical axis was labeled from 420 to 550 to cover the range of the language scores.

After both axes are drawn and labeled, we use *one* dot for each person. The dot is placed so that it is directly over the person's math score, and directly to the right of the person's language score. For example, the dot for Pat's scores is over the 525 on the math axis, and to the right of 550 on the language axis. In Figure 14-1 we have put each person's initial next to his or her mark to help you read the graph, but the initials are usually left off. For convenience, we repeat the math scores and language scores in Table 14-2.

Table 14-2

student	math	language
Pat	525	550
Sue	515	535
Inez	510	535
Arnie	495	520
Gene	430	455
Bob	400	420

Correlation and prediction

You will notice three things about the scattergram.

1. There is one point for each pair of scores, 6 points in all.
2. The points are arranged approximately in a straight line. When this happens, we say that there is good **linear correlation** between the two variables (in this case math skill and language skill).
3. The higher numbers in the math column of the table correspond to the higher numbers in the language column. This causes the line to slope up to the right. This is called **positive correlation.**

MATH SCORES VERSUS THEATER-INTEREST SCORES

Let us now compare the relationship between math scores and theater-interest scores. We repeat these scores in Table 14-3 and graph them in Figure 14-2.

Table 14-3

student	math	theater
Pat	525	30
Sue	515	60
Inez	510	90
Arnie	495	50
Gene	430	30
Bob	400	90

Figure 14-2

Scattergram

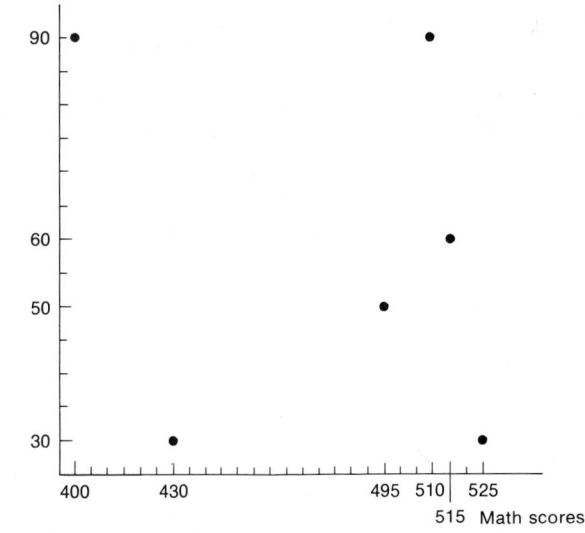

In Figure 14-2 you notice that there is no special tendency for the points to appear in a straight line. We say that there is *little or* **no correlation** between the math scores and the theater-interest scores. Also note that it is not necessary for both variables to be scored on the same scale, since the correlation coefficient describes the *pattern* of the scores, not the actual values.

MATH SCORES VERSUS RETAILING-INTEREST SCORES

In Table 14-4 and Figure 14-3 we compare math scores to retailing-interest scores.

Table 14-4

student	math	retailing
Pat	525	51
Sue	515	55
Inez	510	58
Arnie	495	63
Gene	430	85
Bob	400	95

Figure 14-3

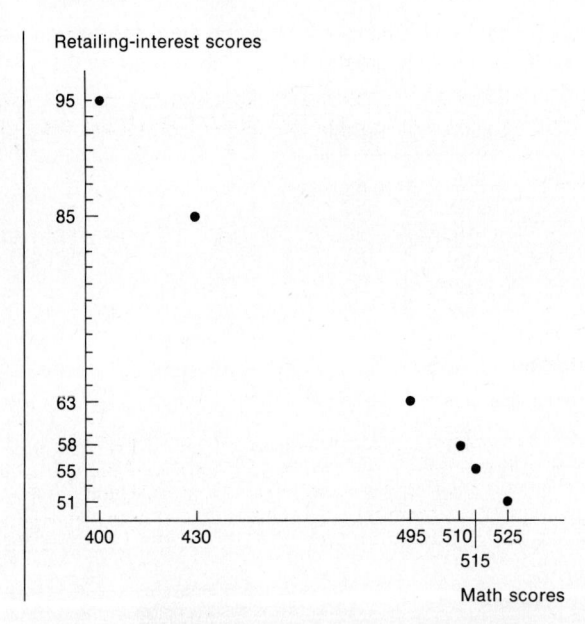

Retailing-interest scores

[Scatter plot with points at approximately (400, 95), (430, 85), (495, 63), (510, 58), (515, 55), (525, 51)]

Math scores

In Figure 14-3 there is a tendency for the points to lie in a line that slopes down to the right. This is called **negative correlation.** This happens because the higher scores in the column for math correspond to the lower scores in the column for retailing interest.

COMPUTATION OF r

Pearson defined r so that the formula for r has a minimum possible value of -1 and a maximum possible value of $+1$. When the sample points lie *exactly* in a line sloping *down to the right,* we say there is *perfect negative correlation*: $r = -1$. When the sample points lie *exactly* in a line sloping *up to the right,* we say there is *perfect positive correlation:* $r = +1$. When there is *no tendency* for the points to lie in a straight line, we say there is *no correlation:* $r = 0$. If r is near $+1$ or -1, we say that we have *high correlation*. If r is near zero, we say that we have *low correlation*.

We would expect that for Figure 14-1, r is near 1; for Figure 14-2, r is near

Correlation and prediction

0; for Figure 14-3, r is near -1. We will show that this is true after we state the formula for r.

The formula for the correlation coefficient r is:

$$r = \frac{n\Sigma XY - (\Sigma X)(\Sigma Y)}{\sqrt{n\Sigma X^2 - (\Sigma X)^2} \sqrt{n\Sigma Y^2 - (\Sigma Y)^2}}$$

or

$$r = \frac{\Sigma(X - \overline{X})(Y - \overline{Y})}{\sqrt{\Sigma(X - \overline{X})^2 \Sigma(Y - \overline{Y})^2}}$$

where X = label for one of the variables
Y = label for the other variable
n = number of *pairs* of scores

All the above notation is familiar except ΣXY. This is found by multiplying the corresponding values of X and Y, and then adding all these products.

ILLUSTRATION OF r COMPUTATIONS

For the correlation depicted in Figure 14-1 we can tabulate the data as follows.

math X	language Y	X^2	Y^2	XY
525	550	275,625	302,500	288,750
515	535	265,225	286,225	275,525
510	535	260,100	286,225	272,850
495	520	245,025	270,400	257,400
430	455	184,900	207,025	195,650
400	420	160,000	176,400	168,000
$\Sigma X = 2875$	$\Sigma Y = 3015$	$\Sigma X^2 =$ 1,390,875	$\Sigma Y =$ 1,528,775	$\Sigma XY =$ 1,458,175

Solving for r,

$$r = \frac{n\Sigma XY - (\Sigma X)(\Sigma Y)}{\sqrt{n\Sigma X^2 - (\Sigma X)^2} \sqrt{n\Sigma Y^2 - (\Sigma Y)^2}}$$

$$= \frac{6(1,458,175) - 2875(3015)}{\sqrt{6(1,390,875) - (2875)^2} \sqrt{6(1,528,775) - (3015)^2}}$$

$$= \frac{80,925}{\sqrt{79,625} \sqrt{82,425}} = \frac{80,925}{282.2(287.1)} = \frac{80,925}{81,019.6} = .999$$

Similarly, for the correlation shown in Figure 14-2 we find that $\Sigma X = 2875$, $\Sigma Y = 350$, $\Sigma X^2 = 1,390,875$, $\Sigma Y^2 = 24,100$, and $\Sigma XY = 166,200$. Thus,

$$r = \frac{n\Sigma XY - (\Sigma X)(\Sigma Y)}{\sqrt{n\Sigma X^2 = -(\Sigma X)^2} \sqrt{n\Sigma Y^2 - (\Sigma Y)^2}}$$

$$= \frac{6(166,200) - 2875(350)}{\sqrt{6(1,390,875) - (2875)^2} \sqrt{6(24,100) - (350)^2}}$$

$$= \frac{997{,}200 - 1{,}006{,}250}{\sqrt{79{,}625}\sqrt{22{,}100}} = \frac{-9050}{282.2(148.7)} = -.22$$

For the correlation shown in Figure 14-3 we get $\Sigma X = 2875$, $\Sigma Y = 407$, $\Sigma X^2 = 1{,}390{,}875$, $\Sigma Y^2 = 29{,}209$, and $\Sigma XY = 190{,}415$. Thus,

$$r = \frac{n\Sigma XY - (\Sigma X)(\Sigma Y)}{\sqrt{n\Sigma X^2 - (\Sigma X)^2}\sqrt{n\Sigma Y^2 - (\Sigma Y)^2}}$$

$$= \frac{6(190{,}415) - 2875(407)}{\sqrt{6(1{,}390{,}875) - (2875)^2}\sqrt{6(29{,}209) - (407)^2}}$$

$$= \frac{1{,}142{,}490 - 1{,}170{,}125}{\sqrt{79{,}625}\sqrt{9605}} = \frac{-27{,}635}{282.2(98.0)} = -.999$$

EXERCISES

14-1 Imagine that you are a school principal and that you have calculated the coefficient of correlation between last year's grades in Señor Oldways' class in Spanish II and this year's grades in Señorita Modern's class in Spanish III. What would it mean if r was close to -1? close to $+1$? close to zero?

14-2 Students taking Professor Springsteen's course in geology compared their grades on the final exam with their final course grades. They calculated $r = .12$. What does this indicate?

14-3 (a) Complete the following table.

X	Y	X^2	Y^2	XY
1	1			
2	1			
3	2			
4	3			
totals				

Find:
(b) n (c) ΣX (d) ΣY
(e) ΣXY (f) ΣX^2 (g) ΣY^2
(h) $(\Sigma X)^2$ (i) $(\Sigma Y)^2$ (j) $n\Sigma XY - (\Sigma X)(\Sigma Y)$
(k) $\sqrt{n\Sigma X^2 - (\Sigma X)^2}\sqrt{n\Sigma Y^2 - (\Sigma Y)^2}$ (l) r

14-4 (a) Complete the following table.
Find (b) to (l) of Exercise 14-3.

X	Y	X^2	Y^2	XY
10	1			
5	9			
3	3			
8	9			
0	0			
totals				

14-5 Follow the instructions in Exercise 14-3, using these data.

X	Y	X^2	Y^2	XY
2	−1			
2	3			
0	−2			
3	4			
4	4			
1	0			
totals				

14-6 Below is a table giving the lengths of some pieces of lumber in feet and in inches.

length, inches	length, feet
12	1
36	3
60	5
48	4
24	2
72	6

(a) Draw a scattergram for the data.
(b) Intuitively guess whether r is near 1, −1, or 0.
(c) Compute r.

14-7 Here is a table showing some people's ages in April 1982, and the years of their births. Guess at a value of r; then compute r.

age in April 1982	year of birth
18	1964
19	1963
20	1961
21	1961
24	1958

14-8 A random sample of men stopped in a shopping center were asked their shoe sizes and the number of ties that they owned. Would you expect there to be any correlation between the two variables? Here are the data. Compute r.

shoe size	number of ties owned
7½	10
9	17
9	17
11	4
8½	10
8	1
13	6

Understanding statistics

14-9 Compute r for these grades of 6 students in Preparatory English and in the first semester of Survey of English Literature.

Prep English	English Literature
50	45
57	57
68	60
75	75
80	84
89	93

14-10 An experiment was done to see if there was any correlation between the volume of water in a fishtank and the average length to which four goldfish grew when they were hatched and raised in that tank. Here are the results. Compute r.

fishtank volume, gallons	fish average length, inches
0.5	1.8
1	2.1
2	2.2
4	2.9
5	3.3

14-11 An experiment was performed where an object was dropped through a certain liquid. The distance it traveled was recorded every second for 6 seconds. Here are the results.

time, seconds	distance object traveled through liquid, feet
0	0
1	1
2	4
3	9
4	16
5	25
6	36

The experimenter recognized that this was not a true linear relationship. She thought that the relationship was nearly linear, though, for the time period from 3 to 6 seconds. Check her intuition by computing r:
(a) For all the data.
(b) For just the data from 3 to 6 seconds.

14-12 Here are some stopping distances for a certain vehicle going at various speeds. Compute r for the data.

speed, miles per hour	stopping distance, feet
30	90
40	150
50	240
60	370
70	530

14-13 For the weather data below sketch a scattergram and compute r.

Correlation and prediction

day	daily high, °F	daily low, °F
Tuesday	70	50
Wednesday	72	50
Thursday	66	48
Friday	73	51
Saturday	67	49

14-14 An investor investigating a possible correlation between two stocks noticed what she thought was a pattern connecting their prices. Sketch a scattergram and compute r for the data to see if there seems to be a linear relationship between their prices.

date	selling price of BQT stock, dollars	selling price of CRV stock, dollars
January 1	47	22
February 1	40	24
March 1	30	26
April 1	15	30

14-15 State two variables for which there would probably be:
(a) Negative correlation.
(b) Zero correlation.
(c) Positive correlation.

TESTING THE SIGNIFICANCE OF r

Suppose that we examined an entire population and computed the correlation coefficient for two variables. If this coefficient equaled zero, we would say that there is no correlation between these 2 variables in this population. Consequently, when we examine a random sample taken from a population, then a *sample* value of r near zero is interpreted as reflecting no correlation between the variables *in the population*. A sample value of r far from zero (near 1 or -1) indicates that there *is* some correlation in the population. The statistician must decide when a sample value of r is far enough from zero to be significant, that is, when it is sufficiently far from zero to reflect correlation in the population.

This test for significance would have to be carried out in different ways, depending on what the distributions of the two variables in the population are. The simplest case results when it can be assumed that both variables are normally distributed.

Strictly speaking, in trying to test a sample value of r for significance, we are actually setting up as a null hypothesis that there is no correlation between the two variables in question. That is,

H_0: correlation coefficient in population = 0

We will be able to reject the null hypothesis when our sample value of r is farther from zero than some critical value. This means that we would be establishing the alternative hypothesis that there is some degree of linear correlation between the two variables in the population. Tables C-9 and C-10 contain lists of critical values of r arranged according to the total number of pairs of scores in the sample, n, the significance level of the test, α, and the number of tails in the test. Values of r are given in this table without signs.

You must determine whether the critical values of r are positive, negative, or both from the alternative hypothesis.

EXAMPLE 14-1 In a study of academic success, a random sample of 30 public school children in a community was taken. For each child the survey recorded verbal aptitude (as determined by a commonly used test) and the annual income of the child's family. The 30 pairs of scores were used to compute r. It was found that $r = .46$. Does this support the hypothesis that there is positive correlation between verbal aptitude and family income in the children of this community? Test at the .05 significance level.

SOLUTION H_0: correlation coefficient in population $= 0$

H_a: correlation coefficient in population > 0 (one-tail test)

$r_{\text{sample data}} = .46$

$r_{\text{critical}} = .31$

Therefore, we conclude that the coefficient of correlation in the population is greater than zero. There is some positive linear correlation in this population between family income and verbal aptitude.

NOTE: The existence of *some* nonzero correlation does not mean that the correlation is *high,* merely that it is not zero. Unless the correlation is far enough from zero it may be of very little practical value. ∎

Caution The fact that the correlation exists does not prove anything about the reasons for the correlation. Researchers must decide independently (1) what causes the correlation, and (2) whether or not the correlation and its causes are of any practical concern.

The significance test may verify the existence of the correlation in the given population; it does not establish cause and effect. For example, it would not be hard to show that there is a positive correlation between the average salary of elementary school teachers in Suffolk County, N.Y., and the number of pizzas sold in Suffolk County over the past 10 years. Can you think of any reasons why this is so?

EXAMPLE 14-2 A research team wondered if there was any correlation between the amount of food a freely feeding rat ate when it was 50 days old and the age at which it died. They planned to experiment with 30 rats who would be left alone to eat as much as they chose. What are their two hypotheses? What values of r would they have to find to establish a correlation between diet and longevity if they use $\alpha = .05$?

SOLUTION H_0: no correlation between food consumption and life span, correlation coefficient in population $= 0$

H_a: some correlation between food consumption and life span, correlation coefficient in population $\neq 0$ (two-tail test)

$r_{\text{critical}} = \pm.36$

If they find that their sample value of r is more than $+.36$ or less than $-.36$, this will be evidence in favor of a correlation between food consumption and life span. A significant positive value of r would indicate an association of eating

Correlation and prediction

■ a lot with living a long time. A significant negative value of r would indicate an association of eating a little with living a long time.

EXERCISES

In the following exercises, assume that both populations are normal.

14-16 A hypothesis test shows that there is high positive correlation between grades at the Police Academy and success "on the street" as a police officer. Yet Sgt. Experience notes that not infrequently a candidate who did well at the Academy fails on the beat. Similarly, he notes that a few officers do well on the job who did not do well at the Academy. Just what information is conveyed by high positive correlation anyhow?

Exercises 14-17 to 14-22

You have already computed r for these exercises. Now we will state what an experimenter was testing for. You should then do the following.
(a) Write out the null and motivated hypotheses.
(b) Decide if the hypotheses call for a one-tail or a two-tail test.
(c) Find the critical values of r from Table C-9 or Table C-10. Test at the .01 significance level.
(d) Decide whether or not there is nonzero correlation.

14-17 (Data from Exercise 14-8.) The experimenter was testing for nonzero correlation.

shoe size	number of ties owned
7½	10
9	17
9	17
11	4
8½	10
8	1
13	6

14-18 (Data from Exercise 14-9.) Test for positive correlation.

Prep English	English Literature
50	45
57	57
68	60
75	75
80	84
89	93

14-19 (Data from Exercise 14-10.) The experimenter was testing for nonzero correlation.

fishtank volume, gallons	fish average length, inches
0.5	1.8
1	2.1
2	2.2
4	2.9
5	3.3

14-20 [Data from Exercise 14-11(a).] Test for positive correlation.

time, seconds	distance object traveled through liquid, feet
0	0
1	1
2	4
3	9
4	16
5	25
6	36

14-21 [Data from Exercise 14-11(b).] Test for positive correlation.

time, seconds	distance object traveled through liquid, feet
3	9
4	16
5	25
6	36

14-22 (Data from Exercise 14-12.) Test for positive correlation.

speed, miles per hour	stopping distance, feet
30	90
40	150
50	240
60	370
70	530

14-23 Suppose you thought that there was some correlation between the length of a male college student's hair and his political beliefs. Imagine that some clever psychology professor has designed a test of political belief. When a person takes this test, the score can run anywhere from 0 (extreme right-wing beliefs) to 200 (extreme left-wing beliefs). You get a random sample of 25 male students. You score each student for the two variables. These are the results.

student number	hair length, inches	test score	student number	hair length, inches	test score
1	0.5	50	14	2.5	100
2	2.0	140	15	4.5	165
3	1.0	60	16	1.5	90
4	2.5	80	17	3.0	105
5	3.0	115	18	2.5	105
6	1.5	75	19	2.0	85
7	4.5	170	20	3.5	140
8	3.5	120	21	5.0	180
9	2.5	95	22	2.5	130
10	3.0	120	23	4.0	150
11	1.0	85	24	3.0	100
12	4.0	160	25	2.0	80
13	2.0	100			

State your hypotheses. Compute r. Test for positive correlation at the .05 significance level.

14-24 An educational testing laboratory is developing a new test to measure computer programming aptitude. They wish to develop two different forms of the test. Theoretically a person should get the same score, no matter which form of the test he or she takes. To determine whether or not both forms give about the same results, they are administered to 30 people. The results were as follows.

candidate number	form A	form B	candidate number	form A	form B
1	99	80	16	67	63
2	97	95	17	67	60
3	97	87	18	65	64
4	90	88	19	65	81
5	89	83	20	65	65
6	83	90	21	63	60
7	80	85	22	62	61
8	80	78	23	61	59
9	75	40	24	60	50
10	70	76	25	59	58
11	69	70	26	50	40
12	69	71	27	43	51
13	68	70	28	40	70
14	68	72	29	20	19
15	68	68	30	3	0

Compute r. State the hypotheses. Test for positive correlation at the .01 significance level.

14-25 The following data were collected at random from Ms. Betty's School for Young People.

student	reading	spelling	math	music
Sam	20	7	100	10
Samantha	15	7	70	...
Toni	25	10	60	3
Anthony	35	8	90	9
Salvatore	30	9	...	20
Sally	50	8	80	15
Pat	40	10	80	5

Find the coefficient of correlation, and perform a hypothesis test at the .05 significance level for each of the following.

(a) Is there positive correlation between the reading scores and the spelling scores?

(b) Is there negative correlation between the spelling scores and the math scores?

(c) Is there nonzero correlation between the spelling scores and the music scores?

PREDICTION BASED ON LINEAR CORRELATION

If the statistician determines that there is high[†] linear correlation between two variables, we can try to represent the correspondence by an ideal line—a line that best represents the linear correspondence. We can then write the formula which determines this line, and use this formula to predict, for instance, which value of the Y variable corresponds ideally to any given value of the X variable.

For example, let us suppose that at State U. grades in English 1 and English 2 have a high positive correlation. Suppose we have found a formula which predicts grades in English 2 from grades in English 1. Given that a grade in English 1 was 85, the formula predicts a grade of 81 for English 2.

Clearly, if 10 students had grades of 85 in English 1, we do not expect all 10 to get an 81 in English 2. In fact, maybe none of them will actually get an 81. The predicting formula really says that our best estimate of their mean grade will be 81. On the other hand, if we do want to predict one student's grade, the best point estimate we can make will be this mean, 81.

In this section we will show how to get the formula for the line which is used to get the best point estimates. The important topic of evaluating the reliability of these estimates will not be discussed. See a more advanced text if you need further discussion of this topic.

EXAMPLE 14-3 For several years a college has been on open enrollment. Because of recent enrollment growth, the dean of admissions now finds it necessary to turn some applicants down. He wants to turn down those applicants who would probably flunk out anyway. A random sample of former students' records are pulled from the files. It is noticed that there is strong positive correlation between student scores on a certain aptitude test and their college grade-point average at the time they leave the college (either graduating or withdrawing). The scattergram for the data is shown in Figure 14-4.

Figure 14-4

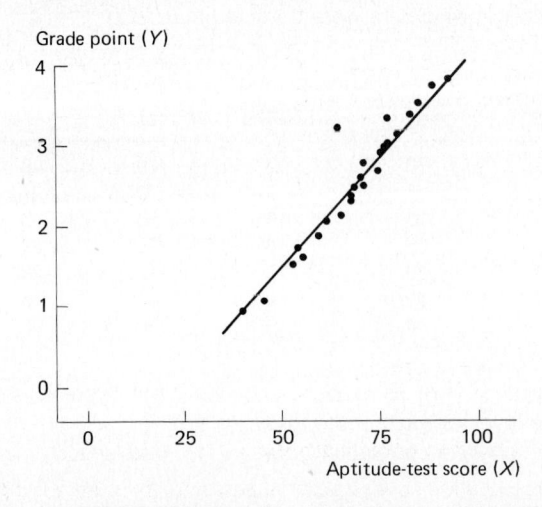

Grade point (Y) vs. Aptitude-test score (X)

To the scattergram has been added the line that best represents a linear correspondence between the two variables. This line shows for any given value

---

[†] What is considered to be a high correlation varies with the field of application.

Correlation and prediction

of X what the corresponding value of Y is according to this ideal line. For example, if someone applies for admission who has scored 75 on the aptitude test, this line predicts a grade-point average of about 2.9. This may be seen by finding 75 on the horizontal axis X, then looking straight up until you hit the predicting line, and then looking directly to the left to see where you hit the vertical axis Y. Follow the dashed arrow on the diagram in Figure 14-5.

Figure 14-5

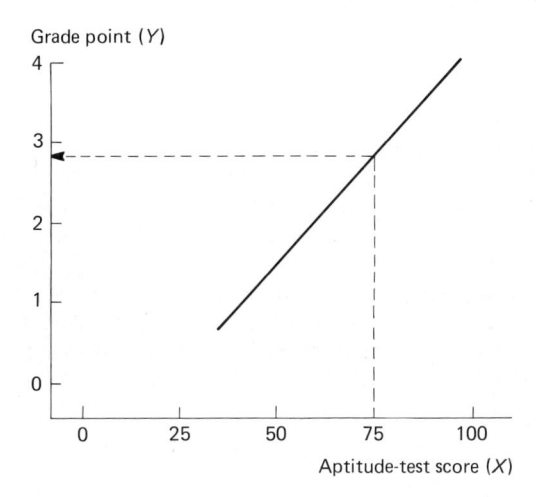

Grade point (Y) vs Aptitude-test score (X)

Similarly if a student scores 62 on the aptitude test, we get a predicted value on the grade-point average of about 2.1. A student with an aptitude score of 50 would have a predicted grade-point average of about 1.4.

NOTE: Realize that the interpretation of the situation says *nothing* about the *reasons* for the correlation, the nature of the test questions, or the intelligence of the students. It merely acknowledges that a pattern exists, and that as long as the population from which their applicants come remains the same, and as long as the college's curriculum and grading system remain the same, then it is likely that the predictions are reasonable.

In this prediction problem, the statistician, after having decided that there is a strong correlation, must then produce the formula for making the predictions. This formula is called the formula of the **best-fitting line** or "the line of regression."

The formula for the best-fitting line is:

$$Y_{predicted} = m_Y + b(X - m_X)$$

where m_X and m_Y are sample means, and

$$b = \frac{n\Sigma XY - (\Sigma X)(\Sigma Y)}{n\Sigma X^2 - (\Sigma X)^2}†$$

n being the number of pairs of scores.

This is the formula to use to predict values of Y. You should choose the labels for your variable so that you assign Y to the variable whose value you want to predict, not to the variable whose value you know.

† Geometrically, b can be interpreted as the slope of the best-fitting line.

Understanding statistics

You can always compute this formula, no matter how the points of the scattergram are arranged, but its accuracy in making predictions declines as r gets closer to zero. An important topic, but one that we omit in this elementary text, is a test for the reliability of the predicted values of Y.[†]

EXAMPLE 14-4 Let us apply the formula to the following set of data (Table 14-5) to try to predict the grade-point average corresponding to aptitude test scores of 60, 70, and 80.

Table 14-5 Sample Data

aptitude test score X	grade-point average Y	XY	X^2
40	0.88	34.20	1600
45	1.02	45.90	2025
53	1.56	82.68	2809
54	1.75	94.50	2916
55	1.63	89.65	3025
60	1.90	114.00	3600
62	2.07	128.34	3844
65	3.21	208.65	4225
66	2.12	139.92	4356
68	2.38	161.84	4624
68	2.40	163.20	4624
69	2.52	173.88	4761
70	2.63	184.10	4900
72	2.52	181.44	5184
72	2.79	200.88	5184
75	2.72	204.00	5625
75	2.90	217.50	5625
76	2.95	224.20	5776
77	3.01	231.77	5929
77	3.35	257.95	5929
80	3.10	248.00	6400
84	3.41	286.44	7056
86	3.52	302.72	7396
89	3.75	333.75	7921
94	3.82	359.08	8836
$\Sigma X = 1732$	$\Sigma Y = 63.91$	$\Sigma XY = 4669.59$	$\Sigma X^2 = 124{,}170$

From the table we compute:

$$n = 25$$

$$m_X = \frac{\Sigma X}{n} = \frac{1732}{25} = 69.28$$

$$m_Y = \frac{\Sigma Y}{n} = \frac{63.91}{25} = 2.56$$

$$b = \frac{n\Sigma XY - (\Sigma X)(\Sigma Y)}{n\Sigma X^2 - (\Sigma X)^2} = \frac{25(4669.59) - 1732(63.91)}{25(124{,}170) - (1732)^2}$$

[†] See W. J. Dixon and F. J. Massey, Jr., *Introduction to Statistical Analysis,* 3d ed., McGraw-Hill Book Company, New York, 1969, for a more complete discussion of this topic.

Correlation and prediction

$$= \frac{116{,}739.75 - 110{,}692.12}{3{,}104{,}250 - 2{,}999{,}824} = \frac{6047.63}{104{,}426} = .06$$

$$Y_{predicted} = m_Y + b(X - m_X)$$
$$= 2.56 + .06(X - 69.28)$$

For $X = 60$, we get

$$Y_{predicted} = 2.56 + .06(60 - 69.28) = 2.56 + .06(-9.28)$$
$$= 2.56 - .56 = 2.00 = 2.0$$

For $X = 70$, we get

$$Y_{predicted} = 2.56 + .06(70 - 69.28) = 2.56 + .06(.72)$$
$$= 2.56 + .04 = 2.60 = 2.6$$

For $X = 80$, we get

$$Y_{predicted} = 2.56 + .06(80 - 69.28) = 2.56 + .06(10.72)$$
$$= 2.56 + .64 = 3.20 = 3.2$$

Therefore for aptitude test scores of 60, 70, and 80 we have predicted final grade-point averages of 2.0, 2.6, and 3.2, respectively. ∎

INTERPRETATION OF b

In the formula for the best-fitting line, the value of b tells you *how much you can expect Y to change when you change X by 1 unit.*

EXAMPLE 14-5 The results of a test where the weight placed on a piston and the pressure it exerted on a fluid are given below.

X, weight, pounds	100	200	300
Y, pressure, pounds per square inch	17	18	19

We can calculate $\overline{X} = 200$, $\overline{Y} = 18$, and $b = .01$ for these data. When X is 200, the predicted value of Y is 18. What will the predicted value of Y be if X is 230?

SOLUTION The change in X is $230 - 200 = 30$. Therefore, the expected change in Y is $b \times 30 = .01(30) = .3$. So $Y_{predicted} = 18 + .3 = 18.3$ pounds per square inch. ∎

INTERPRETING THE MEANING OF THE VALUE OF THE COEFFICIENT OF CORRELATION

Over the past few years, Dr. Giuseppe O'Reilly has found a high correlation, $r = .9$, between the grades of his students in Hebrew III and their grades in Hebrew IV. His regression formula came out to be $Y = .9(X - 75) + 75.5$, where Y equals the Hebrew IV grade and X equals the Hebrew III grade of a particular student. If he lets $X = 90$, he computes $Y = 89$. Of course, this does *not* mean that a student who received a grade of 90 in Hebrew III will definitely get a grade of 89 in Hebrew IV. It *can* be interpreted that all students who get 90 in Hebrew III will receive grades in Hebrew IV whose *average* will be close to 89. Those grades in Hebrew IV that average 89 will have some variation, and this variation can be attributed to many things: increased or decreased interest in the language, response to parental pressures, changed study habits, successes or failures in social life, and previous knowledge of the material such as measured by grades in Hebrew III.

The variation in grades in Hebrew IV can, of course, be measured by the standard deviation (σ or s) and the variance (σ^2 or s^2). If the grades are close together around 89, then the variance will be small. Now it can be shown mathematically that the square of the coefficient of correlation measures what proportion of variance of the Y's can be accounted for by the linear relationship that the Y's have with X. In our example $r = .9$ and $r^2 = .81 = 81$ percent. Thus, we could say that 81 percent *of the variance* of the grades in Hebrew IV is due to the linear relationship between the grades in Hebrew III and the grades in Hebrew IV.

In an informal sense we can say that the closer r^2 is to 1, the more strongly the values of Y "depend" on the values of X, that is, the more likely it is that Y will turn out to be close to what it is predicted to be. In the extreme cases of perfect correlation ($r^2 = 1$), Y is completely determined by X. There is no other source of variation that will cause Y to vary from its predicted value. At the other extreme, that of no correlation ($r = 0$), we can say that X and Y are statistically independent (assuming, as we said earlier, that X and Y are normally distributed).

Note, in this connection, that a coefficient of correlation equal to .8 is not twice as strong as one of .4. If $r_1 = .8$, then $r_1^2 = .64 = 64$ percent, while if $r_2 = .4$, then $r_2^2 = .16 = 16$ percent, and the coefficient of correlation $r_1 = .8$ appears to be *4 times* as strong as the coefficient of correlation $r_2 = .4$.

STUDY AIDS

VOCABULARY

1. Correlation coefficient **2.** Scattergram **3.** Best-fitting line
4. Correlation, linear, positive, negative, zero, perfect

SYMBOLS

1. r **2.** b **3.** $Y_{\text{predicted}}$

FORMULAS

1. $r = \dfrac{n\Sigma XY - (\Sigma X)(\Sigma Y)}{\sqrt{n\Sigma X^2 - (\Sigma X)^2}\sqrt{n\Sigma Y^2 - (\Sigma Y)^2}}$

2. $b = \dfrac{n\Sigma XY - (\Sigma X)(\Sigma Y)}{n\Sigma X^2 - (\Sigma X)^2}$

3. $Y_{\text{predicted}} = m_Y + b(X - m_X)$

Correlation and prediction

EXERCISES

In each of the following exercises, assume that the correlation is high enough to allow for reasonable prediction.

14-26 The graph below appeared in a medical report on hypertension.[†] The two variables are average daily salt intake (in grams) and percentage of people suffering hypertension (high blood pressure). The 5 dots represent 5 different communities.

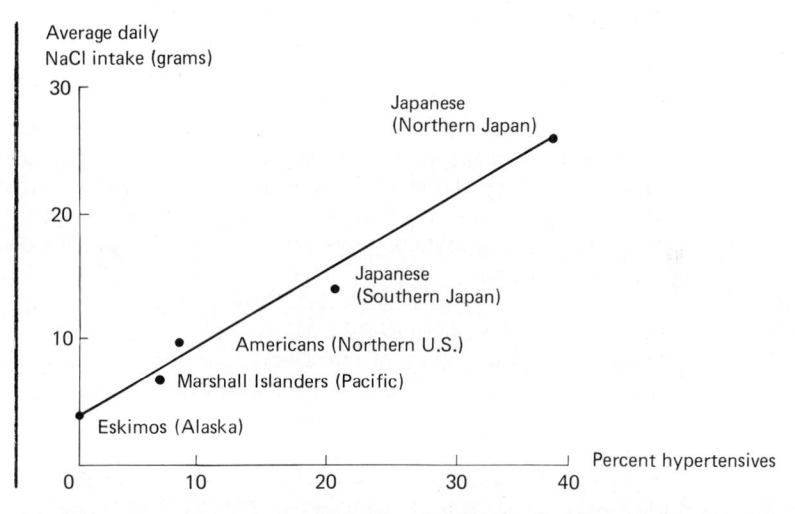

[Graph: Average daily NaCl intake (grams) on y-axis (0 to 30) vs Percent hypertensives on x-axis (0 to 40). Points shown: Eskimos (Alaska), Marshall Islanders (Pacific), Americans (Northern U.S.), Japanese (Southern Japan), Japanese (Northern Japan).]

(a) Does there seem to be a positive correlation between the two variables?
(b) Give at least two different possible explanations for the pattern of results.
(c) If you assume the relationship would hold up for other locations, about what percentage of the population would suffer hypertension if the average daily salt intake were 20 grams of salt?

14-27 A regression formula based on the high correlation between grades in MATH 001, a preparatory mathematics course, and subsequent grades in MATH 112, Introductory College Math, was found to be $Y_{predicted} = 1.5X - 40$.
(a) Using the formula for a student with a grade of 82 in MATH 001, find the predicted grade for MATH 112.
(b) It was noted that of all 47 students who received an 82 in MATH 001 not one scored the predicted 83 in MATH 112. Comment on this fact.

14-28 The following data have been collected from a random sample of 10 families in a community.

total income, in thousands of dollars	13.2	14	15	16	17	17	18.2	20	21	22
percentage spent on food	42	37	33	32	30	27	25	23	22	20

(a) Compute r.
(b) Compute b.
(c) Write the formula for predicting the percentage spent on food when you know the income.

[†] *Hypertension Update*, vol. 1, Health Learning Systems, Bloomfield, N.J., 1979.

Understanding statistics

(d) Predict what percentage a family with an income of $15,000 spends on food.

14-29 Professor Gonzales has a theory that grades on his final exams are related to where the students have been sitting all term. He takes a random sample of 15 students and records their grades and the distance between their seats and his desk.

	distance, feet														
	3	4	5	6	7	8	8	8	9	10	12	12	12	14	15
grade	85	98	93	83	78	70	83	65	80	63	60	57	71	40	90

(a) Compute r.
(b) Compute b.
(c) Write the formula for predicting final grades from seat position.
(d) Predict the grade for a person who sits 11 feet from the professor.

14-30 The secretary of Gamma Gamma Gamma fraternity has done some research on the grading practices of Professor Neumann. By examining a random sample of term papers from Professor Neumann's classes, he has assembled the following data: mean weight of papers = 8.3 ounces, mean grade = 76.6, r = .84, and b = 5.8. Predict the grade for a student who turns in a paper weighing 6 ounces.

14-31 A representative of the fishing industry in a shore community has gathered the following data relating the amount of a certain pollutant in the fishing waters each month and the amount of fish caught: mean amount of pollutant = 36 units per sample of water, mean amount of fish caught = 50 barrels, r = .7, and b = −1.4. Predict the amount of fish caught if the amount of pollutant is 48 units per sample of water.

14-32 The dean of admissions of a particular college found a high correlation between the grade-point average of a student's senior year in high school and the grade-point average for the student's freshman year in college. On a four-point scale she found that the mean high school average was 2.7 and the mean college average was 2.2. If b = 1.2, predict the college average for the following high school averages.
(a) 3.7 (b) 3.0 (c) 2.7 (d) 2.1

14-33 Three researchers were attempting to devise a convenient method for measuring evaporation of water from soils and crops. They discovered a relation between evaporation and what is called "net thermal radiation." Here are their data collected from one field on 8 different days scattered through the year.

day	units of evaporation, calories per square centimeter	units of net thermal radiation, calories per square centimeter
1	17	−87
2	17.5	−86
3	19	−84
4	21	−86
5	55	−62
6	70	−55
7	83	−45
8	90	−41

Correlation and prediction

(a) Draw a scattergram for the data.
(b) Compute r.
(c) Find the best-fitting line.
(d) Predict evaporation if the net thermal radiation is -60.

14-34 If the coefficient of correlation between chemistry grades and math grades is .7, approximately what percentage of the variation in students' chemistry grades is due to the relationship between their chemistry grades and their math grades?

14-35 If the coefficient of correlation between IQ scores and grade-point averages is found to be .3, about what percentage of the variation in students' grade-point averages can be attributed to other factors, such as study habits or personal drive?

14-36 Miss Krobar found that in her third-grade class the weights of the children have a correlation with both their ages and their heights. For the weights and ages she found $r_A = .7$. For the weights and heights she found $r_B = .8$.
(a) What percentage of the variation in weights can be attributed to the relationship between weight and age?
(b) What percentage of the variation in weights can be attributed to the relationship between weight and height?
(c) Discuss the paradox implied by your answers to parts (a) and (b).

14-37 The coefficient of correlation between typing speed and typing errors was found to be .4. The variation in typing errors due to inattention is 4 times as great as the variation due to speed. Find the coefficient of correlation between typing errors and inattention.

CLASS SURVEY

What correlation would you expect to find between the heights of males and their fathers in this class? Compute the coefficient of correlation. Is it near what you expected? Test to see whether or not this is a significant correlation.

FIELD PROJECTS

Compute r for two variables of your choosing. Test the hypothesis that the coefficient of correlation is zero. If it is not zero, you may wish to include the formula for prediction and some pertinent predictions.

Some projects of this type which students have done include investigating correlations between:

1. Speed they drive and time it takes them to get to campus.

2. Number of days left in the month and number of parking tickets placed on cars at a certain location.

3. Length of time a customer spends in a small gift shop and total amount of purchase.

Tests Involving Variance

15

Many important statistical hypotheses are concerned with variability. Two common measures of variability used in hypothesis tests are the *standard deviation* and the square of the standard deviation, the *variance*. You have already worked quite a bit with the standard deviation. One of its advantages is that it measures variability in whatever units the variable is measured. For example, we might say that the average distance from the floor to the doorknobs in a certain apartment house is 36.5 *inches,* with a standard deviation of .25 *inch.* In contrast to this, if you use the variance to measure variability, the variance is not in the original units of the problem and so is not as easy to interpret intuitively. In the example just stated, the variance is $(1/4)^2$, or 1/16, but the unit is "inches squared," which makes no intuitive sense. In this chapter, however, we will see that certain hypotheses should be tested by looking at sample *variances* rather than sample standard deviations, because statistics based on the sample variances can be meaningfully compared with certain well-established tables of critical values. We do not have equivalent tables for statistics based on the sample standard deviations. Note that where we use the letter s to represent standard deviations calculated from samples, we will use s^2 to represent variances calculated from samples. Our basic formulas are:

$$s = \sqrt{\frac{\Sigma X^2 - \frac{(\Sigma X)^2}{n}}{n-1}} \quad \text{and} \quad s^2 = \frac{\Sigma X^2 - \frac{(\Sigma X)^2}{n}}{n-1}$$

In this chapter we are going to illustrate three types of problems.

1. The test of a claim about the variance of a population. For example, a claim might say that the heights in inches of 6-year-old children in the Los Angeles

Tests involving variance

public school system have a variance equal to 4. We would have H_0: variance = 4. In symbols, this is H_0: $\sigma^2 = 4$. This is equivalent to claiming that the standard deviation is 2 inches. This type of claim will be tested by computing a statistic and comparing it with a critical value from a chi-square table. (This is a new application of the chi-square table.) We will call these tests *one-sample tests of variance*.

2. A test to compare the variances of *two* populations. An example might be to test the claim that in sea-farming lobsters, one diet produces more *erratic* size lobsters than another diet. The null hypothesis will be H_0: $\sigma_1^2 = \sigma_2^2$. The computations will involve computing the variances of *two* samples, one from each population, then using these to compute a statistic which can be compared to critical values in a new table, called the *F* table. We will call these tests *two-sample comparisons of variance*.

3. We use the *F* table and sample variances to analyze the hypothesis that the *means* of several populations are equal. This type of test is an extension of the two-sample mean tests already developed in Chapter 11. We will call these tests *comparison of the means of several populations*, or **analysis of variance**.

ONE-SAMPLE TESTS OF VARIANCE

The Deep Dark Device Department of the Kynda Klever Kamera Company manufactures darkroom timers—"the kind you wind." Sample Sam, the quality control man, always tests every timer they make to see if the assembly process is working properly. Sam does the test by setting each timer to the "30 seconds" setting, and then timing it electronically to see how long it actually takes to ring. He repeats this 11 times for each timer. He knows from past experience that if he would test any particular timer that is working properly *thousands* of times and make a frequency chart and histogram of his results, he would see an approximately normal distribution of the times with a mean of 30 seconds (Figure 15-1).

Figure 15-1 Graph for one particular timer when tried thousands of times.

30   Seconds until timer rings

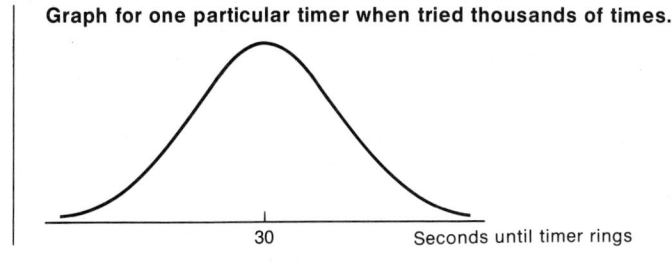

Sam's concern is: does this distribution have a comparatively small variance? This means that this particular timer behaves consistently and is therefore more reliable in the darkroom. For example, if for two particular timers his tests resulted in the two graphs shown in Figure 15-2, this would indicate that timer A is more reliable than timer B.

Sam has decided to reject any timer that appears to have a standard deviation of more than ½ second or, equivalently, a variance of more than $(½)^2$, or ¼. Of course, Sam cannot afford to test each timer thousands of times.

Figure 15-2

Timer A

30   Seconds until timer rings

Timer B

30   Seconds until timer rings

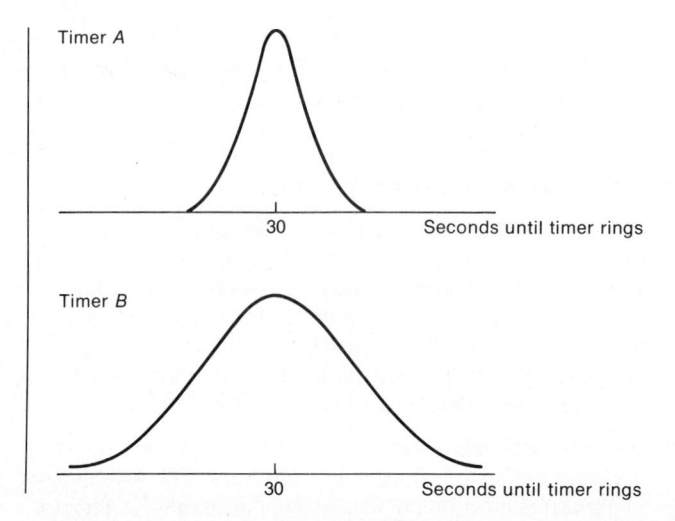

Recall that he tests each timer only 11 times, and he must judge on the basis of these 11 trials what might happen if he were to continue thousands of times. He therefore is in a hypothesis testing situation. His motivated hypothesis, the one that will cause him to reject a timer, is H_a: variance is more than ¼, $\sigma^2 > $ ¼. His null hypothesis, therefore, is H_0: $\sigma^2 = $ ¼. You might guess the general idea of how this test is actually carried out. The first step is to get an estimate of σ^2 from the variability among the 11 test readings. This would be s^2. The second step is to see if s^2 is significantly larger than ¼.

It turns out that if you try to compare the sample value of s^2 with the theoretical value σ^2 by *subtracting* σ^2 from s^2 (as we did with sample means in Chapters 10 and 11), there is no convenient set of critical values to use as a decision rule. However, when it is reasonable to assume that the variable whose variance you are checking is distributed not too differently from normal, then it has been shown that if you *divide* s^2 by $\sigma^2/(n-1)$, you can use the table of critical values for a *chi-square distribution* to get your decision rule. You use the critical values for $n-1$ degrees of freedom. In short, your test statistic is

$$\frac{s^2}{\sigma^2/(n-1)}$$

or, what is the same,

$$\frac{(n-1)s^2}{\sigma^2}$$

Since we will check this against a chi-square table at $n-1$ degrees of freedom, we can call this statistic X^2. That is, we compute

$$X^2 = \frac{(n-1)s^2}{\sigma^2} \quad \text{with degrees of freedom} = n-1$$

NOTE: In the examples that follow it is helpful to not use X for the name of the variable since we will use X^2 for our statistic.

Tests involving variance

EXAMPLE 15-1 In the above problem of darkroom timers our hypotheses are

H_0: variance equals 1/4, $\sigma^2 = 1/4$
H_a: variance is more than 1/4, $\sigma^2 > 1/4$ (one-tail test on the right)

Our sample data can be listed as follows.

	test number										
	1	2	3	4	5	6	7	8	9	10	11
Y = number of seconds until timer rang	29.2	29.5	30.0	30.0	30.0	30.3	28.6	30.0	29.8	29.7	30.2

SOLUTION We assume that the distribution of Y is near normal. We have $n = 11$, and therefore $n - 1 = 10$ degrees of freedom. Using critical values of X^2 for 10 degrees of freedom at $\alpha = .05$, we find $X_c^2 = 18.31$. Therefore, we will reject H_0 if X^2 is greater than 18.31. This is illustrated in Figure 15-3.

Figure 15-3

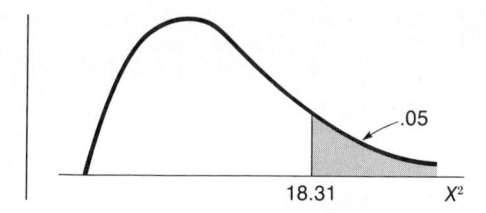

Our data are now rearranged as in Table 15-1.

Table 15-1

test number	Y	Y^2
1	29.2	852.64
2	29.5	870.25
3	30.2	912.04
4	30.0	900.00
5	30.0	900.00
6	30.3	918.09
7	28.6	817.96
8	30.0	900.00
9	29.8	888.04
10	29.7	882.09
11	30.2	912.04
totals	327.5	9753.15

We now solve for s^2 and X^2:

$$s^2 = \frac{\Sigma Y^2 - \frac{(\Sigma Y)^2}{n}}{n-1} = \frac{\Sigma Y^2 - \frac{(\Sigma Y)^2}{11}}{10}$$

$$= \frac{9753.15 - \frac{(327.5)^2}{11}}{10} = \frac{9753.15 - 9750.57}{10} = \frac{2.58}{10} = .258$$

$$X^2 = \frac{(n-1)s^2}{\sigma^2}$$

$$= \frac{10(.258)}{1/4} = \frac{2.58}{.25} = 10.32$$

Sam's value of $X^2 = 10.32$ is not above 18.31. Therefore, Sam would fail to reject the hypothesis. The timer passes his inspection.

EXAMPLE 15-2 A seed packaging company has a type of bean seed from which on the average the plant grows its first leaves 15 days after being planted. The variance for the number of days until first leaves is 5. A new way of selecting the seeds to be packaged is supposed to reduce this variance. A random sample of 61 seeds selected the new way is found to have $s^2 = 3.2$. Is this sufficient to indicate at $\alpha = .05$ that the new selection method *reduces* variance?

SOLUTION *Assumption* The variable "number of days until first leaves" is close to being normally distributed. Our hypotheses will be:

H_0: variance is 5, $\sigma^2 = 5$
H_a: variance is less than 5, $\sigma^2 < 5$ (one-tail test on the left)

Since our degrees of freedom $= n - 1 = 61 - 1 = 60$, we find in Table C-7 that $X_c^2 = 43.19$.† This situation is illustrated in Figure 15-4. If X^2 turns out to be *less* than 43.19, our decision will be to reject H_0, and we will have evidence that the variance has been reduced.

Figure 15-4

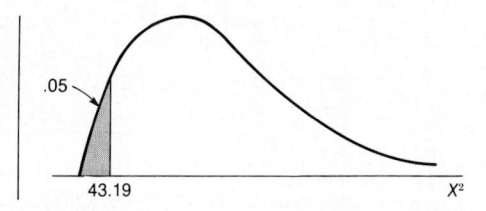

$$X^2 = \frac{(n-1)s^2}{\sigma^2}$$

$$= \frac{60(3.2)}{5} = 12(3.2) = 38.4$$

Since 38.4 *is less* than 43.19, we can reject the null hypothesis. We have evidence that the new packaging procedure reduces variance.

If we have a two-tail test with an alternative hypothesis that the variance is *not equal* to some given variance, then we need to find two critical values of X^2. We will reject the null hypothesis if our value of X^2 is not between these two critical values.

---

† This is the first time we have used the chi-square table for a critical value *on the left*. Be careful in reading the table.

TWO-SAMPLE COMPARISONS OF VARIANCE

Sometimes it is important for an experimenter to know if two populations have the *same* variance. For example, in a *t* test which compares the means of two samples, we assumed that the variances of both populations involved were approximately equal. This is because the *t* test really answers the question, "could these two samples be from the *same* normal population?" When we reject the null hypothesis, we are saying "no" to this question. There are two reasons for two normal populations to be unequal—they could have different means or they could have different variances. If we can assume that the variances *are* equal, then the *t* test becomes solely a test about the *means* of the populations. This is the way that we have been using it in our two-sample tests of means. All of this implies that if you perform a two-sample *t* test for means and you end up rejecting the null hypothesis, then you should also check to rule out the possibility that your results are not caused by unequal variances in the two populations.

Consider some other cases in which you should check for unequal variance in two populations. Suppose you teach Advanced Knot Tying at the local Institute for Marine and Matrimonial Sciences. You think of two ways to teach slipknot tying, but you suspect that method 1 might produce more erratic results than method 2. You try method 1 on one group of randomly selected students, and method 2 on another. Then you give both groups the same exam. Suppose you get the same *mean* results in both samples. This means that the two methods are equally effective on the average. But suppose the variance in group 1 is significantly larger than the variance in group 2. This would indicate that there is a comparatively wide gap between the best and worst students in group 1. In short, method 1 is better for the good students and worse for the bad ones. You may decide not to use method 1 if you want to keep the class "together." Let's give an example of this type.

ONE-TAILED TESTS

In these one-tailed tests we always name the population that we expect to have the larger variance population 1. Hence, our motivated hypothesis is always $\sigma_1^2 > \sigma_2^2$. Our test statistic will be the ratio

$$F = \frac{s_1^2}{s_2^2}$$

where we put the sample variance that we expect to be larger on top. We therefore expect this fraction to be greater than 1. The question is, is it significantly greater than 1? To decide this, we find a critical value F_c in Table C-11, C-12, C-13, or C-14.

The letter F is used to honor Sir Ronald Fisher, who pioneered much of the work discussed in this chapter in the 1920s. The F was first used by the statistician Snedecor in the 1930s.

EXAMPLE 15-3 We have two methods of teaching slipknot tying. We want to see if method A produces exam scores with a greater variance than does method B. The scores on the slipknot final for method A are 75, 80, 90, 100, 110, 120, 125. For

Understanding statistics

method B they are 90, 95, 95, 100, 100, 100, 105, 105, 110. We assume that both methods produce exam scores that are normally distributed.

SOLUTION Because we are testing to see if the variance of the scores for method A is greater, we name the scores of method A population 1. We calculate from the above data:

$n_1 = 7$, $\quad m_1 = 100$, $\quad s_1^2 = 375$

$n_2 = 9$, $\quad m_2 = 100$, $\quad s_2^2 = 37.5$

Our hypotheses are

$H_0: \sigma_1^2 = \sigma_2^2$

$H_a: \sigma_1^2 > \sigma_2^2$ $\quad$ (one-tail test)

Our test statistic is

$$F = \frac{s_1^2}{s_2^2} = \frac{375}{37.5} = 10$$

Now to find the critical value of F at $\alpha = .05$ we find the number of degrees of freedom for both the numerator and denominator:

Degrees of freedom (numerator) $= n_1 - 1 = 7 - 1 = 6$

Degrees of freedom (denominator) $= n_2 - 1 = 9 - 1 = 8$

Using $\alpha = .05$ we look at the table entry corresponding to these degrees of freedom, and we find that $F_c = 3.58$. This is illustrated in Figure 15-5.

Figure 15-5

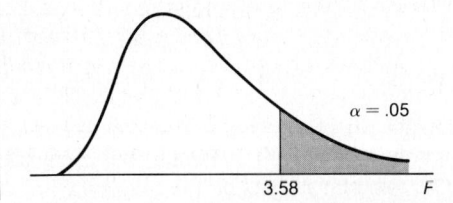

We see that 10, our sample value of F, is well beyond the critical value. This indicates that there is a significant difference in variability between the results in group 1 and the results in group 2. The method used to teach knot tying in group 1 produces a wider split between the best and worst students than the method used in group 2. ∎

EXAMPLE 15-4 In a district court in Big City, the average sentence given for larceny by two judges is about the same. An investigator for a prison reform group has reason to believe that Justice Herthaykum, however, is much more erratic in sentencing the convicted than is Justice Tharthaygoe. The investigator checks random samples of 50 sentences given for this crime by each judge. The sentences given by Justice Herthaykum had $m = 18.2$ months with $s = 6.1$ months. The sentences given by Justice Tharthaygoe had $m = 18.5$ months with $s = 3.4$ months. Is this a significant difference in variability? Use $\alpha = .05$.

Tests involving variance

SOLUTION Because the investigator suspects that the variance of the sentences imposed by Justice Herthaykum is larger, we name these sentences population 1.

H_0: variance is same for both judges, $\sigma_1^2 = \sigma_2^2$

H_a: variance for Herthaykum is larger, $\sigma_1^2 > \sigma_2^2$ (one-tail test)

Thus,

$$F = \frac{s_1^2}{s_2^2} = \frac{6.1^2}{3.4^2} = 3.22$$

For $n_1 = 50$ we have degrees of freedom $= n_1 - 1 = 49$. For $n_2 = 50$ we have degrees of freedom $= n_2 - 1 = 49$. We find from the tables that $F_c = 1.60$ (using the closest entry in the table). This is shown in Figure 15-6. Since F is larger than F_c, we have good evidence that Justice Herthaykum is more erratic than Justice Tharthaygoe. ∎

Figure 15-6

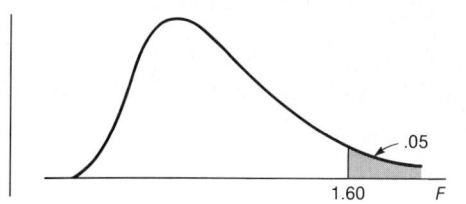

TWO-TAILED TESTS

In both of the last two examples, the alternative hypothesis was *one-sided* with H_a: $\sigma_1^2 > \sigma_2^2$. Therefore, we knew ahead of time that s_1^2 should be larger than s_2^2 if there was to be any support for H_a. So it was clear that we had a one-tail test on the right with s_1^2/s_2^2 as our statistic. In some problems, however, the alternative is two-sided, namely, $\sigma_1^2 \neq \sigma_2^2$. We do not know ahead of time whether s_1^2 or s_2^2 will be larger. At the .05 significance level, for example, we should have a two-tail test as illustrated in Figure 15-7, and either a very low F or a very high F will cause us to reject H_0.

Figure 15-7

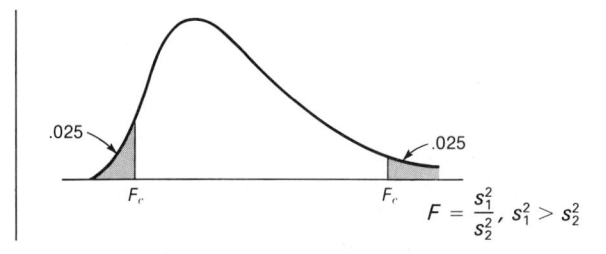

$$F = \frac{s_1^2}{s_2^2}, \quad s_1^2 > s_2^2$$

It is convenient to always put the larger sample variance on the top of fraction F and hence do all tests on the right, maintaining only *one-half of α* (in this case .025). We have supplied separate F tables for these two-tailed tests.

Understanding statistics

EXAMPLE 15-5 Is the variability in temperature the same in the two resort cities of Harvey-Cedars and Claremont? On the first day of every month the highest temperature of the day is recorded at Harvey-Cedars and Claremont. This is done for 1 year. The data collected showed that Harvey-Cedars had a mean high temperature of 70.2 °F, with $s = 10.8$, and that Claremont had a mean high temperature of 74.1 °F, with $s = 15.3$. Test the hypothesis that the high temperatures at Harvey-Cedars and Claremont are equally variable. Use $\alpha = .05$.

SOLUTION Because $s = 15.3$ at Claremont and $s = 10.8$ at Harvey-Cedars, we choose the larger one for s_1, so that the daily high temperatures at Claremont become population 1.

$H_0: \sigma_1^2 = \sigma_2^2$
$H_a: \sigma_1^2 \neq \sigma_2^2$ (two-tailed test)

Our degrees of freedom for $n_1 = 12$ and $n_2 = 12$ will be

Degrees of freedom $= n_1 - 1 = 11$
Degrees of freedom $= n_2 - 1 = 11$

At $\alpha = .05$ on a *two-tailed test* we get $F_c = 3.48$. Our value of F is

$$F = \frac{s_1^2}{s_2^2} = \frac{(15.3)^2}{(10.8)^2} = \frac{234.09}{116.64} = 2.01$$

Since our F statistic is smaller than the critical F value, we cannot reject the hypothesis of equal variance. The variability of high temperatures in the two cities may be the same.

USING SAMPLE VARIANCES TO MAKE INFERENCES ABOUT MEANS
(An introduction to analysis of variance.)

In this section we want to show how analyzing sample variances can help answer questions about population means. Let us consider the kind of problem we usually solve by using a two-sample t test for comparing means. In this kind of problem we assume that the two populations are both approximately normal and that they have the same variance. Then we test the null hypothesis that says they have equal means.

If they do have equal means, then the two populations are essentially the *same* for statistical purposes: they are both normal and they have the same mean and the same variance. We will estimate the numerical value of this common variance. We will do this in two distinct ways. If the null hypothesis is true (the means are equal), then, according to statistical theory, these two distinct estimates of the variance should be "close" to one another as determined by an *F ratio test*. We will take an obvious case for illustrative purposes.

EXAMPLE 15-6 Given two brands of brandy, Brand D and Brand E. Suppose we are measuring how many ounces test subjects must drink until a certain reaction occurs. And suppose we can reasonably assume that to begin with the subjects are about equally susceptible to the effects of the brandies. There are 5 subjects on each brand. The following data show that the two brandies are very different from each other.

Tests involving variance

ounces of Brand D	ounces of Brand E
8, 7, 8, 7, 8	13, 12.5, 12, 13, 12

From these data we compute that $m_D = 7.6$ with $s_D^2 = .30$, and $m_E = 12.5$, with $s_E^2 = .25$. Clearly, Brand D is much more potent than Brand E. The average amount of E needed to elicit the desired reaction is much more than the average amount of D. We want to show how this difference can also be seen indirectly by looking at variances. Recall that we are assuming that the two populations have the same variance. We are going to try to estimate this variance by two methods.

SOLUTION *Method 1: Variability within the Samples* We have two distinct samples of the same size from populations with the same variance. Recall that s^2 from a sample is a good estimate for the variance of a population. Therefore, we have two good estimates of the variance, namely, $s_D^2 = .30$ and $s_E^2 = .25$, and assuming that $\sigma_1^2 = \sigma_2^2$, it is reasonable to use the *average* of s_D^2 and s_E^2 as a good estimate of the value of the population variance. Therefore, we estimate the value of the variance by

$$\frac{s_D^2 + s_E^2}{2} = \frac{.30 + .25}{2} = .275$$

You will remember from our original definition of variance that the variance is the *mean* of a set of *squared* deviations, and is often called a *mean square*. In most textbooks the variance we just estimated is called **mean square within groups** because each value of s^2 was based on the variability *within* the individual sample from which it was calculated. s_D^2 measures variability within sample D, and s_E^2 measures variability within sample E.

We will write s_W^2 for this mean square within groups, where the letter W indicates "within":

$$s_W^2 = \frac{s_D^2 + s_E^2}{2}$$

In our case $s_W^2 = .275$.

In general s_W^2 is

$$s_W^2 = \frac{\Sigma s^2}{N}$$

where N is the number of samples.

We have seen that s_W^2 is a good estimate of the variance of the two populations. Furthermore, it is true that s_W^2 is a good estimate of this variance *whether or not* the populations *also* have equal *means*.

Method 2: Variability among the Sample Means There is another way to estimate the variance of populations which have equal variance, but this way only gives an accurate estimate if the populations also have equal means. When the populations have equal means, then this method gives approximately the same value as s_W^2. But when the populations have different means, then this second method will give a value significantly larger than s_W^2.

If our two normal populations have the same variance *and* the same mean, then for statistical purposes they are considered to be *one* large population. Our two samples can be considered to be two samples of size 5 from the same population. When we have several samples of the *same size* from a population, we can apply what we know about the relationship between the variance of a population and the variance of the means of samples taken from that population.

Recall from the central limit theorem given in Chapter 10 that

$$\sigma_m = \frac{\sigma_{\text{pop}}}{\sqrt{n}}$$

This is equivalent to

$$\sigma_{\text{pop}}^2 = n\sigma_m^2$$

where n is the size of each sample.

NOTE: N is the number of samples; n is the size of each sample.

In words this says that you can find the variance of a population by multiplying n by the variance of the sample means. Now to get the exact value of σ_m^2 we would need an infinite set of sample means from the population. We can estimate σ_m^2 by calculating s_m^2 for as many sample means as we do have.

In our example we have taken *two* samples, so we have *two* sample means, $m_D = 7.6$ and $m_E = 12.5$. We compute s_m^2 from these two values:

m	m^2
7.6	57.76
12.5	156.25
$\Sigma m = 20.1$	$\Sigma m^2 = 214.01$

$$s_m^2 = \frac{\Sigma m^2 - \frac{(\Sigma m)^2}{N}}{N - 1}$$

$$= \frac{214.01 - \frac{(20.1)^2}{2}}{1} = \frac{214.01 - 202.01}{1} = 12.00$$

Now that we have s_m^2 we can apply the above idea that $\sigma_{\text{pop}}^2 = ns_m^2$ and estimate σ_{pop}^2 by ns_m^2, getting $5(12.00) = 60$.

We call this second estimate of σ_{pop}^2 the **mean square *among* groups** because it measures the variability *among* the sample means. We write $s_A^2 = 60$, where the letter A indicates "among." Our formula then is

$$s_A^2 = ns_m^2$$

Thus, s_A^2 is our second estimate of the population variance. We now need to compare s_A^2 and s_W^2, our two estimates of the population variance.

Comparison of the Two Estimates As you might have guessed, the best way to compare two values of s^2 is by the F ratio. We look at the ratio

Tests involving variance

$$F = \frac{\text{mean square } among \text{ groups}}{\text{mean square } within \text{ groups}} = \frac{s_A^2}{s_W^2} = \frac{60}{.275} = 218.2$$

and compare it to a critical F value. We stated above that if the means are equal, then $\sigma_A^2 = \sigma_W^2$; but if the means are not equal, then σ_A^2 is *larger than* σ_W^2. Our hypotheses are:

H_0: variance determined by s_A^2 equals variance determined by s_W^2, the means are equal

H_a: variance determined by s_A^2 *is larger than* variance determined by s_W^2, the means are not equal   (this is always a *one*-tail test of the *variances*)

To use the F table we need to know the degrees of freedom for numerator and denominator. The degrees of freedom in the numerator is $N - 1 = 2 - 1 = 1$ because s_A^2 was computed from a list of 2 means. The degrees of freedom in the denominator is 8 because we pooled two values of s^2, each with degrees of freedom = 4. We use degrees of freedom = $(n_1 - 1) + (n_2 - 1)$. For these two values of degrees of freedom at $\alpha = .05$ we get $F_c = 5.32$.

You can see that F is much larger than the critical value, which is what we expected, since the two samples are not at all alike. A large value of F indicates that there is great variability among the sample means—the means of the samples are not alike in value; that is, a large value of F indicates that the samples probably came from populations with different means.

■

EXAMPLE 15-7 We repeat the brandy problem, but this time we make up sample values to show that the two brands are practically the same in effectiveness. Once again we assume both variables are near normally distributed with equal variances.

ounces of Brand D	ounces of Brand E
8, 7, 8, 8, 7	7, 8, 8, 9, 9

We now have $m_D = 7.6$ with $s_D^2 = .30$ and $m_E = 8.2$ with $s_E^2 = .70$, and $N = 2$ since we have 2 samples with $n_1 = 5$ and $n_2 = 5$.

SOLUTION *Method 1: Mean Square within Groups*

$$s_W^2 = \frac{\Sigma s^2}{N} = \frac{s_D^2 + s_E^2}{2}$$

$$= \frac{.30 + .70}{2} = .50$$

Method 2: Mean Square among Groups

m	m^2
7.6	57.76
8.2	67.24
$\Sigma m = 15.8$	$\Sigma m^2 = 125.00$

$$s_m^2 = \frac{\Sigma m^2 - \frac{(\Sigma m)^2}{N}}{N-1}$$

$$= \frac{125.00 - \frac{15.8^2}{2}}{1} = \frac{125.00 - 124.82}{1} = .16$$

$$s_A^2 = ns_m^2$$
$$= 5(.16) = .80$$

Degrees of freedom (numerator) $= N - 1 = 2 - 1 = 1$
Degrees of freedom (denominator) $= (n_1 - 1) + (n_2 - 1) = 4 + 4 = 8$
$F_c = 5.32$

Comparison of Estimates

$$F = \frac{s_A^2}{s_W^2} = \frac{.80}{.50} = 1.6$$

You see that F is now smaller than F_c. This is because s_A^2, the mean square among groups, decreased from 60 to .80, making the value in the top of the F fraction smaller. Thus, the numerator of F decreases when the sample means become more alike, and increases when they become more variable. If you find that the sample means are variable enough to cause F to be larger than F_c, this is a good indication that the samples came from populations with *different means*. Therefore, if you get F larger than F_c, you can reject a null hypothesis which says that the population means are equal. These F tests are essentially one-tailed. The hypotheses are

H_0: variance as determined by s_A^2 *equals* variance as determined by s_W^2
H_a: variance as determined by s_A^2 *is larger than* variance as determined by s_W^2

We have seen that s_A^2 will be significantly larger than s_W^2 when the sample means are not alike. Therefore, the two hypotheses above are equivalent to:

H_0: population means are equal
H_a: population means are not equal

■ So a *two-tailed* test about *means* can be done as a *one-tailed F* test comparing s_A^2 and s_W^2.

This method of comparing means by looking at variances is the simplest version of a statistical approach called **analysis of variance** (or **anova**, for short). One advantage it has over t tests is that it is not limited to comparing only two samples. See the following example for a test asking: are *three* means equal?

EXAMPLE 15-8   Suppose you are a professional rainmaker, and you run the Seedy Cloud Company. You have thought of three different ways to drop the chemicals into the clouds, and you want to know if there is any difference in the results for these three methods. Using anova, you can test the null hypothesis, $H_0: \mu_1 = \mu_2 = \mu_3$. A large value of F will cause you to reject this hypothesis—you will

Tests involving variance

have evidence that not all the methods are equal. (Be careful! This does not mean they are all different, only that *at least one* of the methods is different from the rest.) Here are your results after having tried each method 6 times. We record the rainfall in millimeters.

	method 1	method 2	method 3
	12	10	8
	12	11	10
	13	10	11
	13	12	12
	12	12	10
	13	10	10
mean	12.5	10.83	10.17
variance	.30	.97	1.77

From these data we find that $m_1 = 12.5$ with $s_1^2 = .30$, $m_2 = 10.83$ with $s_2^2 = .97$, and $m_3 = 10.17$ with $s_3^2 = 1.77$. Using $\alpha = .05$, test the hypothesis that all the methods produce the same average rainfall.

SOLUTION   *Assumption*   All three populations are normal with the same variance.

H_0: $\mu_1 = \mu_2 = \mu_3$
H_a: not all the μ's are equal   (always a *one*-tail test)

Method 1:   Mean Square within Groups

$$s_W^2 = \frac{\Sigma s^2}{N} = \frac{.30 + .97 + 1.77}{3} = 1.01$$

Method 2:   Mean Square among Groups

	m	m^2
	12.5	156.25
	10.83	117.29
	10.17	103.43
$\Sigma m = 33.50$		$\Sigma m^2 = 376.97$

$$s_m^2 = \frac{\Sigma m^2 - \frac{(\Sigma m)^2}{N}}{N-1} = \frac{376.97 - \frac{(33.50)^2}{3}}{2} = 1.44$$

$$s_A^2 = n s_m^2$$
$$= 6(1.44) = 8.64$$

Comparison of Estimates

$$F = \frac{8.64}{1.01} = 8.55$$

Degrees of freedom (numerator) $= N - 1 = 3 - 1 = 2$
Degrees of freedom (denominator) $= (n_1 - 1) + (n_2 - 1) + (n_3 - 1)$
$= 5 + 5 + 5 = 15$

Understanding statistics

$F_c = 3.68$

■ Therefore, we can reject the null hypothesis. The methods of producing rain are not all the same.

SOME COMMENTS

This last section, using variances to test hypotheses about population means, is one aspect of a widely used approach, generally called analysis of variance. We have illustrated only the case where all samples are the same size, and where we assume that there is only one factor responsible for any differences we observe. For example, in the brandy example we assume that differences in reaction time are due to the type of brandy and not, say, to the time of day the test was done or the music playing in the background, etc. The type of approach we have illustrated is often called *one-way* analysis of variance. But procedures have been developed for taking into account unequal sample sizes and multiple factors at work. These more advanced features of anova are important to experimenters and are described in more comprehensive texts.[†]

STUDY AIDS

VOCABULARY

1. Analysis of variance
2. F ratio test
3. Mean square within groups
4. Mean square among groups
5. Anova

SYMBOLS

1. F_c
2. s_W^2
3. s_A^2
4. n
5. N

FORMULAS

1. $X^2 = \dfrac{(n-1)s^2}{\sigma^2}$

2. $F = \dfrac{s_1^2}{s_2^2}, \quad s_1^2 > s_2^2$

3. $s_W^2 = \dfrac{\Sigma s^2}{N}$, where N is the number of samples

4. $\sigma_{\text{pop}}^2 = n\sigma_m^2$, where n is the sample size

5. $s_m^2 = \dfrac{\Sigma m^2 - \dfrac{(\Sigma m)^2}{N}}{N - 1}$

6. $s_A^2 = ns_m^2$ 7. $F = \dfrac{s_A^2}{s_W^2}$

[†] For example, you might see W. J. Dixon and F. J. Massey, Jr., *Introduction to Statistical Analysis*, 3d ed., McGraw-Hill Book Company, New York, 1969.

EXERCISES

15-1 In Chapter 11 we made two assumptions about the populations when we used the two-sample t test. Now we can use an F test to test one of these assumptions. What is the assumption?

15-2 Use $\alpha = .05$.
(a) Test the hypothesis that the weights of six boxes of Soggy Morning Cereal come from a distribution with variance equal to .30.

weight, grams
453.2, 453.0, 452.8, 452.9, 453.2, 452.7

(b) Assume that the weights are distributed normally and test the hypothesis that the mean weight is 453.1 grams.

15-3 Every day for the month of July, Donna Shore marks the high-tide point on a certain beach. She puts a stick in the sand where the water comes farthest onto the beach directly in front of a permanent flagpole. Then she measures the distance from the flagpole to the stick.
(a) At $\alpha = .05$, does she have good evidence that the standard deviation of high-tide marks on her beach is more than 5 feet if her results averaged to $m = 150$ feet with $s = 8$ feet?
(b) Assuming that the distances are distributed normally, test the hypothesis that the mean distance is 125 feet. Use $\alpha = .05$.

15-4 Dr. Lazar Beame, a pediatrician in a large maternity hospital, tells his son to record the number of hours, in the first 24 hours of a baby's life, that a baby sleeps. Here are the results that son Beame recorded for 10 babies: 18.6, 16.2, 22.4, 17.9, 18.9, 20.1, 17.0, 18.8, 19.0, 21.4. Assume such measurements are distributed normally and test the following hypotheses. Use $\alpha = .01$.
(a) Variance equals 4
(b) $\mu = 18$ hours

15-5 A series of measurements were made by the noted anthropologist, Dr. Sanford Q. Krotche.
(a) He measured the length of a certain bone in the limbs of 10 adult males that he dug up. Are Sandy's measurements consistent with his hypothesis that the standard deviation of these bone lengths is 10 mm? Test at $\alpha = .05$ if the lengths of bone, in mm, are 150, 160, 152, 157, 159, 148, 152, 153, 156, and 138.
(b) Sandy then found that the length of a similar bone from an unknown primate was 162 mm. If you assume that the lengths of these bones are normally distributed with standard deviation 10 mm and mean given by m from part (a), what is the probability of finding a bone as large as 162 mm (that is, 162 mm or larger)?

15-6 Reread Exercise 11-4. With $\alpha = .05$, test the claim that the standard deviation is less than 2 mm.

15-7 Reread Exercise 11-5. Is it reasonable at the .05 significance level to assume that σ is less than 10?

15-8 Reread Exercise 11-6. Is it reasonable to assume at $\alpha = .05$ that σ is more than 15?

15-9 Reread Exercise 11-7. It was also claimed that the police officers are very similar to one another in height, with a standard deviation of not more than 1 inch. Is this supported by the data given? Use $\alpha = .05$.

15-10 Reread Exercise 11-8. Using $\alpha = .01$, test the hypothesis that the variance of the psychologists' scores is the same as the variance for the general public. This variance is 64.

15-11 For these following exercises from Chapter 11 check the assumption that the variances of the two populations sampled are equal. Use $\alpha = .05$.
(a) 11-15 (b) 11-16 (c) 11-17 (d) 11-18

15-12 Two typists, Martin and Marvin, type on the average about 60 words per minute. They were each tested 10 times, and their scores are shown below. Using $\alpha = .05$, test the hypothesis that their test scores are equally variable.

Martin	Marvin
50	52
55	57
60	50
65	64
62	65
50	66
51	56
60	61
62	67
65	66

15-13 A chemical was applied to the roots of 20 corn plants and 20 oat plants. The concentration of the chemical in the plant stem was measured 24 hours later. Using $\alpha = .01$, test the hypothesis that the chemical acts more erratically in the oat plants than it does in the corn plants. It was found that the mean concentration of chemical in corn plants was 440 units with $s = 40$ units, and the mean concentration of chemical in oat plants was 620 units with $s = 100$ units.

15-14 Doctors Robin and Jay Byrde were studying nectar production in the trumpet creeper flower. They discovered that the flower produces two distinct types of nectar, one that attracts ants, and one that attracts hummingbirds. In the course of their study they had to measure the heights of the petioles and corollas of the flowers. Robin measured 120 petioles and Jay measured 200 corollas. Are they justified in claiming that the two are equally variable in height if $s = 20$ units for the petioles and $s = 25$ units for the corollas? Use $\alpha = .05$.

15-15 Three scientists were investigating the possibility of developing a pill to cure "jet lag." Working with laboratory rats, they found that one drug, theophylline, would cause about an 18-hour forward shift in the rats' body rhythm. Another drug, phenobarbital, caused about a 12-hour backward shift. Each drug was given to 10 rats after the rats had been adjusted to a precise schedule. Using $\alpha = .01$, test the hypothesis that these pills are equally variable in their effects if the results showed that the theophylline had a mean average shift of $+18.2$ hours with a variability of 1.3 hours, and phenobarbital had a mean average shift of -12.1 hours with a variability of 2.4 hours.

15-16 In recent years there has been much medical interest in the effect of extended exercise on women's menstrual cycles. One study compared 3 groups of women who were similar except for the amount and intensity of regular running they did. They were grouped into runners, joggers, and control (no running at all). Each woman contributed data for 1 year, including the number of menstrual periods which she had. The results were as follows.[†]

[†] Data are based on a study by Dale et al., "Menstrual Dysfunction in Distance Runners," as mentioned by S. Glantz in *Primer of Biostatistics*, McGraw-Hill Book Company, New York, 1981.

Tests involving variance

	runners	joggers	control
mean number of menses m	9.1	10.1	11.5
standard deviation	2.4	2.1	1.3
n	26	26	26

Is this evidence strong enough at $\alpha = .01$ to indicate that running affects the frequency of menstruation?

15-17 An ad agency designed 3 different sample displays for a product. To compare their effectiveness, they were tried in 15 similar retail stores: 5 stores got display 1, 5 got display 2, and 5 got display 3. The displays were left up in each store for 1 week and the number of sales of the product in each store were recorded. Use the results below to decide at $\alpha = .05$ whether there is a statistically significant difference in mean sales according to which display was used.

display 1	display 2	display 3
47	41	46
47	45	46
52	47	54
50	40	49
49	47	46
$m = 49$	$m = 44$	$m = 48.2$
$s = 2.12$	$s = 3.3$	$s = 3.5$

15-18 A study was done to see whether migrating birds are sensitive to magnetic fields. A large circular cage was built around a radio transmitter and antenna system. A few at a time, 51 birds were put into the cage and left alone when the transmitter was off. After 5 minutes in the cage the direction each bird was facing was recorded with reference to a compass. It was found that the mean direction was 151° with a standard deviation of 65°. Then the transmitter was turned on, creating an unnatural magnetic field. The birds were replaced in the cage and the measurements taken again. This time the mean direction was 189° with $s = 80°$. Is this difference in mean direction significant at $\alpha = .05$?

(a) Do this test as a two-sample test of means.
(b) Do this test using anova techniques.

15-19 Some scientists were studying the process by which a type of clam egg usually managed to be fertilized by only one sperm when in the presence of many sperm. They placed an equal suspension of clam eggs in each of 15 test tubes containing seawater. Then they prepared two concentrations of clam-sperm solution, which they called the low concentration and the high concentration.

To each tube they added the low concentration of clam sperm. Then 5 seconds later they added the high concentration to tubes 1, 2, 3, 4, and 5, and 5 seconds after this they added the high concentration to tubes 6, 7, 8, 9, and 10. Then 5 seconds later they added the high concentration to tubes 11, 12, 13, 14, and 15. For each tube they recorded the number of eggs that were fertilized by more than one sperm. The results are shown in the following table. Use the techniques of this chapter to show that there is less than a 1 percent probability that the differences among these three groups are just due to chance. Can you suggest an explanation of what is happening?

high concentration added in first 5 seconds	high concentration added in next 5 seconds	high concentration added in last 5 seconds
800	700	40
820	704	32
798	685	31
808	680	39
790	720	46

15-20 The Persian biologist Dr. Isara Kaz thought that certain strains of laboratory mice have genetically determined preferences for alcohol. Shown below are her results for four strains of mice.

strain 1	strain 2	strain 3	strain 4
$n_1 = 7$	$n_2 = 7$	$n_3 = 7$	$n_4 = 7$
$m_1 = .13$	$m_2 = .11$	$m_3 = .09$	$m_4 = .07$
$s_1 = .02$	$s_2 = .02$	$s_3 = .01$	$s_4 = .01$

Is there a significant difference in the mean amounts of alcohol consumed at the .05 significance level? The experiment is conducted by having two containers, one for water and one for alcohol, in the cage, and then measuring how much of each the different strains of mice drink each day. All values of m and s are recorded in milliliters per gram. For example, when we say $m_1 = .13$ milliliters per gram, this means that for the mice of strain 1, the average consumption is .13 milliliters of alcohol per day for each gram of body weight.

15-21 In a winegrowing experiment conducted by the Julius Caesar Vineyards, Inc., three different, but similar, grape varieties were used to make a certain type of wine. The acidity of the finished wine was measured. Using $\alpha = .05$, test the hypothesis that there is no significant difference in acidity. The results for each type of grape are as follows.

type I	type II	type III
$m_I = 322$ units	$m_{II} = 350$ units	$m_{III} = 352$ units
$s_I = 20$ units	$s_{II} = 23$ units	$s_{III} = 18$ units
$n_I = 15$ samples	$n_{II} = 15$ samples	$n_{III} = 15$ samples

CLASS SURVEY

Test to see whether the variance in the heights of the males on your campus is equal to the variance in the heights of the females.

FIELD PROJECT

Gather two samples of similar data from two populations and test to see whether the variances in these populations are the same or not.

Gather three samples and test with anova for equal population means.

Nonparametric Tests

16

In much of this book we have stressed that in order to carry out hypothesis tests we need to make certain specific assumptions about the types of distributions from which we were sampling. For example, to do t tests we needed to assume that the populations involved were approximately normal. In the two-sample t test we needed to make the more specific assumption that the variances of the two populations were equal. An important part of statistics deals with tests for which we do not need to make such specific assumptions. These tests are called **nonparametric** or **distribution-free** tests.

These tests would ordinarily be used if a parametric test were not appropriate. This might happen, for instance, if you were working with a nonnormal distribution, or a distribution whose shape was not yet evident. It might also happen that you are working with some special type of data for which there was no appropriate parametric test.

One type of data, called *ordinal data,* is the use of numbers to put something in **rank order.** Many consumer surveys ask people to rank something or other. For example, you might give various people a sample of three brands of chewing gum to put in order from their most favorite to their least favorite. Your data then will be a collection of 1s, 2s, and 3s. These numbers should not be treated like ordinary measurements, because we cannot interpret ranks as precisely as measurements. For example, if Brand A is ranked as number 1 (favorite) and Brand B as number 2 (next favorite), we cannot tell from this whether someone thought A was a little better than B or a lot better than B.

We have indicated in the preceding paragraphs that there are times when the standard parametric approaches are not the correct, or best, ones to use. There are many special nonparametric tests which have been developed to meet these various situations. In this chapter we are going to mention only three simple cases, but if you are interested in more, see a text about nonparametric statistics.

One point should be made. In general, if you *do* have data for which there is an appropriate parametric test, then that is the test you should use. It will be a more powerful test because it does take into account certain features about the distribution of your data (say, that it is normal in shape) which the nonparametric test would ignore.

Some students ask, "what happens when I use a parametric test even though my data do not really meet the proper requirements?" Currently, much research goes on to see how well various statistical tests work when the populations do not meet the theoretical requirements. For example, in our introduction to the analysis of variance in Chapter 15 we assumed that our populations were approximately normal. Therefore, anova is a parametric test. But research has indicated that many times its results are still reliable even if the populations are not too close to normal. The idea of measuring how well a statistical test works when we violate some of its theoretical assumptions is associated with the concept of robustness. A test is **robust** if it still works well when the theoretical assumptions are violated.

There are many nonparametric tests in fairly common use, and we have picked just three as examples. We hope that you will get the idea of how such tests work. They are derived from basic probability theory by analyzing all possible patterns of experimental results. We will illustrate the following tests.

1. A two-sample *sign* test for matched pairs of data

2. A *runs* test, useful for testing whether certain patterns of experimental outcomes are random

3. The Mann-Whitney *U* test for comparing two samples

THE TWO-SAMPLE SIGN TEST FOR MATCHED PAIRS OF DATA

EXAMPLE 16-1

Patricia McZirk, the dean of a large school of theology, is thinking of instituting a course designed to increase reading speed and comprehension. She picks 13 of the students at random to try this course. They completed the course and were tested on the day they finished. They were tested 1 month later to see if they maintained the level they had achieved. Dean McZirk expected a drop in their scores. The results were as shown in Table 16-1.

Now suppose we have reason to believe that the two distributions of reading scores are not normal. Then we should not perform a t test on the paired differences of the means. However, no matter what the shapes of the distributions are, we can reason as follows. The null hypothesis assumes that the two samples are taken from the *same* population. Therefore, if we ignored any pair of data in which the two test scores are the same, it is just as likely for X to be greater than Y as it is for X to be less than Y. If we replace each pair of values in which X is greater than Y by a plus sign and each pair of values in which X is less than Y by a minus sign, the null hypothesis leads us to expect about the same number of each type of sign.

Nonparametric tests

Table 16-1

student	X score at course ending	Y score after 1 month
Abraham	50	52
Balaam	48	51
Cain	46	46
David	50	49
Esther	62	50
Felix	80	70
Gideon	23	21
Hosea	30	33
Israel	45	46
Job	53	53
Keturah	49	48
Laban	51	48
Moses	46	48

Thus we have changed the problem into a binomial hypothesis test about *signs* where $p = P(\text{a plus sign}) = .5$. Our data can now be shown as in Table 16-2.

Table 16-2

student	sign	student	sign
Abraham	−	Hosea	−
Balaam	−	Israel	−
Cain	0	Job	0
David	+	Keturah	+
Esther	+	Laban	+
Felix	+	Moses	−
Gideon	+		

Note that if $X = Y$, we write 0 for the sign.

Ignoring the pairs of scores corresponding to Job and Cain because they have a zero, we have a binomial hypothesis test with $p = .5$ and $n = 11$. We proceed as follows.

SOLUTION Let $p = P(\text{a plus sign})$.

H_0: number of plus signs is the same as number of minus signs, $p = .5$
H_a: there will be more plus than minus signs, $p > .5$ (one-tail test)
$\mu = np = 11(.5) = 5.5$
$\sigma = \sqrt{npq} = \sqrt{11(.5)(.5)} = 1.66$

Since $np = nq = 5.5 > 5$, we have a normal distribution of successes, and for $\alpha = .01$ the critical value is $z_c = +2.33$. Therefore,

$S_c = \mu + z\sigma = 5.5 + 2.33(1.66) = 8.4$

Our decision rule will be to reject H_0 if we get more than 8.4 plus signs. Since our outcome is 6 plus signs, we fail to reject H_0. We have failed to show that reading skills have decreased after 1 month.

Understanding statistics

EXAMPLE 16-2 A class was given a test under careful supervision. Although cheating was not specifically mentioned, it was almost certain that nobody could or did cheat. The next day the class was *told* that because some of them cheated, an equivalent test would be given with several proctors in the room. The marks for both tests were as shown in Table 16-3.

Table 16-3

student	X, 1st day	Y, 2nd day	student	X, 1st day	Y, 2nd day
Ellington	70	66	Mancini	84	84
Lombardo	43	39	Alpert	78	70
Prima	91	85	Severinsen	92	69
Dorsey	89	92	Welk	83	84
Basie	73	72	Floren	75	74
James	64	63	Cugat	73	72
Kostelanetz	51	40	Duchin	89	83
Bach	83	88			

Did the announcement lower the test results? Use $\alpha = .05$.

SOLUTION Let $p = P(X > Y)$.

H_0: announcement did not affect grades, $p = .5$
H_a: announcement lowered grades, $p > .5$ (one-tail test)

Our data can now be shown as in Table 16-4.

Table 16-4

student	sign	student	sign
Ellington	+	Mancini	0
Lombardo	+	Alpert	+
Prima	+	Severinsen	+
Dorsey	−	Welk	−
Basie	+	Floren	+
James	+	Cugat	+
Kostelanetz	+	Duchin	+
Bach	−		

Ignoring the pair of scores for Mancini, which were the same, we have $n = 14$, $np = nq = 7 > 5$, $\mu = 7$, $\sigma = \sqrt{7(.5)} = 1.87$. Therefore, $S_c = 7 + 1.65(1.87) = 10.1$. Our decision rule will be to reject the null hypothesis if our outcome is more than 10.1 plus signs. Our outcome is 11 plus signs, and we can reject the null hypothesis. The evidence indicates that the announcement lowered the test grades. ■

EXERCISES

16-1 If we are given data from 2 samples, sometimes it is appropriate to perform
    (1) a two-sample sign test
    (2) a two-sample test of comparison of means
    (3) a matched-pairs t test
Why did statisticians develop these three different tests? How do you decide which one to use?

16-2 A test was done to compare the efficacy of fertilizer A with that of fertilizer B. One season a dozen pear trees in a controlled-environment hothouse were treated with Brand A, and the next season they were given Brand B. The number of baskets of marketable pears for each tree were as follows.

tree number	first season	second season
1	1.5	2.0
2	2.0	2.5
3	1.5	1.5
4	3.0	2.5
5	2.5	2.0
6	2.0	1.5
7	2.0	2.5
8	2.5	2.0
9	3.0	(died)
10	1.5	1.0
11	1.0	1.5
12	2.0	1.5

Perform a two-sample sign test for paired data on this pear data. Use $\alpha = .05$.

16-3 All freshmen are weighed by the medical office when school opens in September. Prof. Haman Deggs teaches a class in nutrition. At the end of the semester in December he asks his students to be weighed again. Testing at the .05 significance level, are the variations in weight shown random?

student number	September weight	December weight	student number	September weight	December weight
1	140	137	11	75	81
2	112	110	12	210	212
3	176	210	13	193	189
4	98	98	14	145	140
5	180	165	15	144	142
6	140	145	16	139	139
7	154	150	17	180	164
8	193	185	18	165	159
9	128	129	19	98	97
10	102	101	20	141	136

(a) Test by the two-sample sign test.
(b) Assume that weights are distributed normally and test by a matched-pairs t test.

16-4 Do premed students at Leach College score better in physics than in math? The final exam scores for a random sample of these students are given below. Use $\alpha = .05$.

student	math	physics	student	math	physics
Korn, A.	90	95	Lash, I.	78	78
Tropey, N.	78	76	Ective, F.	90	60
Shorr, C.	80	83	Sera, K.	87	94
Frost, D.	81	82	Bow, L.	78	83
Lope, E.	94	90	Knott, Y.	72	48
Kupp, T.	30	31	Cleaf, O.	99	98
Nee, G.	63	60	Kann, P.	70	80
Bohr, R.	70	78	Kneeaform, Q.	62	78

(a) Test using the two-sample sign test.
(b) Assume the grades are normally distributed and use the matched-pairs t test.

16-5 Of Mr. Abel Riemann's pupils in geometry, 70 were in his algebra classes last year. To compare their grades in these two different subjects, he converted their grades to z scores and then found the differences of their algebra grades minus their geometry grades. He got 40 plus signs, 25 minus signs, and 5 zeros. Test at the .05 significance level.

16-6 Given that 13 Plutans have more arms than legs, 32 have more legs than arms, and 3 have the same number of arms and legs, test at $\alpha = .01$ whether the Plutan population splits equally into those with more arms and those with more legs.

16-7 Since half of the scores fall above the median and half below, we can use reasoning similar to that used in a two-sample sign test to test the hypothesis that the median of a population is some given number. A plus sign corresponds to a score above the median and a minus sign corresponds to a score below the median. Use this approach to solve the following problem.

Eve Wormwood packages apples with a median of 8.5 apples per box. Wormwood samples some packages of her competitor, Adam Upright. She finds the following numbers of apples per box: 10, 12, 8, 7, 15, 9, 8, 7, 12, 8, 8, 9, 9, 9, 9, 10, 12, 7, 8, 15, 14, 18, 12, 6, 8, 9, 9, and 8. Should Wormwood be convinced that Upright's median is greater than hers? Use $\alpha = .05$.

16-8 Is the median age of teachers at Gray University 62 years? Test at $\alpha = .05$ if a sample of teachers had the following ages: 25, 47, 53, 53, 58, 59, 61, 62, 62, 65, 66, 66, 66, 73, 81, 85, and 94.

16-9 Lloyd is the attendant on a drawbridge from 10 P.M. to 6 A.M. His job is not exciting. He notices that when he raises the bridge it seldom interferes with any traffic, even though he may raise the bridge several times per night. He wonders if the median of the number of times in a *week* that any traffic stops for the raised bridge is less than 1. In the past 60 weeks he noticed that in 30 of the weeks no traffic was stopped, in 20 of the weeks traffic was stopped once, and in the remaining 10 weeks traffic was stopped more than once. Test at the .01 significance level.

16-10 Given a matched pairs sign test with 20 nonzero pairs. In a one-tail test on the right, what is the smallest number of plus signs that will cause a rejection of H_0 if $\alpha = .05$?

THE RUNS TEST FOR RANDOMNESS

Ida Noh, a safety expert, has been monitoring her radar set behind a billboard. Each time a car passes doing the speed limit or less she writes S for slow. Each time a car passes doing more than the speed limit she writes F for fast. The results after 40 cars were:

S S F F F F S S S S S S S S S F S S F F
F F S S S S S S S F F F F S S S S F S S

Ida wants to know whether or not speeders and nonspeeders occur randomly. That is, do speeders tend to come bunched together? She breaks the series of outcomes into **runs** of S's and **runs** of F's as follows:

S S   F F F F   S S S S S S S S   F   S S   F F F F
S S S S S S S   F F F F   S S S S   F   S S

Nonparametric tests

She has 11 runs. If we let n_1 equal the number of S's, n_2 equal the number of F's and R equal the number of runs, then $n_1 = 26$, $n_2 = 14$, and $R = 11$.

This is just one possible way that her string of 26 S's and 14 F's might have turned out. Some other arrangement would result, perhaps, in a different number of runs. So we can talk about R, the number of runs, as a random variable. If we repeatedly got random arrangements of 26 S's and 14 F's we would end up with a *distribution* of R's.

The *mean* number of runs, when you *randomly* arrange n_1 items of one kind and n_2 items of another kind, is

$$\mu_R = \frac{2n_1 n_2}{n_1 + n_2} + 1$$

In this case

$$\mu_R = \frac{2(26)(14)}{26 + 14} + 1 = 19.2$$

The standard deviation of the number of runs is

$$\sigma_R = \sqrt{\frac{2n_1 n_2 (2n_1 n_2 - n_1 - n_2)}{(n_1 + n_2)^2 (n_1 + n_2 - 1)}}$$

In this case

$$\sigma_R = \sqrt{\frac{2(26)(14)[2(26)(14) - 26 - 14]}{(26 + 14)^2 (26 + 14 - 1)}} = \sqrt{8.03} = 2.83$$

Further, the distribution of R's is approximately normal if both n_1 and n_2 are greater than 10. Therefore, Ms. Noh has

$ND(\mu_R = 19.2, \sigma_R = 2.83)$
H_0: fast and slow cars arrive randomly
H_a: fast and slow cars do not arrive randomly

This is a two-tail test, since there could be either too many runs or too few runs if they are not random. Testing at the .05 significance level, we have

$$R_c = \mu_R + z_c \sigma_R = 19.2 \pm 1.96(2.83) = 13.6 \quad \text{and} \quad 24.8$$

Her decision rule will be to reject the null hypothesis if her outcome is either less than 13.6 or greater than 24.8 runs. If she gets less than 13.6 runs, then she will conclude that fast and slow cars do not arrive randomly but seem to come in bunches. If she gets more than 24.8 runs, she will conclude that fast and slow cars do not arrive randomly but tend to alternate. Her outcome is $R = 11$ runs, and Ms. Noh can reject the null hypothesis. Fast cars and slow cars arrive in bunches and do not arrive randomly.

EXAMPLE 16-3 Sal, Jean, and Pat leave work together every day and walk to the bus stop. Sal takes the Q43 bus home, Pat rides the Q36, and Jean can take either bus home. If the Q36 bus arrives first, Pat and Jean get on and Sal goes home alone. If the Q43 bus arrives first, Sal and Jean get on and Pat goes home alone. If we let P indicate that Pat went home alone and S that Sal went home alone, for the past 25 days we have:

P P S S P P P P S S P S P P P S S S S P S S S S S

Testing with $\alpha = .05$, do the buses arrive randomly?

SOLUTION H_0: buses do arrive randomly
H_a: buses do not arrive randomly   (two-tail test)

Let n_1 = the number of P's and let n_2 = the number of S's. We have $n_1 = 11 > 10$ and $n_2 = 15 > 10$. Therefore, $z_c = \pm 1.96$. We find

$$\mu_R = \frac{2(11)(15)}{11 + 15} + 1 = 13.7$$

$$\sigma_R = \sqrt{\frac{2(11)(15)[2(11)(15) - 11 - 15]}{(11 + 15)^2(11 + 15 - 1)}} = 2.44$$

$R_c = 13.7 \pm 1.96(2.44) = 13.7 \pm 4.8 = 8.9$ and   18.5

Our decision rule will be to reject the null hypothesis if our outcome is less than 8.9 runs or more than 18.5 runs. Our outcome is $R = 10$ runs, and so we fail to reject the null hypothesis. The buses may arrive randomly.

EXERCISES

16-11 To test a coin for fairness we could do a one-sample binomial hypothesis test to see if $P(\text{heads}) = \frac{1}{2}$. If our sample data were 10 heads followed by 10 tails, we would not claim that the coin was biased. However, these same data used in a runs test would lead us to reject the null hypothesis. Comment. What is the difference between the null hypotheses in the two tests?

16-12 Find n_1, n_2, R, μ_R and σ_R for the following runs.
(a) M M M F F M F M M F M
(b) T T T T F F F T T T F F T
(c) S F F S S S F F F F S S S S S F F F F F F
(d) X X Y X X Y X X Y Y X X X Y X
(e) N O N O N O O N N O N N N O O N O N

16-13 G. Ringo, a tourist, is gambling in a resort in Acapulco. On the last 25 games he won and lost in the following order: WWWLW LWWLL LLWWL LWLLL WWWWL. Are these wins and losses randomly distributed at $\alpha = .05$?

16-14 A person, asked to give a "random" string of A's and B's, says: A B A A B B A B B A B B B A A A B A A B B A. Does this seem to be random if at $\alpha = .05$?

16-15 Confined to bed by a broken leg and bored to tears, Xerxes has been watching people go past his window. He has clocked the time a person remains in his view and found the median time to be 6.3 minutes. He wonders if slower persons and faster persons appear randomly. He records the following list of data on the next 25 people: S F F F S S F S S S S S S F F F F S S F F S S. Test with $\alpha = .05$.

16-16 Carol Louis inspects wabes for Brillig Brothers Inc. The median wabe is 15.3 toves. Every ½ hour she takes one wabe from the assembly line and determines its toves. Her results are 14.7, 14.9, 15.1, 15.5, 15.6, 15.6, 15.2,

Nonparametric tests

15.7, 15.6, 14.9, 14.7, 14.8, 15.2, 15.4, 15.2, 14.8, 15.0, 16.0, 15.8, 15.5, 15.9, 15.4, 15.4. Are the variations from the median random at the .05 significance level?

16-17 In a series of five true and false questions, one answer is true and four are false. There are five different arrangements of one true and four false answers.
(a) List the five arrangements and the number of runs in each arrangement. Find the mean of these numbers of runs.
(b) Justify the formula for the mean number of runs by finding the mean using the formula for μ_R and comparing your answer with part (a).
(c) Justify the formula for the standard deviation of the number of runs by finding it both ways.

THE MANN-WHITNEY U TEST FOR COMPARISON OF TWO GROUPS

ONE-TAIL TEST

This test is used to help decide whether the numbers in the two populations tend to be the same. It is a distribution-free test, because we can use it for almost any two populations. In particular we do not need to assume that both populations are normal.

It is used when items in the population are ranked rather than measured. Let us examine a typical application, one where a new treatment of some kind is compared to a standard or "control" treatment.

EXAMPLE 16-4 There are 14 laboratory rats cloned from the same parent. They are split randomly into two groups of size 7. Their diets and living conditions are identical, except for one vitamin which is added to the diet in group 1. The motivated hypothesis is that the vitamin will make the rats in group 1 smarter. Group 1 is considered the treatment group, and group 2 is the control group. During the experiment, two rats in group 1 escape.

After 1 month the rats (on the basis of rat IQ tests) are ranked from smartest (rank = 1) to most stupid (rank = 12). Here are the results.

treatment group ranks	1	2	4	5	8			$n_1 = 5$
control group ranks	3	6	7	9	10	11	12	$n_2 = 7$

$$N = n_1 + n_2 = 12$$

You can see that the smarter rats tend to be in the treatment group. This is reflected by the fact that there are *more numerically lower ranks* in the treatment group.

NOTE: In this book we always assign rank 1 to the "best" outcome, and Table C-15 is written from this point of view.

SOLUTION The idea of this statistical test is to summarize this relationship by a single statistic. One way to do this is as follows. Since we are trying to express the fact that the treatment rats tend to have lower ranks than the control ranks, we can count the number of times that a rat in the treatment group does better (that is, gets a lower rank) than a rat in the other group.

Understanding statistics

We will compare in turn each rat in the treatment group to all of the rats in the control group. A rat in the treatment group scores +1 for each rat in the other group that he outranks. The total number we get for all rats in the treatment group will be our statistic. We show the computation using the data above.

Computation

(a) The first rat in the treatment group had a rank of 1. This was better than all 7 rats in the other group. So this rat scores 7.

(b) The second rat in the treatment group had a rank of 2. This was also better than all 7 rats in the other group. So this rat scores 7.

(c) The third rat in the treatment group had a rank of 4. This is better (lower) than 6 of the rats in the other group. So this rat scores 6.

(d) Similarly, the fourth rat scores 6.

(e) Finally, the last rat in the treatment group had a rank of 8. This was better than 4 rats in the other group. This rat scores 4.

The total score for the treatment group then is 7 + 7 + 6 + 6 + 4 = 30. This statistic is one way to summarize the amount by which the treatment group had better ranks than the control group. This total is often labeled U and is called the **Mann-Whitney U statistic.** If it is bigger than a tabled critical value, we conclude that the population represented by the treatment group is smarter than the population represented by the control group. In short, we conclude that the diet is effective.

For our particular example we have

H_0: There is no difference in IQs of rats who get the vitamins and rats who do not. (Technically, we are testing that there is *no difference in the mean rank* of rats in the two populations.)

H_a: Rats which receive the vitamin have higher IQs than the rats who do not. (One-tail test.)

We refer to Table C-15 for one-tail tests, and we find that for samples of size 5 and 7 and $\alpha = .05$ the critical value of U is 29. Since our sample value is 30, we have evidence that the diet is effective.

SUMMARY MANN-WHITNEY U TEST

1. Arrange the ranks in both samples from smallest to largest.

2. Compute the sum U for the group that the motivated hypothesis expects to have the lower ranks.

3. Compare U with the tabled critical value found in Table C-15. If U is bigger than the critical value, you can reject the null hypothesis.

EXAMPLE 16-5 Sometimes the experimental data are not ranks to begin with, but the investigator is willing to treat them as such because he suspects that the population from which they came is not normal. Here is such a case.

Patients with a certain disease are treated with two different medicines. The

Nonparametric tests

physicians evaluate the patients in terms of "speed of recovery" in days. The motivated hypothesis is that the new medicine will lead to speedier recovery. We will test by the Mann-Whitney U test at $\alpha = .05$. Suppose these are the results.

	speed of recovery, days
old medicine	13 10 12 14 16 18 20
new medicine	8 17 9 11 15

SOLUTION (a) Put the outcomes in each group in increasing order.

	speed of recovery, days
old medicine	10 12 13 14 16 18 20
new medicine	8 9 11 15 17

(b) Compute U as we did in Example 16-4. We get a score for each patient on new medicine by counting the number of patients on the old medicine who had a slower recovery. Notice that we are treating the recovery times as ranks because we only consider whether one time is shorter than another and *not how much* shorter it is. The value of U for the new medicine group is $7 + 7 + 6 + 3 + 2 = 25$.

(c) The two hypotheses are

H_0: the recovery time is the same for both medicines (there is no difference in the mean ranks of the two groups of patient recovery times)

H_a: the new medicine decreases the recovery time (one-tail test; recall that we are always seeking $U > U_c$)

(d) The critical value is found from Table C-15 as $U_c = 29$.

(e) Since the sample value is not bigger than the critical value, our conclusion is that we have not clearly established the superiority of the new medicine. ■

TWO-TAIL TEST

In the case of a two-sided test we must calculate U for both groups. If *either* one is bigger than the appropriate critical value in Table C-16, we can reject the hypothesis of equality.

EXAMPLE 16-6 A young student of the occult has developed two love potions. To compare them she selects 18 subjects at random and gives half potion 1 and half potion 2. Then she watches and ranks the subjects from most affected (1) to least affected (18). Analyze the results and test for equality of potions. Use $\alpha = .05$.

	subject
	1 2 3 4 5 6 7 8 9 10 11 12 13 14 15 16 17 18
potion	1 1 1 1 1 1 1 1 1 2 2 2 2 2 2 2 2 2
rank	4 10 3 12 6 1 11 5 2 16 17 9 15 7 18 14 13 8

SOLUTION

	ranks		
potion 1	1 2 3 4 5 6 10 11 12	sum = 54	$n_1 = 9$
potion 2	7 8 9 13 14 15 16 17 18	sum = 117	$n_2 = 9$
			$N = 18$

Note that the sum $U_1 + U_2$ always equals $n_1 n_2$. We compute U_1 by counting.

$U_1 = 9 + 9 + 9 + 9 + 9 + 9 + 6 + 6 + 6 = 72$

We compute U_2 as

$U_2 = n_1 n_2 - U_1$
$= 9(9) - 72 = 81 - 72 = 9$

The hypotheses are

H_0: the effects of two potions are the same (mean rank for potion 1 equals mean rank for potion 2)

H_a: effects of two potions are different

Since this is a two-tail test, we refer to Table C-16. The critical value for $\alpha = .05$ with $n_1 = n_2 = 9$ is $U_c = 64$. Thus if either U_1 or U_2 exceeds 64, we reject H_0.

Since $72 > 64$, we reject H_0. We have evidence that potion 1 is more effective than potion 2.

LARGE-SAMPLE TESTS (BOTH n_1 AND n_2 GREATER THAN 10)

When both sample sizes are greater than 10, it can be shown that the distribution of sample values of U is approximately normal. This means that we can compute the critical values of U using the critical values of normal z scores. We will not have to use special tables.

The mean and the standard deviation of the normal curve which describes the distribution of U when the null hypothesis is true are:

$$\mu_U = \frac{n_1 n_2}{2}$$

$$\sigma_U = \sqrt{\frac{n_1 n_2 (N + 1)}{12}}$$

EXAMPLE 16-7

The director of a laboratory can purchase equipment from two suppliers. Over a period of time she has ranked shipments according to quality as shown below. Rank 1 equals highest quality.

	ranks		
supplier A	1 2 6 8 10 11 14 15 17 18 20 23	sum = 145	$n_1 = 12$
supplier B	3 4 5 7 9 12 13 16 19 21 22	sum = 131	$n_2 = 11$
			$N = 12 + 11 = 23$

Nonparametric tests

Test for significant difference in quality. Use $\alpha = .05$. This is a two-tail test.

SOLUTION The hypotheses are

H_0: equipment from both suppliers is of the same quality (mean ranks are equal)

H_a: equipment from both suppliers is not of the same quality (two-tail test)

Since both samples are large (greater than 10), we use the normal approximation.

$$\mu_U = \frac{n_1 n_2}{2} = \frac{12(11)}{2} = 66$$

$$\sigma_U = \sqrt{\frac{n_1 n_2 (N+1)}{12}} = \sqrt{\frac{12(11)(24)}{12}} = \sqrt{264} = 16.2$$

The critical values of U are

$$U_c = \mu_U \pm z_c \sigma_U$$
$$= 66 \pm 1.96(16.2) = 66 \pm 31.8 = 34.2 \quad \text{and} \quad 97.8$$

It would take an observed value of U either less than 34.2 or more than 97.8 to reject the hypothesis of equality.

$$U_A = 11 + 11 + 8 + 7 + 6 + 6 + 4 + 4 + 3 + 3 + 2 = 65$$
$$U_B = n_1 n_2 - U_A = 12(11) - 65 = 132 - 65 = 67$$

Since neither sample value of U is extreme enough to allow us to reject the hypothesis of equality, we have not established that there is a significant difference in quality between the two labs.

THE CASE OF TIES (SMALL SAMPLES)

Very often ranked observations result in ties. When this happens, certain adjustments in the calculations must be made.

EXAMPLE 16-8 Suppose our speed of recovery data from Example 16-5 had turned out like this.

	days to recover
old medicine	10 13 14 16 18 18 20
new medicine	8 8 11 13 18

SOLUTION We are interested in U_{new}, so we get a score for each person in the new medicine group. Since ties are present, it may turn out that someone in the old medicine group may be tied with someone in the new medicine group. When this happens, the person in the new medicine group scores ½. (This is a compromise between scoring 1 for being better and scoring 0 for being worse.)

The first person in the new medicine group did better than all 7 of the people on the old medicine, and scores 7. Obviously, the second person also scores a 7. The next person's time of 11 days was better than 6 of the others. He or she scores 6.

The next person's time of 13 days is better than 5 people in the other group and tied with one of them. He scores $5 + \frac{1}{2} = 5.5$.

The last person's time is better than 1 person in the other group and tied with two of them. He scores $1 + \frac{1}{2} + \frac{1}{2} = 2$. This gives a total U score for the new medicine group of

$$7 + 7 + 6 + 5.5 + 2 = 27.5$$

The hypotheses are

H_0: there is no difference in the mean ranks of the two groups of patient recovery times

H_a: the new medicine decreases the recovery time (one-tail test)

From Table C-15, for $\alpha = .05$, with $n_1 = 7$ and $n_2 = 5$, we find $U_c = 29$.

Since our sample value is not larger than the critical value, we do not have conclusive evidence that the new medicine is superior.

NOTE: Strictly speaking, when ties occur in the ranks, we should not use this table because the tabled critical values were computed assuming no ties. But for most purposes if there are not very many ties, the tabled results are approximately correct.

LARGE SAMPLE WITH TIES

This is the situation that results with many surveys where people are asked to rate something by placing it in one of a few categories (such as good, fair, or bad). We will illustrate this shortly. When there are ties, the formula for the standard deviation of the U distribution must be adjusted. This is because the presence of ties reduces the number of different values U can take on, and so reduces the variability in U. We still have a normal distribution. The mean is still given by

$$\mu_U = \frac{n_1 n_2}{2}$$

But we modify the standard deviation formula.

The standard deviation formula for large samples with ties is

$$\sigma_U = \sqrt{\frac{n_1 n_2 (N + 1)}{12} - C}$$

where C is the correction factor for ties. C is calculated by

$$C = \frac{n_1 n_2}{12 N(N - 1)} \Sigma(T^3 - T)$$

where T counts the number of scores *tied* at each rank.

Nonparametric tests

We illustrate with a typical example.

EXAMPLE 16-9 In a study of executive ambition, a group of 80 middle-level executives in a large corporation were split randomly into a control group of 40 who got the usual pep talk and incentives. An experimental group of the other 40 got a newfangled psychodynamic treatment, complete with hypnosis and special diet. At the end of the study, each executive is evaluated as super, good, fair, or poor. Here are the results.

	super	good	fair	poor	
special treatment	12	16	7	5	$n_1 = 40$
control group	9	15	9	7	$n_2 = 40$
					$N = 80$

Do these data indicate that the special treatment is producing more ambitious executives? Use $\alpha = .05$. This is a one-tail test.

SOLUTION We have large samples so we can use a normal approximation to find the critical values for U. There were 12 people in the special treatment group who were ranked "super." Each of them tied with 9 people in the control group and did better than $15 + 9 + 7 = 31$ people in the control group.

Hence the score for each one of the 12 people is

$9(½) + 15 + 9 + 7 = 35.5$

Since 12 people have this score, we get

$12(35.5) = 426$

Similarly, the 16 people in the special treatment group who were rated "good" score

$16[15(½) + 9 + 7] = 376$

The 7 people rated "fair" score $7[9(½) + 7] = 80.5$.

The 5 persons rated "poor" score $5[7(½)] = 17.5$.

The sum of these numbers, $426 + 376 + 80.5 + 17.5$, is 900. Therefore, for the special treatment group $U = 900$.

The hypotheses are

H_0: mean ratings for both treatments are the same
H_a: mean rating in the special treatment group is better

Since we are using the normal approximation, we now compute the mean and the standard deviation. Recall the formulas:

$$\mu_U = \frac{n_1 n_2}{2}$$

$$\sigma_U = \sqrt{\frac{n_1 n_2 (N+1)}{12} - C}$$

where

$$C = \frac{n_1 n_2}{12N(N-1)} \Sigma(T^3 - T)$$

and T is the number of ties at each rank. We compute C with the help of the following table.

	special treatment group	control group	number of ties T	T^3	$T^3 - T$
super	12	9	12 + 9 = 21	9261	9240
good	16	15	16 + 15 = 31	29,791	29,760
fair	7	9	7 + 9 = 16	4096	4080
poor	5	7	5 + 7 = 12	1728	1716

$$\Sigma(T^3 - T) = 44{,}796$$

Then,

$$\mu_U = \frac{n_1 n_2}{2} = \frac{40(40)}{2} = 800$$

Since

$$C = \frac{n_1 n_2}{12N(N-1)} \Sigma(T^3 - T)$$
$$= \frac{40(40)}{12(80)(79)} (44{,}796) = 945.1$$

we have

$$\sigma_U = \sqrt{\frac{n_1 n_2 (N+1)}{12} - C}$$
$$= \sqrt{\frac{40(40)(81)}{12} - 945.1} = \sqrt{9854.9} = 99.27$$

Thus the critical value of U is:

$$U_c = \mu + z_c \sigma_U \quad \text{(one-tail test)}$$
$$= 800 + 1.65(99.27) = 800 + 163.8 = 963.8$$

Since our sample value of $U = 900$ is not greater than the critical value, $U_c = 963.8$, we do not have sufficient evidence to establish that the special treatment does a better job than the usual one in producing ambitious corporate executives.

STUDY AIDS

VOCABULARY

1. Nonparametric statistic
2. Distribution-free test
3. Rank order
4. Robust
5. Run
6. Mann-Whitney U statistic
7. Ordinal data

Nonparametric tests

SYMBOLS

1. R **2.** μ_R **3.** σ_R **4.** U **5.** C **6.** T

FORMULAS

1. $\mu_R = \dfrac{2n_1 n_2}{n_1 + n_2} + 1$

2. $\sigma_R = \sqrt{\dfrac{2n_1 n_2 (2n_1 n_2 - n_1 - n_2)}{(n_1 + n_2)^2 (n_1 + n_2 - 1)}}$

3. $N = n_1 + n_2$

4. $U_1 + U_2 = n_1 n_2$

5. $\mu_U = \dfrac{n_1 n_2}{2}$

6. $\sigma_U = \sqrt{\dfrac{n_1 n_2 (N + 1)}{12}}$

7. $\sigma_U = \sqrt{\dfrac{n_1 n_2 (N + 1)}{12} - C}$

8. $C = \dfrac{n_1 n_2}{12 N(N - 1)} \Sigma(T^3 - T)$

EXERCISES

16-18 For the data shown calculate U_A.

(a)

sample from population A	sample from population B
1	6
2	6
3	0
7	5
8	10

(b)

sample from population A	sample from population B
2	2
3	2
4	5
5	6
	6

16-19 The following are the scores on the National Tie Manufacturers' annual scholarship exam.

	score
*Debra Draw	50
Stan Stalemate	48
*Stu Standoff	52
Dan Deadlock	48
*Norma Knott	50
Ben Bind	48
Carl Cravat	46

Those marked with asterisks were tutored before the test by Dr. Noah Vail. Test to see whether they ranked higher than the others. Use $\alpha = .05$.

16-20 John Smith eats in two fast-food places. To find out if one is faster than the other he times how long it takes to be served a drink, burger, and fries. The results of his last 13 visits were as follows.

Understanding statistics

McBurgers	2 5 13 8 7 12 4 6
King Donalds	9 3 11 1 10

Test at $\alpha = .05$, using the Mann-Whitney U test.

16-21 Wanda the witch buys packages of bat wings for her brews. Recently she has switched suppliers because she was getting too many substandard wings from her former supplier. To be fair she decides to compare the two suppliers. Her data are as follows.

number of substandard wings per package of 13 wings

former supplier	current supplier
3	0
2	0
1	0
3	2
0	2
0	1
3	0
1	3
2	

Using the Mann-Whitney U test at $\alpha = .05$, decide whether there is a difference in quality between the two suppliers.

16-22 A waiter at Mount Cupid Honeymoon Lodge kept tabs on the number of meals missed during one week by newly married couples. Of the couples that he served during the first week in June he noted the following. The number of meals missed by couples on the first week of their honeymoon were 7, 5, 11, 8, 7, 11, and 17; while the number of meals missed by couples on the second week of their honeymoon were 0, 13, 3, 8, 10, 6, 10, and 6. Rank the 15 numbers above from 1 to 15. By using the rank sum test, test the hypothesis that there is no difference between the number of meals missed by first- and second-week honeymooners. Use $\alpha = .05$.

16-23 Two groups of married people were asked to compare how much they were "in love with" their spouse today with how they felt on their wedding day. The data were as follows.

	married less than 7 years	married at least 7 years
a lot less	30	10
a little less	20	20
the same	10	30
a little more	20	20
a lot more	30	10

Calculate μ_U, the correction factor for ties, and the standard deviation.

16-24 In the interest of science, students who studied under two professors were asked to rank their jokes. The data were as follows.

Nonparametric tests

	funny	average	rank	
Prof. Laurel	20	80	100	200
Prof. Hardy	40	80	80	200

Use $\alpha = .05$. Do this as a two-tail U test.

16-25 Espionage agents in two different countries are trained to endure pain. Then at the Fourth Annual International Espionage Agent Pain Endurance Competition, 11 agents from each country enter the contest. Here are the results.

name of agent	country	volts endured
001	A	100
002	A	143
003	A	128
004	A	118
005	A	89
006	A	118
007	A	132
008	A	107
009	A	141
010	A	93
011	A	101
001	B	130
002	B	135
003	B	142
004	B	120
005	B	60
006	B	140
007	B	95
008	B	120
009	B	60
010	B	97
011	B	102

Does this indicate at the .05 significance level that either country produces a superior agent?

16-26 The vice president of Air Languid wants to measure the results of a recent change in flight scheduling aimed at reducing the number of empty seats per flight. The number of empty seats on the last 14 flights under the old schedule is to be compared with the first 14 flights under the new schedule. The data are as follows.

old schedule	15 17 9 17 7 21 19 25 21 9 27 23 23 27
new schedule	14 14 16 20 24 10 18 12 20 6 16 8 16 14

Using $\alpha = .05$, decide if the new schedule reduces the number of empty seats.

16-27 A chemical refinery has developed a new type of gasohol. To see whether it is better than their old gasohol, 50 drivers used the new type in their automobiles for 1000 miles, and then they used the old type for 1000 miles. The drivers reported their average number of miles per gallon. The data were accumulated as follows.

	10–15 miles per gallon	15–20 miles per gallon	20–25 miles per gallon	25–30 miles per gallon	
old	10	20	15	5	50
new	0	10	15	25	50

Test at $\alpha = .05$.

CLASS SURVEY

Compare the heights of the females in the class with the heights of their mothers.
1. Use the sign test for matched pairs. This tests whether or not a daughter tends to be taller than her mother.
2. Use the Mann-Whitney U test. This tests whether one generation tends to be taller than the other.

FIELD PROJECT

Using one of the tables in Appendix C in the back of this book, check to see if the *second* digits are randomly odd or even, or perform some other nonparametric project of your own choosing.

SAMPLE TEST FOR CHAPTERS 14, 15, AND 16

1. Would you expect the following to be positively or negatively correlated? Explain.
(a) Height and weight in a population of male college freshmen.
(b) Gas mileage (miles per gallon) and weight of automobiles in a population of American-made automobiles.
(c) Number of cigarettes smoked per day and days of work missed because of illness in a population of female factory workers.
(d) Number of polo ponies owned and amount spent on manure removal.
(e) The price of gold and the value of the U.S. dollar.
2. What would you say if you were told that someone had calculated r between height and weight for newborn infants to be 2.38?
3. For the following data find: (a) r, (b) b, (c) the formula for the best-fitting line.

city	X elevation	Y mean high temperature for January 1978
Bakersfield, CA	475	62.7
Dallas–Ft. Worth, TX	551	42.3
Denver, CO	5283	37.5
Los Angeles, CA	97	65.3
San Francisco, CA	8	58.3
Seattle, WA	400	48.7

(d) Show the scattergram with the best-fitting line superimposed.

Nonparametric tests

4. A production line in a factory is supposed to fill containers with at least 10 pounds of detergent powder. The standard deviation is supposed to be less than one-tenth of a pound. These are the weights of 10 packages pulled at random from a lot: 10.09, 10.02, 10.02, 10.03, 10.08, 9.95, 10.01, 10.07, 10.03, 10.00 pounds. Is there evidence at $\alpha = .05$ that the line is working erratically?

5. Two different kinds of plastic are being tested for use in an artificial heart valve. Ten samples of each are tried in laboratory animals. Among the questions to be answered is, "are the two materials equally variable in their working life span?" Answer the above question with $\alpha = .05$, based on the following data.

	life span of valve in days
material 1	459 458 459 493 415 471 494 465 459 433
material 2	491 430 426 493 401 469 470 433 401 442

6. A refinery produces 4 grades of gasoline. The gasolines are tested in a sample motor, and certain exhaust pollutants are measured. Is there evidence at $\alpha = .05$ that the grades of gasoline differ in average amount of exhaust pollutants?

	units of pollutant
grade A	100 110 120 113
grade B	105 115 125 100
grade C	130 140 145 140
grade D	135 140 160 150

7. People in Disasterville, N.J., were called at home and asked if they regularly watched the 6 o'clock news. If they said yes, they were asked whether they watched on channel 0 or on channel 1. The results for 118 calls were as follows:

37 watched channel 0

28 watched channel 1

53 did not usually watch the 6 o'clock news

Analyze these data by the sign test with $\alpha = .05$. Among the regular watchers, does the sign test indicate that channel 0 is more popular in the population? Use the .05 significance level.

APPENDIXES

Arithmetic Review Appendix

A

It is advisable for you to know what arithmetic skills are needed for the material in this book. The following exercises are a sample of the types of problems you will be asked to do. If you have trouble with these, consult a text on arithmetic or see your instructor for suggestions. The answers to this review are given at the end of the text.

EXERCISES

A-1 *Decimal Review*   Evaluate the following.
(a) 4.1 + 5.03 + 14 + .6   (b) 5 − .03
(c) .05 − .004   (d) 6.01(.2)

(e) 5.12 + 1.96(2.3)   (f) $\dfrac{.3(.04)}{10}$

(g) 17/.8   (h) 6.1/12.2

(i) $\dfrac{.18}{9}$ (j) $\dfrac{0}{4.3}$

A-2 *Size of Decimals*
(a) Which is larger, −2.58 or −2.33?
List in order with the smallest first.
(b) 4.7, .41, .081, .6, and 4.51
(c) .41, −.273, and .273

A-3 *Symbols*   Answer true or false.
(a) 5 < 8   (b) 5 ≤ 8
(c) 8 > 9   (d) 6 ≤ 6

Appendix A

If X is any whole number from 0 to 6 inclusive, list which values of X satisfy the following conditions.
(e) $X > 4$ (f) $X \geq 4$
(g) $X < 5$ (h) $X \leq 5$
(i) $2 < X < 5$ (j) $2 \leq X \leq 5$

A-4 *Percent Review*
(a) Change .05 to a percentage.
(b) Change .003 to a percentage.
(c) Change 37 percent to a decimal.
(d) Change 3.2 percent to a decimal.
(e) Change 3/8 to a decimal and to a percentage.
(f) Change 5/19 to a decimal and to a percentage.
(g) Find 23 percent of 50.
(h) Find 4 percent of 200.
(i) 15 is what percentage of 50?
(j) 27 is what percentage of 108?

A-5 *Signed Number Review* Evaluate the following.
(a) $4 + (-7) + (13) + (-25)$
(b) $-3.07(-5)$
(c) $-1.65(10)$
(d) $-16/8$
(e) $10/(-2)$

A-6 *Formulas* Evaluate the following.
Given $x = 7.3$, $y = 1.02$, $\sigma = .1$, $\mu = 11.4$, and $z = -2.33$, find:

(a) xy (b) $\dfrac{x - \mu}{\sigma}$ (c) $\mu + z\sigma$

(d) Given $\mu = 3$, $\sigma = 2$, and $z = \pm 1.96$, find $\mu + z\sigma$.

(e) True or false? $\sqrt{pq/n} = \sqrt{\dfrac{pq}{n}}$

A-7 *Exponents* Evaluate the following.
(a) $(-4)^2$ (b) $(.7)^3$
(c) $(½)^4$ (d) $10(.7)^3(.3)^2$

A-8 *Square Roots* Compute the following.
(a) Given $n = 10$, $p = .4$, and $q = .6$, find $\sqrt{npq}$.

(b) $\sqrt{\dfrac{50 - (10)^2/8}{7}}$ (c) $\sqrt{\dfrac{(3)^2}{5} + \dfrac{(6)^2}{2}}$

Probability Appendix

B

This appendix assumes the knowledge of Chapter 4. We are going to consider probability in a little more detail. In Chapter 4 in order to compute a probability we listed the total number of possible outcomes of an event and found the proportion of outcomes that were favorable to that event. Now we will use this method from Chapter 4 and then show another approach to obtain the same results.

INDEPENDENT AND DEPENDENT EVENTS

Two events are said to be **independent** if the occurrence or nonoccurrence of one has no effect on the occurrence or nonoccurrence of the other.

EXAMPLE B-1

Let us consider two machines which operate independently of each other. This means that if one machine fails to function, it will have no effect on whether the other machine works or fails. Let us suppose that each machine fails half of the days, so that $P(A) = P$(machine A fails) $= 1/2$ and $P(B) = P$(machine B fails) $= 1/2$. Let us find the probability that both machines fail on the same day.

SOLUTION 1

Using the method of Chapter 4, since the machines operate independently, if machine A works, then machine B may work or fail, and similarly if machine A fails, machine B may work or fail. Since there are two possible outcomes for machine A and two possible outcomes for machine B, there are $2 \times 2 = 4$ possible outcomes. They are listed in the table.

outcome	machine A	machine B
1	work	work
2	work	fail
3	fail	work
4	fail	fail

Since each of the above possibilities is equally likely, we can see that the probability of both failing is 1/4. We write:

P(machine A fails and machine B fails) $= P(A$ and $B) = 1/4$

SOLUTION 2 We could have reached this result by multiplying the probability that machine A fails times the probability that machine B fails. That is,

$P(A$ and $B) = P(A)P(B) = 1/2 \cdot 1/2 = 1/4$

When two events are independent, we can always multiply their probabilities together in order to find the probability that they will occur simultaneously. In fact, many authors use this as the definition of independence. They say that two events are independent if the probability of their occurring together equals the product of the probabilities of each occurring separately. In symbols we write: events A and B are independent if and only if

■ $P(A$ and $B) = P(A)P(B)$

We cannot use this rule for events that are dependent. Let us consider a second example.

EXAMPLE B-2 Let us suppose that someone decides to play Russian roulette with a six-shooter in which the chambers are numbered 1 through 6. If he places a bullet in one chamber, spins the cylinder, and then pulls the trigger twice in a row, find the probability that he lives.

SOLUTION 1 By the method of Chapter 4, if the cylinder lands so that chamber 3 will be fired first, then chamber 4 will be fired next. Similarly, if chamber 6 is fired first, then chamber 1 is fired next. Thus, there are six possible outcomes.

first try	second try
chamber 1	chamber 2
chamber 2	chamber 3
chamber 3	chamber 4
chamber 4	chamber 5
chamber 5	chamber 6
chamber 6	chamber 1

Notice that each chamber is listed twice. The bullet is in only one chamber. Therefore, there are four outcomes of the six in which both chambers are empty. If we let E_1 stand for the event that the first chamber tried is empty, and E_2 stand for the event that the second chamber tried is empty, we have

P(both chambers are empty) $= P(E_1$ and $E_2) = 4/6 = 2/3$

Understanding statistics

SOLUTION 2

Let us use E_1 and E_2 as defined above and calculate the probability that the person will live by computing the probabilities associated with each pull of the trigger separately. The probability that the first chamber is empty is 5/6 since there is only one bullet. We write:

P(first chamber tried is empty) = $P(E_1)$ = 5/6

The probability that the second chamber is empty depends on what happened when we pulled the trigger the first time. If we got the bullet on the first try, we know for sure that the next chamber is empty, but we probably would not care. On the other hand, if the first chamber selected were empty, then of the remaining 5 chambers, 4 are empty. Thus, we write:

P(second chamber is empty given first chamber is empty)
= $P(E_2$ given $E_1)$ = 4/5

Thus, we can say that event E_2 is **dependent** on event E_1, since the occurrence or nonoccurrence of E_1 will affect the occurrence or nonoccurrence of E_2.

To find the probability that both chambers are empty, we multiply the probability that E_1 occurs times the probability that E_2 occurs given that E_1 has already occurred:

$P(E_1$ and $E_2)$ = $P(E_1)P(E_2$ given $E_1)$
           = 5/6 · 4/5 = 4/6 = 2/3

■ Note that this is the same result we obtained in Solution 1.

We would also like to mention that we could have solved the problem by considering the second chamber. The probability that the second chamber is empty is 5/6, since there is only one bullet. We write:

P(second chamber is empty) = $P(E_2)$ = 5/6

Given that the second chamber is empty, of the 5 remaining chambers only 1 has a bullet, and so the probability that our first chamber is empty is 4/5. We write:

P(first chamber is empty given second chamber is empty)
= $P(E_1$ given $E_2)$ = 4/5

Therefore,

$P(E_1$ and $E_2)$ = $P(E_2)P(E_1$ given $E_2)$
           = 5/6 · 4/5 = 4/6 = 2/3

Here we see that E_1 now depends upon the outcome of E_2.

In general, if A and B are dependent events, the probability that they occur together is found by

$P(A$ and $B)$ = $P(A)P(B$ given $A)$
or $P(A$ and $B)$ = $P(B)P(A$ given $B)$

Appendix B

REPLACEMENT AND NONREPLACEMENT

Let us consider another type of problem. Suppose we have an urn which contains three white balls and 2 red balls. What is the probability that we randomly select 2 red balls in a row? The answer to the question depends upon whether or not we replace the first ball before drawing the second ball.

EXAMPLE B-3 **Replacement** Suppose that after we select the first ball we replace it and then select another ball. Since the outcome of the first selection has no effect on the outcome of the second selection, the events are independent. Thus, we can use the formula for independent events. Since 2 of the 5 balls are red, the probability of selecting a red is $P(\text{red}) = 2/5$. Thus, the probability of selecting 2 reds is

■ $P(\text{red and red}) = P(\text{red})P(\text{red}) = 2/5 \cdot 2/5 = 4/25$

EXAMPLE B-4 **Nonreplacement** Let us reconsider the same problem of getting 2 reds from the urn with 3 white and 2 red balls, but this time we will not replace the first ball selected before we randomly select the second ball. The probability of getting a red on the first selection, $P(\text{red})$, is still 2/5, but the probability of getting a red on the second selection depends on what occurred on the first selection.

SOLUTION If a red ball were selected first, there would be 4 balls remaining of which only one was a red. Thus, the probability of a red given a red on the first selection is one chance in 4; we write $P(\text{red}_2 \text{ given red}_1) = 1/4$. Thus, according to the rule for dependent events, the probability of selecting 2 reds will be

$P(\text{red}_1 \text{ and red}_2) = P(\text{red}_1)P(\text{red}_2 \text{ given red}_1)$

■ $\qquad = 2/5 \cdot 1/4 = 2/20 = 1/10$

Let us compare our answers in Examples B-3 and B-4. We found that the probability of selecting 2 reds was $4/25 = 16/100 = .16$ when we replaced the first ball, and that the probability of 2 reds was $1/10 = .10$ when we did not replace the first ball. You can see that there is a difference of $.06 = 6/100$ between the two methods. As we increase the number of balls in the urn, the difference becomes less pronounced. In fact, for a large number of balls the difference is negligible, as shown in Example B-5:

EXAMPLE B-5 Suppose our urn contains 5000 balls of which 3000 are white and 2000 are red. Find the probability of randomly selecting 2 red balls.

SOLUTION 1 *Replacement* Since $P(\text{red}) = 2000/5000 = 2/5$ and each selection is independent, we have

$P(\text{red and red}) = 2/5 \cdot 2/5 = 4/25 = .16$

SOLUTION 2 *Nonreplacement* On the first selection $P(\text{red}) = 2000/5000 = 2/5$, and the probability of getting a red on the second selection given a red on the first selection is $P(\text{red given red}) = 1999/4999$ since there is one less red ball in our urn. Thus

$P(\text{red and red}) = P(\text{red})P(\text{red given red})$
$= 2/5 \cdot 1999/4999$
$= 3998/24{,}995 = .15995$

You can see that the difference in our answers using the two methods is .00005 or 1/20,000.

■

The fact that two answers are close to each other becomes important when we take random samples from large populations because in theory we assume replacement (independence), but in practice we often use nonreplacement. For example, if one were asking people whom they planned to vote for, one would not usually ask the same person twice, and so we are not placing this person back into the population.

MUTUALLY EXCLUSIVE EVENTS

Let us consider a regular deck of 52 playing cards. Since there are 4 kings and 4 queens, we can write $P(\text{randomly selecting a king}) = P(K) = 4/52 = 1/13$ and $P(\text{randomly selecting a queen}) = P(Q) = 4/52 = 1/13$. If we wanted to know the probability of selecting a king or a queen, we would count 8 in all so that $P(\text{randomly selecting a king or a queen}) = P(K \text{ or } Q) = 8/52 = 2/13$. (This problem asks the probability of a king *or* a queen on one draw, and the answer is 2/13. The probability of a king *and* a queen on one draw is zero. Do not confuse the word "and" with the word "or.") Note that we could have obtained this last result by adding the individual probabilities of randomly selecting a king and randomly selecting a queen. Thus,

$P(K \text{ or } Q) = P(K) + P(Q) = 1/13 + 1/13 = 2/13$

We must be careful because this procedure will not always work. Suppose we wanted to know the probability of selecting a heart or a jack. Out of the 52 cards there are 13 hearts, 1 of which is a jack and there are 3 jacks of other suits. Thus there are 16 favorable outcomes. We write $P(\text{randomly selecting a heart or a jack}) = P(H \text{ or } J) = 16/52 = 4/13$. If we consider hearts and jacks separately, we have $P(\text{randomly selecting a heart}) = P(H) = 13/52$ since there are 13 hearts and $P(\text{randomly selecting a jack}) = P(J) = 4/52$ since there are 4 jacks. Note that if we add these results together we do *not* get the right answer of 16/52, since $P(H) + P(J) = 13/52 + 4/52 = 17/52$.

We got the wrong answer because we counted the jack of hearts twice. We counted it once as a heart and once as a jack. In order to count the jack of hearts only once we must subtract the probability of getting the jack of hearts, which is 1/52. We write $P(\text{jack of hearts}) = P(\text{randomly selecting a heart and a jack} = P(H \text{ and } J) = 1/52$. Thus, our formula becomes

$P(H \text{ or } J) = P(H) + P(J) - P(H \text{ and } J) = 13/52 + 4/52 - 1/52 = 16/52$

Let us compare the two situations. In the first problem there was no card that was both a king and a queen. Thus, we say that selecting kings and selecting queens are **mutually exclusive events**.

Formally, we say that two events are mutually exclusive if the probability that they occur simultaneously is zero. We write P(selecting a king and a queen in a single draw) $= P(K$ and $Q) = 0$. Since the probability of simultaneous occurrence is always zero for mutually exclusive events, many authors use this fact as the definition and say that events K and Q are mutually exclusive if and only if $P(K$ and $Q) = 0$.

In the second problem there was a card which is both a heart and a jack, and so we say that the events of selecting a heart and of selecting a jack are not mutually exclusive. Since P(jack of hearts) $= P(J$ and $H) = 1/52 \neq 0$, we say that the events J and H are not mutually exclusive.

We summarize by considering two events A and B. We have seen that

$$P(A \text{ or } B) = P(A) + P(B) - P(A \text{ and } B)$$

If A and B are mutually exclusive, then

$$P(A \text{ and } B) = 0$$

and the formula reduces to

$$P(A \text{ or } B) = P(A) + P(B)$$

STUDY AIDS

VOCABULARY

1. Independent event
2. Dependent event
3. Replacement
4. Nonreplacement
5. Mutually exclusive events

SYMBOLS

1. $P(A$ and $B)$
2. $P(A$ given $B)$
3. $P(A$ or $B)$

FORMULAS

1. $P(A$ and $B) = P(A)P(B)$, when A and B are independent
2. $P(A$ and $B) = P(A)P(B$ given $A)$ or $P(A$ and $B) = P(B)P(A$ given $B)$
3. $P(A$ or $B) = P(A) + P(B) - P(A$ and $B)$
4. $P(A$ or $B) = P(A) + P(B)$, when A and B are mutually exclusive

EXERCISES

B-1 Decide intuitively if the following pairs of events are independent or dependent.
(a) Tossing a penny and tossing a dime.
(b) Spinning a spinner twice.
(c) Selecting two toadstools from a bag containing three mushrooms and two toadstools (without replacement).
(d) Part (c) above but with replacement.

B-2 Decide intuitively whether the following pairs of events are independent or dependent.
(a) Playing Russian roulette twice and spinning the cylinder each time.
(b) Playing Russian roulette twice and spinning the cylinder only the first time.
(c) Dealing two cards from an ordinary deck.
(d) Rolling two dice.
(e) Rolling one die twice.

B-3 $P(A) = 1/2$ and $P(B) = 1/3$, while $P(A$ and $B)$ is not 1/6. Explain how this is possible.

B-4 Are the following outcomes of the given events mutually exclusive or not?
(a) In picking a card from an ordinary deck:
  (1) Getting a jack and a spade.
  (2) Getting a jack and a queen.
(b) In predicting the sex of your next two children:
  (1) Two the same sex and at least one boy.
  (2) Both girls, and at least one boy.

B-5 Are the following outcomes mutually exclusive or not?
(a) Being a surgeon and being a woman.
(b) Being a male and being a mother.
(c) Given that you own one pet, owning a cat and a dog.
(d) Given that you own more than one pet, owning a boa constrictor and a prairie dog.

B-6 If $P(A$ or $B) = 5/7, P(A) = 3/7$, and $P(B) = 2/7$
(a) Find $P(A$ and $B)$.
(b) Are A and B mutually exclusive?

B-7 If $P(A$ or $B) = .62$ and $P(A) = .41$ and $P(B) = .41$:
(a) Find $P(A$ and $B)$.
(b) Are A and B mutually exclusive?

B-8 Mauro prefers to date Elena, but if she turns him down, he calls Jill. The probability that Elena says yes is 0.4. If the probability that both Elena and Jill say no is 0.2, find the probability that Jill says no given that Elena said no.

B-9 Two airplane engines operate independently. The plane can fly if either engine works. If engine 1 fails once every 100 flights and engine 2 fails once every 10,000 flights, find the probability that both engines will fail on the same flight.

B-10 In an assortment of 30 chocolates it is impossible to tell the fruits from the nuts. There are 20 fruits and 10 nuts, and the type is identified under each candy. Two chocolates are picked at random.
Case 1 Replacement:
(a) Find the probability of two nuts.
(b) Find the probability of a fruit first and then a nut.
(c) Find the probability of one of each in any order.
Case 2 Without replacement:
(d) Repeat part (a).
(e) Repeat part (b).
(f) Repeat part (c).

B-11 A deck of playing cards is shuffled, and two cards are randomly selected.

Case 1 Replacement (the first card is returned to the deck prior to the second pick and the deck is reshuffled):
(a) Find the probability of two hearts.
(b) Find the probability of an ace and then a 7.
(c) Find the probability of an ace and a 7 in any order.
(d) Find the probability of two picture cards.
Case 2 Without replacement:
(e) Repeat part (a).
(f) Repeat part (b).
(g) Repeat part (c).
(h) Repeat part (d).

B-12 Two machines work according to the following rules. The probability that machine A will fail is 1/3. The probability that machine B will fail, given that machine A has failed is 1/8.
(a) Find the probability that machines A and B both fail.
(b) If the probability that machine B will fail is 1/4, then using the results from part (a), find the probability that machine A will fail, given that machine B has failed.

B-13 A coin-tossing game is played as follows. To win the game you must toss exactly 2 heads. If the first toss produces heads, then the coin is tossed 2 more times. However, if the first toss produces tails, then the coin is tossed 3 more times.
(a) What is the probability of winning, given that the first toss is heads?
(b) What is the probability of winning, given that the first toss is tails?

B-14 What is the probability of picking 3 cards from a shuffled deck such that the first card is an ace, the second card is the king of the same suit, and the third card is the queen of the same suit?

B-15 What is the probability of picking 2 cards at random without replacement from a deck with an outcome of an ace and a king of the same suit, if either card can be picked first?

B-16 From a shuffled deck you draw one card.
(a) Find P(ace or spade).
(b) Given that it is a picture card, find P(10 or jack).

B-17 If you are in an automobile accident, find the probability that you are injured and do not collect insurance if P(injured) $=$ .4 and P(not collecting, given that you are injured) $=$ .2.

Tables Appendix C

C-1	Pascal's Triangle, $\binom{n}{S}$	
C-2	Binomial Probabilities	
C-3	Areas to the Left of z under the Normal Curve: Short Form	
C-4	Areas to the Left of z under the Normal Curve: Long Form	
C-5	Critical Values of t for a Two-Tail Test	
C-6	Critical Values of t for a One-Tail Test	
C-7	Critical Values of X^2 for a One-Tail Test	
C-8	Critical Values of X^2 for a Two-Tail Test	
C-9	Critical Values of r for a Two-Tail Test	
C-10	Critical Values of r for a One-Tail Test	
C-11	Critical Values of F for $\alpha = .05$ (for a One-Tail Test)	
C-12	Critical Values of F for $\alpha = .025$ (for a Two-Tail Test with $\alpha = .05$)	
C-13	Critical Values of F for $\alpha = .01$ (for a One-Tail Test)	
C-14	Critical Values of F for $\alpha = .005$ (for a Two-Tail Test with $\alpha = .01$)	
C-15	Critical Values of U for a One-Tail Test	
C-16	Critical Values of U for a Two-Tail Test	

Table C-1 Pascal's Triangle, $\binom{n}{S}$

n \ S	0	1	2	3	4	5	6	7	8	9	10	11	12	13	14	15	16	17	18	19	20
0	1																				
1	1	1																			
2	1	2	1																		
3	1	3	3	1																	
4	1	4	6	4	1																
5	1	5	10	10	5	1															
6	1	6	15	20	15	6	1														
7	1	7	21	35	35	21	7	1													
8	1	8	28	56	70	56	28	8	1												
9	1	9	36	84	126	126	84	36	9	1											
10	1	10	45	120	210	252	210	120	45	10	1										
11	1	11	55	165	330	462	462	330	165	55	11	1									
12	1	12	66	220	495	792	924	792	495	220	66	12	1								
13	1	13	78	286	715	1287	1716	1716	1287	715	286	78	13	1							
14	1	14	91	364	1001	2002	3003	3432	3003	2002	1001	364	91	14	1						
15	1	15	105	455	1365	3003	5005	6435	6435	5005	3003	1365	455	105	15	1					
16	1	16	120	560	1820	4368	8008	11,440	12,870	11,440	8008	4368	1820	560	120	16	1				
17	1	17	136	680	2380	6188	12,376	19,448	24,310	24,310	19,448	12,376	6188	2380	680	136	17	1			
18	1	18	153	816	3060	8568	18,564	31,824	43,758	48,620	43,758	31,824	18,564	8568	3060	816	153	18	1		
19	1	19	171	969	3876	11,628	27,132	50,388	75,582	92,378	92,378	75,582	50,388	27,132	11,628	3876	969	171	19	1	
20	1	20	190	1140	4845	15,504	38,760	77,520	125,970	167,960	184,756	167,960	125,970	77,520	38,760	15,504	4845	1140	190	20	1

Understanding statistics

Table C-2 Binomial Probabilities

$n=2$ S\\p	.05	.10	.20	.30	.40	.50	.60	.70	.80	.90	.95
0	.9025	.8100	.6400	.4900	.3600	.2500	.1600	.0900	.0400	.0100	.0025
1	.0950	.1800	.3200	.4200	.4800	.5000	.4800	.4200	.3200	.1800	.0950
2	.0025	.0100	.0400	.0900	.1600	.2500	.3600	.4900	.6400	.8100	.9025

$n=3$ S\\p	.05	.10	.20	.30	.40	.50	.60	.70	.80	.90	.95
0	.8574	.7290	.5120	.3430	.2160	.1250	.0640	.0270	.0080	.0010	.0001
1	.1354	.2430	.3840	.4410	.4320	.3750	.2880	.1890	.0960	.0270	.0071
2	.0071	.0270	.0960	.1890	.2880	.3750	.4320	.4410	.3840	.2430	.1354
3	.0001	.0010	.0080	.0270	.0640	.1250	.2160	.3430	.5120	.7290	.8574

$n=4$ S\\p	.05	.10	.20	.30	.40	.50	.60	.70	.80	.90	.95
0	.8145	.6561	.4096	.2401	.1296	.0625	.0256	.0081	.0016	.0001	.0000
1	.1715	.2916	.4096	.4116	.3456	.2500	.1536	.0756	.0256	.0036	.0005
2	.0135	.0486	.1536	.2646	.3456	.3750	.3456	.2646	.1536	.0486	.0135
3	.0005	.0036	.0256	.0756	.1536	.2500	.3456	.4116	.4096	.2916	.1715
4	.0000	.0001	.0016	.0081	.0256	.0625	.1296	.2401	.4096	.6561	.8145

$n=5$ S\\p	.05	.10	.20	.30	.40	.50	.60	.70	.80	.90	.95
0	.7738	.5905	.3277	.1681	.0778	.0313	.0102	.0024	.0003	.0000	.0000
1	.2036	.3281	.4096	.3602	.2592	.1563	.0768	.0284	.0064	.0005	.0000
2	.0214	.0729	.2048	.3087	.3456	.3125	.2304	.1323	.0512	.0081	.0011
3	.0011	.0081	.0512	.1323	.2304	.3125	.3456	.3087	.2048	.0729	.0214
4	.0000	.0005	.0064	.0284	.0768	.1563	.2592	.3602	.4096	.3281	.2036
5	.0000	.0000	.0003	.0024	.0102	.0313	.0778	.1681	.3277	.5905	.7738

$n=6$ S\\p	.05	.10	.20	.30	.40	.50	.60	.70	.80	.90	.95
0	.7351	.5314	.2621	.1176	.0467	.0156	.0041	.0007	.0001	.0000	.0000
1	.2321	.3543	.3932	.3025	.1866	.0938	.0369	.0102	.0015	.0001	.0000
2	.0305	.0984	.2458	.3241	.3110	.2344	.1382	.0595	.0154	.0012	.0001
3	.0021	.0146	.0819	.1852	.2765	.3125	.2765	.1852	.0819	.0146	.0021
4	.0001	.0012	.0154	.0595	.1382	.2344	.3110	.3241	.2458	.0984	.0305
5	.0000	.0001	.0015	.0102	.0369	.0938	.1866	.3025	.3932	.3543	.2321
6	.0000	.0000	.0001	.0007	.0041	.0156	.0467	.1176	.2621	.5314	.7351

$n=7$ S\\p	.05	.10	.20	.30	.40	.50	.60	.70	.80	.90	.95
0	.6983	.4783	.2097	.0824	.0280	.0078	.0016	.0002	.0000	.0000	.0000
1	.2573	.3720	.3670	.2471	.1306	.0547	.0172	.0036	.0004	.0000	.0000
2	.0406	.1240	.2753	.3177	.2613	.1641	.0774	.0250	.0043	.0002	.0000
3	.0036	.0230	.1147	.2269	.2903	.2734	.1935	.0972	.0287	.0026	.0002
4	.0002	.0026	.0287	.0972	.1935	.2734	.2903	.2269	.1147	.0230	.0036
5	.0000	.0002	.0043	.0250	.0774	.1641	.2613	.3177	.2753	.1240	.0406
6	.0000	.0000	.0004	.0036	.0172	.0547	.1306	.2471	.3670	.3720	.2573
7	.0000	.0000	.0000	.0002	.0016	.0078	.0280	.0824	.2097	.4783	.6983

Appendix C

Table C-2 (*continued*)

$n=8$ S \ p	.05	.10	.20	.30	.40	.50	.60	.70	.80	.90	.95
0	.6634	.4305	.1678	.0576	.0168	.0039	.0007	.0001	.0000	.0000	.0000
1	.2793	.3826	.3355	.1977	.0896	.0313	.0079	.0012	.0001	.0000	.0000
2	.0515	.1488	.2936	.2965	.2090	.1094	.0413	.0100	.0011	.0000	.0000
3	.0054	.0331	.1468	.2541	.2787	.2188	.1239	.0467	.0092	.0004	.0000
4	.0004	.0046	.0459	.1361	.2322	.2734	.2322	.1361	.0459	.0046	.0004
5	.0000	.0004	.0092	.0467	.1239	.2188	.2787	.2541	.1468	.0331	.0054
6	.0000	.0000	.0011	.0100	.0413	.1094	.2090	.2965	.2936	.1488	.0515
7	.0000	.0000	.0001	.0012	.0079	.0313	.0896	.1977	.3355	.3826	.2793
8	.0000	.0000	.0000	.0001	.0007	.0039	.0168	.0576	.1678	.4305	.6634

$n=9$ S \ p	.05	.10	.20	.30	.40	.50	.60	.70	.80	.90	.95
0	.6302	.3874	.1342	.0404	.0101	.0020	.0003	.0000	.0000	.0000	.0000
1	.2985	.3874	.3020	.1556	.0605	.0176	.0035	.0004	.0000	.0000	.0000
2	.0629	.1722	.3020	.2668	.1612	.0703	.0212	.0039	.0003	.0000	.0000
3	.0077	.0446	.1762	.2668	.2508	.1641	.0743	.0210	.0028	.0001	.0000
4	.0006	.0074	.0661	.1715	.2508	.2461	.1672	.0735	.0165	.0008	.0000
5	.0000	.0008	.0165	.0735	.1672	.2461	.2508	.1715	.0661	.0074	.0006
6	.0000	.0001	.0028	.0210	.0743	.1641	.2508	.2668	.1762	.0446	.0077
7	.0000	.0000	.0003	.0039	.0212	.0703	.1612	.2668	.3020	.1722	.0629
8	.0000	.0000	.0000	.0004	.0035	.0176	.0605	.1556	.3020	.3874	.2985
9	.0000	.0000	.0000	.0000	.0003	.0020	.0101	.0404	.1342	.3874	.6302

$n=10$ S \ p	.05	.10	.20	.30	.40	.50	.60	.70	.80	.90	.95
0	.5987	.3487	.1074	.0282	.0060	.0010	.0001	.0000	.0000	.0000	.0000
1	.3151	.3874	.2684	.1211	.0403	.0098	.0016	.0001	.0000	.0000	.0000
2	.0746	.1937	.3020	.2335	.1209	.0439	.0106	.0014	.0001	.0000	.0000
3	.0105	.0574	.2013	.2668	.2150	.1172	.0425	.0090	.0008	.0000	.0000
4	.0010	.0112	.0881	.2001	.2508	.2051	.1115	.0368	.0055	.0001	.0000
5	.0001	.0015	.0264	.1029	.2007	.2461	.2007	.1029	.0264	.0015	.0001
6	.0000	.0001	.0055	.0368	.1115	.2051	.2508	.2001	.0881	.0112	.0010
7	.0000	.0000	.0008	.0090	.0425	.1172	.2150	.2668	.2013	.0574	.0105
8	.0000	.0000	.0001	.0014	.0106	.0439	.1209	.2335	.3020	.1937	.0746
9	.0000	.0000	.0000	.0001	.0016	.0098	.0403	.1211	.2684	.3874	.3151
10	.0000	.0000	.0000	.0000	.0001	.0010	.0060	.0282	.1074	.3487	.5987

$n=11$ S \ p	.05	.10	.20	.30	.40	.50	.60	.70	.80	.90	.95
0	.5688	.3138	.0859	.0198	.0036	.0005	.0000	.0000	.0000	.0000	.0000
1	.3293	.3835	.2362	.0932	.0266	.0054	.0007	.0000	.0000	.0000	.0000
2	.0867	.2131	.2953	.1998	.0887	.0269	.0052	.0005	.0000	.0000	.0000
3	.0137	.0710	.2215	.2568	.1774	.0806	.0234	.0037	.0002	.0000	.0000
4	.0014	.0158	.1107	.2201	.2365	.1611	.0701	.0173	.0017	.0000	.0000
5	.0001	.0025	.0388	.1321	.2207	.2256	.1471	.0566	.0097	.0003	.0000
6	.0000	.0003	.0097	.0566	.1471	.2256	.2207	.1321	.0388	.0025	.0001
7	.0000	.0000	.0017	.0173	.0701	.1611	.2365	.2201	.1107	.0158	.0014
8	.0000	.0000	.0002	.0037	.0234	.0806	.1774	.2568	.2215	.0710	.0137
9	.0000	.0000	.0000	.0005	.0052	.0269	.0887	.1998	.2953	.2131	.0867
10	.0000	.0000	.0000	.0000	.0007	.0054	.0266	.0932	.2362	.3835	.3293
11	.0000	.0000	.0000	.0000	.0000	.0005	.0036	.0198	.0859	.3138	.5688

Understanding statistics

Table C-2 *(continued)*

$n=12$ S \ p	.05	.10	.20	.30	.40	.50	.60	.70	.80	.90	.95
0	.5404	.2824	.0687	.0138	.0022	.0002	.0000	.0000	.0000	.0000	.0000
1	.3413	.3766	.2062	.0712	.0174	.0029	.0003	.0000	.0000	.0000	.0000
2	.0988	.2301	.2835	.1678	.0639	.0161	.0025	.0002	.0000	.0000	.0000
3	.0173	.0852	.2362	.2397	.1419	.0537	.0125	.0015	.0001	.0000	.0000
4	.0021	.0213	.1329	.2311	.2128	.1208	.0420	.0078	.0005	.0000	.0000
5	.0002	.0038	.0532	.1585	.2270	.1934	.1009	.0291	.0033	.0000	.0000
6	.0000	.0005	.0155	.0792	.1766	.2256	.1766	.0792	.0155	.0005	.0000
7	.0000	.0000	.0033	.0291	.1009	.1934	.2270	.1585	.0532	.0038	.0002
8	.0000	.0000	.0005	.0078	.0420	.1208	.2128	.2311	.1329	.0213	.0021
9	.0000	.0000	.0001	.0015	.0125	.0537	.1419	.2397	.2362	.0852	.0173
10	.0000	.0000	.0000	.0002	.0025	.0161	.0639	.1678	.2835	.2301	.0988
11	.0000	.0000	.0000	.0000	.0003	.0029	.0174	.0712	.2062	.3766	.3413
12	.0000	.0000	.0000	.0000	.0000	.0002	.0022	.0138	.0687	.2824	.5404

$n=13$ S \ p	.05	.10	.20	.30	.40	.50	.60	.70	.80	.90	.95
0	.5133	.2542	.0550	.0097	.0013	.0001	.0000	.0000	.0000	.0000	.0000
1	.3512	.3672	.1787	.0540	.0113	.0016	.0001	.0000	.0000	.0000	.0000
2	.1109	.2448	.2680	.1388	.0453	.0095	.0012	.0001	.0000	.0000	.0000
3	.0214	.0997	.2457	.2181	.1107	.0349	.0065	.0006	.0000	.0000	.0000
4	.0028	.0277	.1535	.2337	.1845	.0873	.0243	.0034	.0001	.0000	.0000
5	.0003	.0055	.0691	.1803	.2214	.1571	.0656	.0142	.0011	.0000	.0000
6	.0000	.0008	.0230	.1030	.1968	.2095	.1312	.0442	.0058	.0001	.0000
7	.0000	.0001	.0058	.0442	.1312	.2095	.1968	.1030	.0230	.0008	.0000
8	.0000	.0000	.0011	.0142	.0656	.1571	.2214	.1803	.0691	.0055	.0003
9	.0000	.0000	.0001	.0034	.0243	.0873	.1845	.2337	.1535	.0277	.0028
10	.0000	.0000	.0000	.0006	.0065	.0349	.1107	.2181	.2457	.0997	.0214
11	.0000	.0000	.0000	.0001	.0012	.0095	.0453	.1388	.2680	.2448	.1109
12	.0000	.0000	.0000	.0000	.0001	.0016	.0113	.0540	.1787	.3672	.3512
13	.0000	.0000	.0000	.0000	.0000	.0001	.0013	.0097	.0550	.2542	.5133

$n=14$ S \ p	.05	.10	.20	.30	.40	.50	.60	.70	.80	.90	.95
0	.4877	.2288	.0440	.0068	.0008	.0001	.0000	.0000	.0000	.0000	.0000
1	.3593	.3559	.1539	.0407	.0073	.0009	.0001	.0000	.0000	.0000	.0000
2	.1229	.2570	.2501	.1134	.0317	.0056	.0005	.0000	.0000	.0000	.0000
3	.0259	.1142	.2501	.1943	.0845	.0222	.0033	.0002	.0000	.0000	.0000
4	.0037	.0349	.1720	.2290	.1549	.0611	.0136	.0014	.0000	.0000	.0000
5	.0004	.0078	.0860	.1963	.2066	.1222	.0408	.0066	.0003	.0000	.0000
6	.0000	.0013	.0322	.1262	.2066	.1833	.0918	.0232	.0020	.0000	.0000
7	.0000	.0002	.0092	.0618	.1574	.2095	.1574	.0618	.0092	.0002	.0000
8	.0000	.0000	.0020	.0232	.0918	.1833	.2066	.1262	.0322	.0013	.0000
9	.0000	.0000	.0003	.0066	.0408	.1222	.2066	.1963	.0860	.0078	.0004
10	.0000	.0000	.0000	.0014	.0136	.0611	.1549	.2290	.1720	.0349	.0037
11	.0000	.0000	.0000	.0002	.0033	.0222	.0845	.1943	.2501	.1142	.0259
12	.0000	.0000	.0000	.0000	.0005	.0056	.0317	.1134	.2501	.2570	.1229
13	.0000	.0000	.0000	.0000	.0001	.0009	.0073	.0407	.1539	.3559	.3593
14	.0000	.0000	.0000	.0000	.0000	.0001	.0008	.0068	.0440	.2288	.4877

Appendix C

Table C-2 (continued)

$n = 15$

S\p	.05	.10	.20	.30	.40	.50	.60	.70	.80	.90	.95
0	.4633	.2059	.0352	.0047	.0005	.0000	.0000	.0000	.0000	.0000	.0000
1	.3658	.3432	.1319	.0305	.0047	.0005	.0000	.0000	.0000	.0000	.0000
2	.1348	.2669	.2309	.0916	.0219	.0032	.0003	.0000	.0000	.0000	.0000
3	.0307	.1285	.2501	.1700	.0634	.0139	.0016	.0001	.0000	.0000	.0000
4	.0049	.0428	.1876	.2186	.1268	.0417	.0074	.0006	.0000	.0000	.0000
5	.0006	.0105	.1032	.2061	.1859	.0916	.0245	.0030	.0001	.0000	.0000
6	.0000	.0019	.0430	.1472	.2066	.1527	.0612	.0116	.0007	.0000	.0000
7	.0000	.0003	.0138	.0811	.1771	.1964	.1181	.0348	.0035	.0000	.0000
8	.0000	.0000	.0035	.0348	.1181	.1964	.1771	.0811	.0138	.0003	.0000
9	.0000	.0000	.0007	.0116	.0612	.1527	.2066	.1472	.0430	.0019	.0000
10	.0000	.0000	.0001	.0030	.0245	.0916	.1859	.2061	.1032	.0105	.0006
11	.0000	.0000	.0000	.0006	.0074	.0417	.1268	.2186	.1876	.0428	.0049
12	.0000	.0000	.0000	.0001	.0016	.0139	.0634	.1700	.2501	.1285	.0307
13	.0000	.0000	.0000	.0000	.0003	.0032	.0219	.0916	.2309	.2669	.1348
14	.0000	.0000	.0000	.0000	.0000	.0005	.0047	.0305	.1319	.3432	.3658
15	.0000	.0000	.0000	.0000	.0000	.0000	.0005	.0047	.0352	.2059	.4633

$n = 16$

S\p	.05	.10	.20	.30	.40	.50	.60	.70	.80	.90	.95
0	.4401	.1853	.0281	.0033	.0003	.0000	.0000	.0000	.0000	.0000	.0000
1	.3706	.3294	.1126	.0228	.0030	.0002	.0000	.0000	.0000	.0000	.0000
2	.1463	.2745	.2111	.0732	.0150	.0018	.0001	.0000	.0000	.0000	.0000
3	.0359	.1423	.2463	.1465	.0468	.0085	.0008	.0000	.0000	.0000	.0000
4	.0061	.0514	.2001	.2040	.1014	.0278	.0040	.0002	.0000	.0000	.0000
5	.0008	.0137	.1201	.2099	.1623	.0667	.0142	.0013	.0000	.0000	.0000
6	.0001	.0028	.0550	.1649	.1983	.1222	.0392	.0056	.0002	.0000	.0000
7	.0000	.0004	.0197	.1010	.1889	.1746	.0840	.0185	.0012	.0000	.0000
8	.0000	.0001	.0055	.0487	.1417	.1964	.1417	.0487	.0055	.0001	.0000
9	.0000	.0000	.0012	.0185	.0840	.1746	.1889	.1010	.0197	.0004	.0000
10	.0000	.0000	.0002	.0056	.0392	.1222	.1983	.1649	.0550	.0028	.0001
11	.0000	.0000	.0000	.0013	.0142	.0667	.1623	.2099	.1201	.0137	.0008
12	.0000	.0000	.0000	.0002	.0040	.0278	.1014	.2040	.2001	.0514	.0061
13	.0000	.0000	.0000	.0000	.0008	.0085	.0468	.1465	.2463	.1423	.0359
14	.0000	.0000	.0000	.0000	.0001	.0018	.0150	.0732	.2111	.2745	.1463
15	.0000	.0000	.0000	.0000	.0000	.0002	.0030	.0228	.1126	.3294	.3706
16	.0000	.0000	.0000	.0000	.0000	.0000	.0003	.0033	.0281	.1853	.4401

Understanding statistics

Table C-2 (*continued*)

n=17 S\p	.05	.10	.20	.30	.40	.50	.60	.70	.80	.90	.95
0	.4181	.1668	.0225	.0023	.0002	.0000	.0000	.0000	.0000	.0000	.0000
1	.3741	.3150	.0957	.0169	.0019	.0001	.0000	.0000	.0000	.0000	.0000
2	.1575	.2800	.1914	.0581	.0102	.0010	.0001	.0000	.0000	.0000	.0000
3	.0415	.1556	.2393	.1245	.0341	.0052	.0004	.0000	.0000	.0000	.0000
4	.0076	.0605	.2093	.1868	.0796	.0182	.0021	.0001	.0000	.0000	.0000
5	.0010	.0175	.1361	.2081	.1379	.0472	.0081	.0006	.0000	.0000	.0000
6	.0001	.0039	.0680	.1784	.1839	.0944	.0242	.0026	.0001	.0000	.0000
7	.0000	.0007	.0267	.1201	.1927	.1484	.0571	.0095	.0004	.0000	.0000
8	.0000	.0001	.0084	.0644	.1606	.1855	.1070	.0276	.0021	.0000	.0000
9	.0000	.0000	.0021	.0276	.1070	.1855	.1606	.0644	.0084	.0001	.0000
10	.0000	.0000	.0004	.0095	.0571	.1484	.1927	.1201	.0267	.0007	.0000
11	.0000	.0000	.0001	.0026	.0242	.0944	.1839	.1784	.0680	.0039	.0001
12	.0000	.0000	.0000	.0006	.0081	.0472	.1379	.2081	.1361	.0175	.0010
13	.0000	.0000	.0000	.0001	.0021	.0182	.0796	.1868	.2093	.0605	.0076
14	.0000	.0000	.0000	.0000	.0004	.0052	.0341	.1245	.2393	.1556	.0415
15	.0000	.0000	.0000	.0000	.0001	.0010	.0102	.0581	.1914	.2800	.1575
16	.0000	.0000	.0000	.0000	.0000	.0001	.0019	.0169	.0957	.3150	.3741
17	.0000	.0000	.0000	.0000	.0000	.0000	.0002	.0023	.0225	.1668	.4181

n=18 S\p	.05	.10	.20	.30	.40	.50	.60	.70	.80	.90	.95
0	.3972	.1501	.0180	.0016	.0001	.0000	.0000	.0000	.0000	.0000	.0000
1	.3763	.3002	.0811	.0126	.0012	.0001	.0000	.0000	.0000	.0000	.0000
2	.1683	.2835	.1723	.0458	.0069	.0006	.0000	.0000	.0000	.0000	.0000
3	.0473	.1680	.2297	.1046	.0246	.0031	.0002	.0000	.0000	.0000	.0000
4	.0093	.0700	.2153	.1681	.0614	.0117	.0011	.0000	.0000	.0000	.0000
5	.0014	.0218	.1507	.2017	.1146	.0327	.0045	.0002	.0000	.0000	.0000
6	.0002	.0052	.0816	.1873	.1655	.0708	.0145	.0012	.0000	.0000	.0000
7	.0000	.0010	.0350	.1376	.1892	.1214	.0374	.0046	.0001	.0000	.0000
8	.0000	.0002	.0120	.0811	.1734	.1669	.0771	.0149	.0008	.0000	.0000
9	.0000	.0000	.0033	.0386	.1284	.1855	.1284	.0386	.0033	.0000	.0000
10	.0000	.0000	.0008	.0149	.0771	.1669	.1734	.0811	.0120	.0002	.0000
11	.0000	.0000	.0001	.0046	.0374	.1214	.1892	.1376	.0350	.0010	.0000
12	.0000	.0000	.0000	.0012	.0145	.0708	.1655	.1873	.0816	.0052	.0002
13	.0000	.0000	.0000	.0002	.0045	.0327	.1146	.2017	.1507	.0218	.0014
14	.0000	.0000	.0000	.0000	.0011	.0117	.0614	.1681	.2153	.0700	.0093
15	.0000	.0000	.0000	.0000	.0002	.0031	.0246	.1046	.2297	.1680	.0473
16	.0000	.0000	.0000	.0000	.0000	.0006	.0069	.0458	.1723	.2835	.1683
17	.0000	.0000	.0000	.0000	.0000	.0001	.0012	.0126	.0811	.3002	.3763
18	.0000	.0000	.0000	.0000	.0000	.0000	.0001	.0016	.0180	.1501	.3972

Appendix C

Table C-2 (continued)

$n=19$ $s \backslash p$	.05	.10	.20	.30	.40	.50	.60	.70	.80	.90	.95
0	.3774	.1351	.0144	.0011	.0001	.0000	.0000	.0000	.0000	.0000	.0000
1	.3774	.2852	.0685	.0093	.0008	.0000	.0000	.0000	.0000	.0000	.0000
2	.1787	.2852	.1540	.0358	.0046	.0003	.0000	.0000	.0000	.0000	.0000
3	.0533	.1796	.2182	.0869	.0175	.0018	.0001	.0000	.0000	.0000	.0000
4	.0112	.0798	.2182	.1491	.0467	.0074	.0005	.0000	.0000	.0000	.0000
5	.0018	.0266	.1636	.1916	.0933	.0222	.0024	.0001	.0000	.0000	.0000
6	.0002	.0069	.0955	.1916	.1451	.0518	.0085	.0005	.0000	.0000	.0000
7	.0000	.0014	.0443	.1525	.1797	.0961	.0237	.0022	.0000	.0000	.0000
8	.0000	.0002	.0166	.0981	.1797	.1442	.0532	.0077	.0003	.0000	.0000
9	.0000	.0000	.0051	.0514	.1464	.1762	.0976	.0220	.0013	.0000	.0000
10	.0000	.0000	.0013	.0220	.0976	.1762	.1464	.0514	.0051	.0000	.0000
11	.0000	.0000	.0003	.0077	.0532	.1442	.1797	.0981	.0166	.0002	.0000
12	.0000	.0000	.0000	.0022	.0237	.0961	.1797	.1525	.0443	.0014	.0000
13	.0000	.0000	.0000	.0005	.0085	.0518	.1451	.1916	.0955	.0069	.0002
14	.0000	.0000	.0000	.0001	.0024	.0222	.0933	.1916	.1636	.0266	.0018
15	.0000	.0000	.0000	.0000	.0005	.0074	.0467	.1491	.2182	.0798	.0112
16	.0000	.0000	.0000	.0000	.0001	.0018	.0175	.0869	.2182	.1796	.0533
17	.0000	.0000	.0000	.0000	.0000	.0003	.0046	.0358	.1540	.2852	.1787
18	.0000	.0000	.0000	.0000	.0000	.0000	.0008	.0093	.0685	.2852	.3774
19	.0000	.0000	.0000	.0000	.0000	.0000	.0001	.0011	.0144	.1351	.3774

$n=20$ $s \backslash p$	.05	.10	.20	.30	.40	.50	.60	.70	.80	.90	.95
0	.3585	.1216	.0115	.0008	.0000	.0000	.0000	.0000	.0000	.0000	.0000
1	.3774	.2702	.0576	.0068	.0005	.0000	.0000	.0000	.0000	.0000	.0000
2	.1887	.2852	.1369	.0278	.0031	.0002	.0000	.0000	.0000	.0000	.0000
3	.0596	.1901	.2054	.0716	.0123	.0011	.0000	.0000	.0000	.0000	.0000
4	.0133	.0898	.2182	.1304	.0350	.0046	.0003	.0000	.0000	.0000	.0000
5	.0022	.0319	.1746	.1789	.0746	.0148	.0013	.0000	.0000	.0000	.0000
6	.0003	.0089	.1091	.1916	.1244	.0370	.0049	.0002	.0000	.0000	.0000
7	.0000	.0020	.0545	.1643	.1659	.0739	.0146	.0010	.0000	.0000	.0000
8	.0000	.0004	.0222	.1144	.1797	.1201	.0355	.0039	.0001	.0000	.0000
9	.0000	.0001	.0074	.0654	.1597	.1602	.0710	.0120	.0005	.0000	.0000
10	.0000	.0000	.0020	.0308	.1171	.1762	.1171	.0308	.0020	.0000	.0000
11	.0000	.0000	.0005	.0120	.0710	.1602	.1597	.0654	.0074	.0001	.0000
12	.0000	.0000	.0001	.0039	.0355	.1201	.1797	.1144	.0222	.0004	.0000
13	.0000	.0000	.0000	.0010	.0146	.0739	.1659	.1643	.0545	.0020	.0000
14	.0000	.0000	.0000	.0002	.0049	.0370	.1244	.1916	.1091	.0089	.0003
15	.0000	.0000	.0000	.0000	.0013	.0148	.0746	.1789	.1746	.0319	.0022
16	.0000	.0000	.0000	.0000	.0003	.0046	.0350	.1304	.2182	.0898	.0133
17	.0000	.0000	.0000	.0000	.0000	.0011	.0123	.0716	.2054	.1901	.0596
18	.0000	.0000	.0000	.0000	.0000	.0002	.0031	.0278	.1369	.2852	.1887
19	.0000	.0000	.0000	.0000	.0000	.0000	.0005	.0068	.0576	.2702	.3774
20	.0000	.0000	.0000	.0000	.0000	.0000	.0000	.0008	.0115	.1216	.3585

Table C-3 Areas to the Left of z Under the Normal Curve: Short Form

z score	proportion of area to the left of z
−4	.00003
−3	.0013
−2.58	.0049
−2.33	.0099
−2	.0228
−1.96	.0250
−1.65	.0495
−1	.1587
0	.5000
1	.8413
1.65	.9505
1.96	.9750
2	.9772
2.33	.9901
2.58	.9951
3	.9987
4	.99997

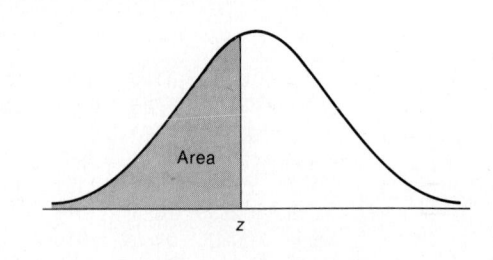

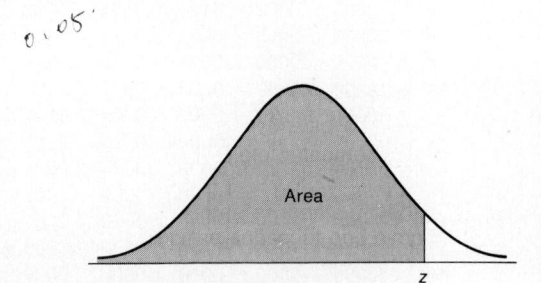

A QUICK REFERENCE TO SOME IMPORTANT z SCORES

Values of z are given without signs. You must determine whether the critical values are positive, negative, or both from the alternative hypothesis.

	$\alpha = .05$	$\alpha = .01$
one-tail	1.65	2.33
two-tail	1.96	2.58

Appendix C

Table C-4 Areas to the Left of z Under the Normal Curve: Long Form

z	area	z	area	z	area
−4	.00003	−2.74	.0031	−2.29	.0110
−3.9	.00005	−2.73	.0032	−2.28	.0113
−3.8	.0001	−2.72	.0033	−2.27	.0116
−3.7	.0001	−2.71	.0034	−2.26	.0119
−3.6	.0002	−2.70	.0035	−2.25	.0122
−3.5	.0002	−2.69	.0036	−2.24	.0125
−3.4	.0003	−2.68	.0037	−2.23	.0129
−3.3	.0005	−2.67	.0038	−2.22	.0132
−3.2	.0007	−2.66	.0039	−2.21	.0136
−3.1	.0010	−2.65	.0040	−2.20	.0139
−3.09	.0010	−2.64	.0041	−2.19	.0143
−3.08	.0010	−2.63	.0043	−2.18	.0146
−3.07	.0011	−2.62	.0044	−2.17	.0150
−3.06	.0011	−2.61	.0045	−2.16	.0154
−3.05	.0011	−2.60	.0047	−2.15	.0158
−3.04	.0012	−2.59	.0048	−2.14	.0162
−3.03	.0012	−2.58	.0049	−2.13	.0166
−3.02	.0013	−2.57	.0051	−2.12	.0170
−3.01	.0013	−2.56	.0052	−2.11	.0174
−3.00	.0013	−2.55	.0054	−2.10	.0179
−2.99	.0014	−2.54	.0055	−2.09	.0183
−2.98	.0014	−2.53	.0057	−2.08	.0188
−2.97	.0015	−2.52	.0059	−2.07	.0192
−2.96	.0015	−2.51	.0060	−2.06	.0197
−2.95	.0016	−2.50	.0062	−2.05	.0202
−2.94	.0016	−2.49	.0064	−2.04	.0207
−2.93	.0017	−2.48	.0066	−2.03	.0212
−2.92	.0017	−2.47	.0068	−2.02	.0217
−2.91	.0018	−2.46	.0069	−2.01	.0222
−2.90	.0019	−2.45	.0071	−2.00	.0228
−2.89	.0019	−2.44	.0073	−1.99	.0233
−2.88	.0020	−2.43	.0075	−1.98	.0239
−2.87	.0021	−2.42	.0078	−1.97	.0244
−2.86	.0021	−2.41	.0080	−1.96	.0250
−2.85	.0022	−2.40	.0082	−1.95	.0256
−2.84	.0023	−2.39	.0084	−1.94	.0262
−2.83	.0023	−2.38	.0087	−1.93	.0268
−2.82	.0024	−2.37	.0089	−1.92	.0274
−2.81	.0025	−2.36	.0091	−1.91	.0281
−2.80	.0026	−2.35	.0094	−1.90	.0287
−2.79	.0026	−2.34	.0096	−1.89	.0294
−2.78	.0027	−2.33	.0099	−1.88	.0301
−2.77	.0028	−2.32	.0102	−1.87	.0307
−2.76	.0029	−2.31	.0104	−1.86	.0314
−2.75	.0030	−2.30	.0107	−1.85	.0322

Table C-4 (Continued)

z	area	z	area	z	area
−1.84	.0329	−1.39	.0823	−.94	.1736
−1.83	.0336	−1.38	.0838	−.93	.1762
−1.82	.0344	−1.37	.0853	−.92	.1788
−1.81	.0352	−1.36	.0869	−.91	.1814
−1.80	.0359	−1.35	.0885	−.90	.1841
−1.79	.0367	−1.34	.0901	−.89	.1867
−1.78	.0375	−1.33	.0918	−.88	.1894
−1.77	.0384	−1.32	.0934	−.87	.1922
−1.76	.0392	−1.31	.0951	−.86	.1949
−1.75	.0401	−1.30	.0968	−.85	.1977
−1.74	.0409	−1.29	.0985	−.84	.2005
−1.73	.0418	−1.28	.1003	−.83	.2033
−1.72	.0427	−1.27	.1020	−.82	.2061
−1.71	.0436	−1.26	.1038	−.81	.2090
−1.70	.0446	−1.25	.1056	−.80	.2119
−1.69	.0455	−1.24	.1075	−.79	.2148
−1.68	.0465	−1.23	.1093	−.78	.2177
−1.67	.0475	−1.22	.1112	−.77	.2206
−1.66	.0485	−1.21	.1131	−.76	.2236
−1.65	.0495	−1.20	.1151	−.75	.2266
−1.64	.0505	−1.19	.1170	−.74	.2296
−1.63	.0516	−1.18	.1190	−.73	.2327
−1.62	.0526	−1.17	.1210	−.72	.2358
−1.61	.0537	−1.16	.1230	−.71	.2389
−1.60	.0548	−1.15	.1251	−.70	.2420
−1.59	.0559	−1.14	.1271	−.69	.2451
−1.58	.0571	−1.13	.1292	−.68	.2483
−1.57	.0582	−1.12	.1314	−.67	.2514
−1.56	.0594	−1.11	.1335	−.66	.2546
−1.55	.0606	−1.10	.1357	−.65	.2578
−1.54	.0618	−1.09	.1379	−.64	.2611
−1.53	.0630	−1.08	.1401	−.63	.2643
−1.52	.0643	−1.07	.1423	−.62	.2676
−1.51	.0655	−1.06	.1446	−.61	.2709
−1.50	.0668	−1.05	.1469	−.60	.2743
−1.49	.0681	−1.04	.1492	−.59	.2776
−1.48	.0694	−1.03	.1515	−.58	.2810
−1.47	.0708	−1.02	.1539	−.57	.2843
−1.46	.0722	−1.01	.1562	−.56	.2877
−1.45	.0735	−1.00	.1587	−.55	.2912
−1.44	.0749	−.99	.1611	−.54	.2946
−1.43	.0764	−.98	.1635	−.53	.2981
−1.42	.0778	−.97	.1660	−.52	.3015
−1.41	.0793	−.96	.1685	−.51	.3050
−1.40	.0808	−.95	.1711	−.50	.3085

Appendix C

Table C-4 (Continued)

z	area	z	area	z	area
−.49	.3121	−.04	.4840	.41	.6591
−.48	.3156	−.03	.4880	.42	.6628
−.47	.3192	−.02	.4920	.43	.6664
−.46	.3228	−.01	.4960	.44	.6700
−.45	.3264	.00	.5000	.45	.6736
−.44	.3300	.01	.5040	.46	.6772
−.43	.3336	.02	.5080	.47	.6808
−.42	.3372	.03	.5120	.48	.6844
−.41	.3409	.04	.5160	.49	.6849
−.40	.3446	.05	.5199	.50	.6915
−.39	.3483	.06	.5239	.51	.6950
−.38	.3520	.07	.5279	.52	.6985
−.37	.3557	.08	.5319	.53	.7019
−.36	.3594	.09	.5359	.54	.7054
−.35	.3632	.10	.5398	.55	.7088
−.34	.3669	.11	.5438	.56	.7123
−.33	.3707	.12	.5478	.57	.7157
−.32	.3745	.13	.5517	.58	.7190
−.31	.3783	.14	.5557	.59	.7224
−.30	.3821	.15	.5596	.60	.7257
−.29	.3859	.16	.5636	.61	.7291
−.28	.3897	.17	.5675	.62	.7324
−.27	.3936	.18	.5714	.63	.7357
−.26	.3974	.19	.5753	.64	.7389
−.25	.4013	.20	.5793	.65	.7422
−.24	.4052	.21	.5832	.66	.7454
−.23	.4090	.22	.5871	.67	.7486
−.22	.4129	.23	.5910	.68	.7517
−.21	.4168	.24	.5948	.69	.7549
−.20	.4207	.25	.5987	.70	.7580
−.19	.4247	.26	.6026	.71	.7611
−.18	.4286	.27	.6064	.72	.7642
−.17	.4325	.28	.6103	.73	.7673
−.16	.4364	.29	.6141	.74	.7704
−.15	.4404	.30	.6179	.75	.7734
−.14	.4443	.31	.6217	.76	.7764
−.13	.4483	.32	.6255	.77	.7794
−.12	.4522	.33	.6293	.78	.7283
−.11	.4562	.34	.6331	.79	.7852
−.10	.4602	.35	.6368	.80	.7881
−.09	.4641	.36	.6406	.81	.7910
−.08	.4681	.37	.6443	.82	.7939
−.07	.4721	.38	.6480	.83	.7967
−.06	.4761	.39	.6517	.84	.7995
−.05	.4801	.40	.6554	.85	.8023

Table C-4 (*Continued*)

z	area	z	area	z	area
.86	.8051	1.31	.9049	1.76	.9608
.87	.8078	1.32	.9066	1.77	.9616
.88	.8106	1.33	.9082	1.78	.9625
.89	.8133	1.34	.9099	1.79	.9633
.90	.8159	1.35	.9115	1.80	.9641
.91	.8186	1.36	.9131	1.81	.9649
.92	.8212	1.37	.9147	1.82	.9656
.93	.8238	1.38	.9162	1.83	.9664
.94	.8264	1.39	.9177	1.84	.9671
.95	.8289	1.40	.9192	1.85	.9678
.96	.8315	1.41	.9207	1.86	.9686
.97	.8340	1.42	.9222	1.87	.9693
.98	.8365	1.43	.9236	1.88	.9699
.99	.8389	1.44	.9251	1.89	.9706
1.00	.8413	1.45	.9265	1.90	.9713
1.01	.8438	1.46	.9278	1.91	.9719
1.02	.8461	1.47	.9292	1.92	.9726
1.03	.8485	1.48	.9306	1.93	.9732
1.04	.8508	1.49	.9319	1.94	.9738
1.05	.8531	1.50	.9332	1.95	.9744
1.06	.8554	1.51	.9345	1.96	.9750
1.07	.8577	1.52	.9357	1.97	.9756
1.08	.8599	1.53	.9370	1.98	.9761
1.09	.8621	1.54	.9382	1.99	.9767
1.10	.8643	1.55	.9394	2.00	.9772
1.11	.8665	1.56	.9406	2.01	.9778
1.12	.8686	1.57	.9418	2.02	.9783
1.13	.8708	1.58	.9429	2.03	.9788
1.14	.8729	1.59	.9441	2.04	.9793
1.15	.8749	1.60	.9452	2.05	.9798
1.16	.8770	1.61	.9463	2.06	.9803
1.17	.8790	1.62	.9474	2.07	.9808
1.18	.8810	1.63	.9484	2.08	.9812
1.19	.8830	1.64	.9495	2.09	.9817
1.20	.8849	1.65	.9505	2.10	.9821
1.21	.8869	1.66	.9515	2.11	.9826
1.22	.8888	1.67	.9525	2.12	.9830
1.23	.8907	1.68	.9535	2.13	.9834
1.24	.8925	1.69	.9545	2.14	.9838
1.25	.8944	1.70	.9554	2.15	.9842
1.26	.8962	1.71	.9564	2.16	.9846
1.27	.8980	1.72	.9573	2.17	.9850
1.28	.8997	1.73	.9582	2.18	.9854
1.29	.9015	1.74	.9591	2.19	.9857
1.30	.9032	1.75	.9599	2.20	.9861

Appendix C

Table C-4 (*Continued*)

z	area	z	area	z	area
2.21	.9864	2.66	.9961	3.2	.9993
2.22	.9868	2.67	.9962	3.3	.9995
2.23	.9871	2.68	.9963	3.4	.9997
2.24	.9875	2.69	.9964	3.5	.9998
2.25	.9878	2.70	.9965	3.6	.9998
2.26	.9881	2.71	.9966	3.7	.9999
2.27	.9884	2.72	.9967	3.8	.9999
2.28	.9887	2.73	.9968	3.9	.99995
2.29	.9890	2.74	.9969	4.0	.99997
2.30	.9893	2.75	.9970		
2.31	.9896	2.76	.9971		
2.32	.9898	2.77	.9972		
2.33	.9901	2.78	.9973		
2.34	.9904	2.79	.9974		
2.35	.9906	2.80	.9974		
2.36	.9909	2.81	.9975		
2.37	.9911	2.82	.9976		
2.38	.9913	2.83	.9977		
2.39	.9916	2.84	.9977		
2.40	.9918	2.85	.9978		
2.41	.9920	2.86	.9979		
2.42	.9922	2.87	.9979		
2.43	.9925	2.88	.9980		
2.44	.9927	2.89	.9981		
2.45	.9929	2.90	.9981		
2.46	.9931	2.91	.9982		
2.47	.9932	2.92	.9982		
2.48	.9934	2.93	.9983		
2.49	.9936	2.94	.9984		
2.50	.9938	2.95	.9984		
2.51	.9940	2.96	.9985		
2.52	.9941	2.97	.9985		
2.53	.9943	2.98	.9986		
2.54	.9945	2.99	.9986		
2.55	.9946	3.00	.9987		
2.56	.9948	3.01	.9987		
2.57	.9949	3.02	.9987		
2.58	.9951	3.03	.9988		
2.59	.9952	3.04	.9988		
2.60	.9953	3.05	.9989		
2.61	.9955	3.06	.9989		
2.62	.9956	3.07	.9989		
2.63	.9957	3.08	.9990		
2.64	.9959	3.09	.9990		
2.65	.9960	3.10	.9990		

Understanding statistics

Table C-5
Critical Values of t for a Two-Tail Test

(Values of t_c in this table are given without signs. All values are both positive and negative, that is, $t_c = \pm 12.71$.)

degrees of freedom σ	t_c for $\alpha = .05$	t_c for $\alpha = .01$
1	12.71	63.66
2	4.30	9.92
3	3.18	5.84
4	2.78	4.60
5	2.57	4.03
6	2.45	3.71
7	2.36	3.50
8	2.31	3.36
9	2.26	3.25
10	2.23	3.17
11	2.20	3.11
12	2.18	3.06
13	2.16	3.01
14	2.14	2.98
15	2.13	2.95
16	2.12	2.92
17	2.11	2.90
18	2.10	2.88
19	2.09	2.86
20	2.09	2.84
21	2.08	2.83
22	2.07	2.82
23	2.07	2.81
24	2.06	2.80
25	2.06	2.79
26	2.06	2.78
27	2.05	2.77
28	2.05	2.76
29	2.04	2.76
30	2.04	2.75
40	2.02	2.70
50	2.01	2.68
60	2.00	2.66
80	1.99	2.64
100	1.98	2.63
120	1.98	2.62
200	1.97	2.60
500	1.96	2.59
infinity	1.96	2.58

Table C-6
Critical Values of t for a One-Tail Test

(Values of t_c are given without signs. You must determine whether t_c is positive or negative from the alternative hypothesis.)

degrees of freedom σ	t_c for $\alpha = .05$	t_c for $\alpha = .01$
1	6.31	31.82
2	2.92	6.96
3	2.35	4.54
4	2.13	3.75
5	2.02	3.36
6	1.94	3.14
7	1.90	3.00
8	1.86	2.90
9	1.83	2.82
10	1.81	2.76
11	1.80	2.72
12	1.78	2.68
13	1.77	2.65
14	1.76	2.62
15	1.75	2.60
16	1.75	2.58
17	1.74	2.57
18	1.73	2.55
19	1.73	2.54
20	1.72	2.53
21	1.72	2.52
22	1.72	2.51
23	1.71	2.50
24	1.71	2.49
25	1.71	2.48
26	1.71	2.48
27	1.70	2.47
28	1.70	2.47
29	1.70	2.46
30	1.70	2.46
40	1.68	2.42
50	1.68	2.40
60	1.67	2.40
80	1.66	2.37
100	1.66	2.36
120	1.66	2.36
200	1.65	2.34
500	1.65	2.33
infinity	1.65	2.33

Appendix C

Table C-7   Critical Values of X^2 for a One-Tail Test

(For contingency table tests use one tail on the right.)

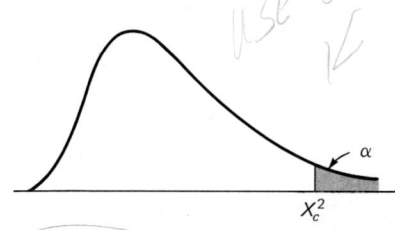

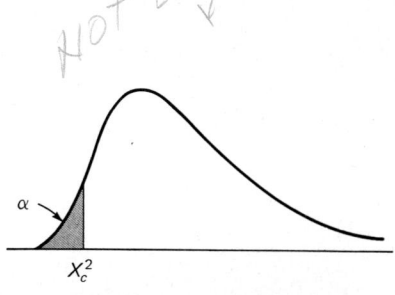

	one tail on the right		one tail on the left	
degrees of freedom	X_c^2 for $\alpha = .05$	X_c^2 for $\alpha = .01$	X_c^2 for $\alpha = .05$	X_c^2 for $\alpha = .01$
1	3.84	6.63	.0039	.00016
2	5.99	9.21	.1026	.0201
3	7.81	11.34	.352	.115
4	9.49	13.28	.711	.297
5	11.07	15.09	1.15	.554
6	12.59	16.81	1.64	.872
7	14.07	18.48	2.17	1.24
8	15.51	20.09	2.73	1.65
9	16.92	21.67	3.33	2.09
10	18.31	23.21	3.94	2.56
11	19.68	24.73	4.57	3.05
12	21.03	26.22	5.23	3.57
13	22.36	27.69	5.89	4.11
14	23.68	29.14	6.57	4.66
15	25.00	30.58	7.26	5.23
16	26.30	32.00	7.96	5.81
18	28.87	34.81	9.39	7.01
20	31.41	37.57	10.85	8.26
24	36.42	42.98	13.85	10.86
30	43.77	50.89	18.49	14.95
40	55.76	63.69	26.51	22.16
60	79.08	88.38	43.19	37.48
120	146.57	158.95	95.70	86.92

For critical values of X^2 whose degrees of freedom exceed 30 we can approximate X_c^2 by the formula

$$X_c^2 = \frac{1}{2}\left(z_c + \sqrt{2d - 1}\right)^2$$

where d is the degrees of freedom and z_c is either ± 1.65 or ± 2.33.

Table C-8

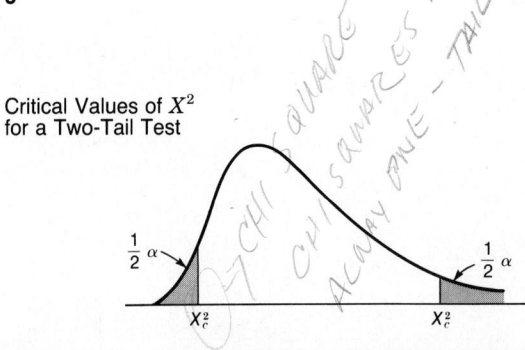

Critical Values of X^2 for a Two-Tail Test

degrees of freedom	X_c^2 for $\alpha = .05$		X_c^2 for $\alpha = .01$	
1	.00098 and	5.02	.000039 and	7.88
2	.0506	7.38	.0100	10.60
3	.216	9.35	.0717	12.84
4	.484	11.14	.207	14.86
5	.831	12.83	.412	16.75
6	1.24	14.45	.676	18.55
7	1.69	16.01	.989	20.28
8	2.18	17.53	1.34	21.96
9	2.70	19.02	1.73	23.59
10	3.25	20.48	2.16	25.19
11	3.82	21.92	2.60	26.76
12	4.40	23.34	3.07	28.30
13	5.01	24.74	3.57	29.82
14	5.63	26.12	4.07	31.32
15	6.26	27.49	4.60	32.80
16	6.91	28.85	5.14	34.27
18	8.23	31.53	6.26	37.16
20	9.59	34.17	7.43	40.00
24	12.40	39.36	9.89	45.56
30	16.79	46.98	13.79	53.67
40	24.43	59.34	20.71	66.77
60	40.48	83.30	35.53	91.95
120	91.58	152.21	83.85	163.64

For critical values of X^2 whose degrees of freedom exceed 30 we can approximate X_c^2 by the formula

$$X_c^2 = \frac{1}{2}\left(z_c + \sqrt{2d - 1}\right)^2$$

where d is the degrees of freedom and z_c is ± 1.96 or ± 2.58.

Appendix C

Table C-9
Critical Values of r for a
Two-Tail Test

(Values of r are given without signs. All values are both positive and negative, that is, $r_c = \pm 1.00$.)

n	r_c for $\alpha = .05$	r_c for $\alpha = .01$
3	1.00	1.00
4	.95	.99
5	.88	.96
6	.81	.92
7	.75	.87
8	.71	.83
9	.67	.80
10	.63	.76
11	.60	.73
12	.58	.71
13	.53	.68
14	.53	.66
15	.51	.64
16	.50	.61
17	.48	.61
18	.47	.59
19	.46	.58
20	.44	.56
21	.43	.55
22	.42	.54
23	.41	.53
24	.40	.52
25	.40	.51
26	.39	.50
27	.38	.49
28	.37	.48
29	.37	.47
30	.36	.46

For values of r_c, when n is greater than 30, use

$$r_c = \frac{t_c}{\sqrt{t_c^2 + (n-2)}}$$

where t_c is the corresponding critical value of t for $(n-2)$ degrees of freedom in Table C-5.

Table C-10
Critical Values of r for a
One-Tail Test

(Values of r are given without signs. You must determine whether r_c is positive or negative from the alternative hypothesis.)

n	r_c for $\alpha = .05$	r_c for $\alpha = .01$
3	.99	1.00
4	.90	.98
5	.81	.93
6	.73	.88
7	.67	.83
8	.62	.79
9	.58	.75
10	.54	.72
11	.52	.69
12	.50	.66
13	.48	.63
14	.46	.61
15	.44	.59
16	.42	.57
17	.41	.56
18	.40	.54
19	.39	.53
20	.38	.52
21	.37	.50
22	.36	.49
23	.35	.48
24	.34	.47
25	.34	.46
26	.33	.45
27	.32	.45
28	.32	.44
29	.31	.43
30	.31	.42

For values of r_c, when n is greater than 30, use

$$r_c = \frac{t_c}{\sqrt{t_c^2 + (n-2)}}$$

where t_c is the corresponding critical value of t for $(n-2)$ degrees of freedom in Table C-6.

Table C-11
Critical Values of F
for $\alpha = .05$
(for a One-Tail Test)

Degrees of freedom for numerator

	1	2	3	4	5	6	7	8	9	10	12	15	20	24	30	40	50	∞
1	161	200	216	225	230	234	237	239	241	242	244	246	248	249	250	251	252	254
2	18.5	19.0	19.2	19.2	19.3	19.3	19.4	19.4	19.4	19.4	19.4	19.4	19.4	19.5	19.5	19.5	19.5	19.5
3	10.1	9.55	9.28	9.12	9.01	8.94	8.89	8.85	8.81	8.79	8.74	8.70	8.66	8.64	8.62	8.59	8.58	8.53
4	7.71	6.94	6.59	6.39	6.26	6.16	6.09	6.04	6.00	5.96	5.91	5.86	5.80	5.77	5.75	5.72	5.70	5.63
5	6.61	5.79	5.41	5.19	5.05	4.95	4.88	4.82	4.77	4.74	4.68	4.62	4.56	4.53	4.50	4.46	4.44	4.37
6	5.99	5.14	4.76	4.53	4.39	4.28	4.21	4.15	4.10	4.06	4.00	3.94	3.87	3.84	3.81	3.77	3.75	3.67
7	5.59	4.74	4.35	4.12	3.97	3.87	3.79	3.73	3.68	3.64	3.57	3.51	3.44	3.41	3.38	3.34	3.32	3.23
8	5.32	4.46	4.07	3.84	3.69	3.58	3.50	3.44	3.39	3.35	3.28	3.22	3.15	3.12	3.08	3.04	3.03	2.93
9	5.12	4.26	3.86	3.63	3.48	3.37	3.29	3.23	3.18	3.14	3.07	3.01	2.94	2.90	2.86	2.83	2.80	2.71
10	4.96	4.10	3.71	3.48	3.33	3.22	3.14	3.07	3.02	2.98	2.91	2.85	2.77	2.74	2.70	2.66	2.64	2.54
11	4.84	3.98	3.59	3.36	3.20	3.09	3.01	2.95	2.90	2.85	2.79	2.72	2.65	2.61	2.57	2.53	2.50	2.40
12	4.75	3.89	3.49	3.26	3.11	3.00	2.91	2.85	2.80	2.75	2.69	2.62	2.54	2.51	2.47	2.43	2.40	2.30
13	4.67	3.81	3.41	3.18	3.03	2.92	2.83	2.77	2.71	2.67	2.60	2.53	2.46	2.42	2.38	2.34	2.32	2.21
14	4.60	3.74	3.34	3.11	2.96	2.85	2.76	2.70	2.65	2.60	2.53	2.46	2.39	2.35	2.31	2.27	2.24	2.13
15	4.54	3.68	3.29	3.06	2.90	2.79	2.71	2.64	2.59	2.54	2.48	2.40	2.33	2.29	2.25	2.20	2.18	2.07
16	4.49	3.63	3.24	3.01	2.85	2.74	2.66	2.59	2.54	2.49	2.42	2.35	2.28	2.24	2.19	2.15	2.13	2.01
17	4.45	3.59	3.20	2.96	2.81	2.70	2.61	2.55	2.49	2.45	2.38	2.31	2.23	2.19	2.15	2.10	2.08	1.96
18	4.41	3.55	3.16	2.93	2.77	2.66	2.58	2.51	2.46	2.41	2.34	2.27	2.19	2.15	2.11	2.06	2.04	1.92
19	4.38	3.52	3.13	2.90	2.74	2.63	2.54	2.48	2.42	2.38	2.31	2.23	2.16	2.11	2.07	2.03	2.00	1.88
20	4.35	3.49	3.10	2.87	2.71	2.60	2.51	2.45	2.39	2.35	2.28	2.20	2.12	2.08	2.04	1.99	1.96	1.84
25	4.24	3.39	2.99	2.76	2.60	2.49	2.40	2.34	2.28	2.24	2.16	2.09	2.01	1.96	1.92	1.87	1.84	1.71
30	4.17	3.32	2.92	2.69	2.53	2.42	2.33	2.27	2.21	2.16	2.09	2.01	1.93	1.89	1.84	1.79	1.76	1.62
40	4.08	3.23	2.84	2.61	2.45	2.34	2.25	2.18	2.12	2.08	2.00	1.92	1.84	1.79	1.74	1.69	1.66	1.51
50	4.03	3.18	2.79	2.56	2.40	2.29	2.20	2.13	2.07	2.02	1.95	1.87	1.78	1.74	1.69	1.63	1.60	1.44
∞	3.84	3.00	2.60	2.37	2.21	2.10	2.01	1.94	1.88	1.83	1.75	1.67	1.57	1.52	1.46	1.39	1.35	1.00

Degrees of freedom for denominator

Table C-12
Critical Values of F
for $\alpha = .025$
(for a Two-Tail Test with $\alpha = .05$)

	1	2	3	4	5	6	7	8	9	10	11	12	15	20	24	30	40	50	∞
1	648	800	864	900	922	937	948	957	963	969	973	977	985	993	997	1000	1010	1010	1020
2	38.5	39.0	39.2	39.2	39.3	39.3	39.4	39.4	39.4	39.4	39.4	39.4	39.4	39.4	39.5	39.5	39.5	39.5	39.5
3	17.4	16.0	15.4	15.1	14.9	14.7	14.6	14.5	14.5	14.4	14.3	14.3	14.3	14.2	14.1	14.1	14.0	14.0	13.9
4	12.2	10.6	9.98	9.60	9.36	9.20	9.07	8.98	8.90	8.84	8.79	8.75	8.66	8.56	8.51	8.46	8.41	8.38	8.26
5	10.0	8.43	7.76	7.39	7.15	6.98	6.85	6.76	6.68	6.62	6.57	6.52	6.43	6.33	6.28	6.23	6.18	6.14	6.02
6	8.81	7.26	6.60	6.23	5.99	5.82	5.70	5.60	5.52	5.46	5.41	5.37	5.27	5.17	5.12	5.07	5.01	4.98	4.85
7	8.07	6.54	5.89	5.52	5.29	5.12	4.99	4.90	4.82	4.76	4.71	4.67	4.57	4.47	4.42	4.36	4.31	4.27	4.14
8	7.57	6.06	5.42	5.05	4.82	4.65	4.53	4.43	4.36	4.30	4.25	4.20	4.10	4.00	3.95	3.89	3.84	3.80	3.67
9	7.21	5.71	5.08	4.72	4.48	4.32	4.20	4.10	4.03	3.96	3.91	3.87	3.77	3.67	3.61	3.56	3.51	3.47	3.33
10	6.94	5.46	4.83	4.47	4.24	4.07	3.95	3.85	3.78	3.72	3.67	3.62	3.52	3.42	3.37	3.31	3.26	3.22	3.08
11	6.72	5.26	4.63	4.28	4.04	3.88	3.76	3.66	3.59	3.53	3.48	3.43	3.33	3.23	3.17	3.12	3.06	3.02	2.88
12	6.55	5.10	4.47	4.12	3.89	3.73	3.61	3.51	3.44	3.37	3.32	3.28	3.18	3.07	3.02	2.96	2.91	2.87	2.72
13	6.41	4.97	4.35	4.10	3.77	3.60	3.48	3.39	3.31	3.25	3.20	3.15	3.05	2.95	2.89	2.84	2.78	2.74	2.60
14	6.30	4.86	4.24	3.89	3.66	3.50	3.38	3.29	3.20	3.15	3.10	3.05	2.95	2.84	2.79	2.73	2.67	2.64	2.49
15	6.20	4.77	4.15	3.80	3.58	3.41	3.29	3.20	3.12	3.06	3.01	2.96	2.86	2.76	2.70	2.64	2.59	2.55	2.40
16	6.12	4.69	4.08	3.73	3.50	3.34	3.22	3.12	3.05	2.99	2.93	2.89	2.79	2.68	2.63	2.57	2.51	2.47	2.32
17	6.04	4.62	4.01	3.66	3.44	3.28	3.16	3.06	2.98	2.92	2.87	2.82	2.72	2.62	2.56	2.50	2.44	2.41	2.25
18	5.98	4.56	3.95	3.61	3.38	3.22	3.10	3.01	2.93	2.87	2.81	2.77	2.67	2.56	2.50	2.44	2.38	2.35	2.19
19	5.92	4.51	3.90	3.56	3.33	3.17	3.05	2.96	2.88	2.82	2.76	2.72	2.62	2.51	2.45	2.39	2.33	2.30	2.13
20	5.87	4.46	3.86	3.51	3.29	3.13	3.01	2.91	2.84	2.77	2.72	2.68	2.57	2.46	2.41	2.35	2.29	2.25	2.09
25	5.69	4.29	3.69	3.35	3.13	2.97	2.85	2.75	2.68	2.61	2.56	2.51	2.41	2.30	2.24	2.18	2.12	2.08	1.91
30	5.57	4.18	3.59	3.25	3.03	2.87	2.75	2.65	2.57	2.51	2.46	2.41	2.31	2.20	2.14	2.07	2.01	1.97	1.79
40	5.42	4.05	3.46	3.13	2.90	2.74	2.62	2.53	2.45	2.39	2.33	2.29	2.18	2.07	2.01	1.94	1.88	1.83	1.64

Degrees of freedom for numerator

Degrees of freedom for denominator

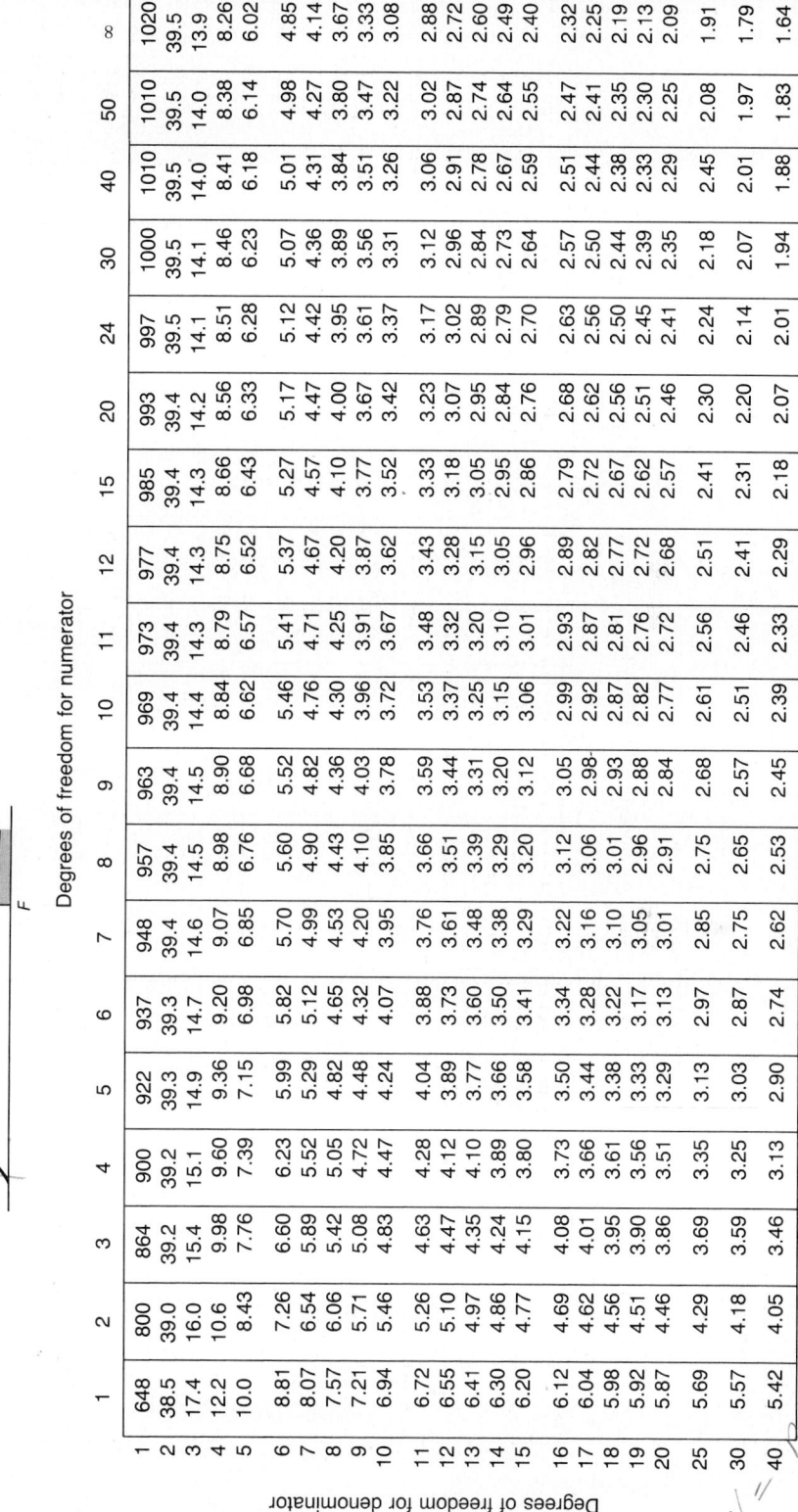

Table C-13
Critical Values of F
for $\alpha = .01$
(for a One-Tail Test)

	1	2	3	4	5	6	7	8	9	10	12	15	20	24	30	40	50	∞
1	4052	5000	5403	5625	5764	5859	5928	5982	6023	6056	6106	6157	6209	6235	6261	6287	6302	6366
2	98.5	99.0	99.2	99.2	99.3	99.3	99.4	99.4	99.4	99.4	99.4	99.4	99.4	99.5	99.5	99.5	99.5	99.5
3	34.1	30.8	29.5	28.7	28.2	27.9	27.7	27.5	27.3	27.2	27.1	26.9	26.7	26.6	26.5	26.4	26.4	26.1
4	21.2	18.0	16.7	16.0	15.5	15.2	15.0	14.8	14.7	14.5	14.4	14.2	14.0	13.9	13.8	13.7	13.7	13.5
5	16.3	13.3	12.1	11.4	11.0	10.7	10.5	10.3	10.2	10.1	9.89	9.72	9.55	9.47	9.38	9.29	9.24	9.02
6	13.7	10.9	9.78	9.15	8.75	8.47	8.26	8.10	7.98	7.87	7.72	7.56	7.40	7.31	7.23	7.14	7.09	6.88
7	12.2	9.55	8.45	7.85	7.46	7.19	6.99	6.84	6.72	6.62	6.47	6.31	6.16	6.07	5.99	5.91	5.85	5.65
8	11.3	8.65	7.59	7.01	6.63	6.37	6.18	6.03	5.91	5.81	5.67	5.52	5.36	5.28	5.20	5.12	5.06	4.86
9	10.6	8.02	6.99	6.42	6.06	5.80	5.61	5.47	5.35	5.26	5.11	4.96	4.81	4.73	4.65	4.57	4.51	4.31
10	10.0	7.56	6.55	5.99	5.64	5.39	5.20	5.06	4.94	4.85	4.71	4.56	4.41	4.33	4.25	4.17	4.12	3.91
11	9.65	7.21	6.22	5.67	5.32	5.07	4.89	4.74	4.63	4.54	4.40	4.25	4.10	4.02	3.94	3.86	3.80	3.60
12	9.33	6.93	5.95	5.41	5.06	4.82	4.64	4.50	4.39	4.30	4.16	4.01	3.86	3.78	3.70	3.62	3.56	3.36
13	9.07	6.70	5.74	5.21	4.86	4.62	4.44	4.30	4.19	4.10	3.96	3.82	3.66	3.59	3.51	3.43	3.37	3.17
14	8.86	6.51	5.56	5.04	4.70	4.46	4.28	4.14	4.03	3.94	3.80	3.66	3.51	3.43	3.35	3.27	3.21	3.00
15	8.68	6.36	5.42	4.89	4.56	4.32	4.14	4.00	3.89	3.80	3.67	3.52	3.37	3.29	3.21	3.13	3.07	2.87
16	8.53	6.23	5.29	4.77	4.44	4.20	4.03	3.89	3.78	3.69	3.55	3.41	3.26	3.18	3.10	3.02	2.96	2.75
17	8.40	6.11	5.19	4.67	4.34	4.10	3.93	3.79	3.68	3.59	3.46	3.31	3.16	3.08	3.00	2.92	2.86	2.65
18	8.29	6.01	5.09	4.58	4.25	4.01	3.84	3.71	3.60	3.51	3.37	3.23	3.08	3.00	2.92	2.84	2.78	2.57
19	8.19	5.93	5.01	4.50	4.17	3.94	3.77	3.63	3.52	3.43	3.30	3.15	3.00	2.92	2.84	2.76	2.70	2.49
20	8.10	5.85	4.94	4.43	4.10	3.87	3.70	3.56	3.46	3.37	3.23	3.09	2.94	2.86	2.78	2.69	2.63	2.42
25	7.77	5.57	4.68	4.18	3.86	3.63	3.46	3.32	3.22	3.13	2.99	2.85	2.70	2.62	2.53	2.45	2.40	2.17
30	7.56	5.39	4.51	4.02	3.70	3.47	3.30	3.17	3.07	2.98	2.84	2.70	2.55	2.47	2.39	2.30	2.24	2.01
40	7.31	5.18	4.31	3.83	3.51	3.29	3.12	2.99	2.89	2.80	2.66	2.52	2.37	2.29	2.20	2.11	2.05	1.80
50	7.17	5.06	4.20	3.72	3.41	3.18	3.02	2.88	2.78	2.70	2.56	2.42	2.26	2.18	2.10	2.00	1.94	1.69
∞	6.63	4.61	3.78	3.32	3.02	2.80	2.64	2.51	2.41	2.32	2.18	2.04	1.88	1.79	1.70	1.59	1.52	1.00

Degrees of freedom for numerator

Degrees of freedom for denominator

Table C-14
Critical Values of F
for $\alpha = .005$
(for a Two-Tail Test with $\alpha = .01$)

Degrees of freedom for numerator

	1	2	3	4	5	6	7	8	9	10	11	12	15	20	24	30	40	50	∞
1	16,200	20,000	21,600	22,500	23,100	23,400	23,700	23,900	24,100	24,200	24,300	24,400	24,630	24,836	24,940	25,040	25,148	25,211	25,465
2	198	199	199	199	199	199	199	199	199	199	199	199	199	199	199	199	199	199	199
3	55.6	49.8	47.5	46.2	45.4	44.8	44.4	44.1	43.9	43.7	43.5	43.4	43.1	42.8	42.6	42.5	42.3	42.2	41.8
4	31.3	26.3	24.3	23.2	22.4	22.0	21.6	21.4	21.1	21.0	20.8	20.7	20.4	20.2	20.0	19.9	19.8	19.7	19.3
5	22.8	18.3	16.5	15.6	14.9	14.5	14.2	14.0	13.8	13.6	13.5	13.4	13.1	12.9	12.8	12.7	12.5	12.4	12.1
6	18.6	14.5	12.9	12.0	11.5	11.1	10.8	10.6	10.4	10.3	10.1	10.0	9.81	9.59	9.47	9.36	9.24	9.17	8.88
7	16.2	12.4	10.9	10.1	9.52	9.16	8.89	8.68	8.52	8.38	8.27	8.18	7.97	7.75	7.65	7.53	7.42	7.35	7.08
8	14.7	11.0	9.60	8.81	8.30	7.95	7.69	7.50	7.34	7.21	7.10	7.01	6.81	6.61	6.50	6.40	6.29	6.22	5.95
9	13.6	10.1	8.72	7.96	7.47	7.13	6.88	6.69	6.54	6.42	6.32	6.23	6.03	5.83	5.73	5.62	5.52	5.45	5.19
10	12.8	9.43	8.08	7.34	6.87	6.54	6.30	6.12	5.97	5.85	5.75	5.66	5.47	5.27	5.17	5.07	4.97	4.90	4.64
11	12.2	8.91	7.60	6.88	6.42	6.10	5.86	5.68	5.54	5.42	5.32	5.24	5.05	4.86	4.76	4.65	4.55	4.49	4.23
12	11.8	8.51	7.23	6.52	6.07	5.76	5.52	5.35	5.20	5.09	4.99	4.91	4.72	4.53	4.43	4.33	4.23	4.16	3.90
13	11.4	8.19	6.93	6.23	5.79	5.48	5.25	5.08	4.94	4.82	4.73	4.64	4.46	4.27	4.17	4.07	3.97	3.91	3.65
14	11.1	7.92	6.68	6.00	5.53	5.26	5.03	4.86	4.72	4.60	4.51	4.43	4.25	4.06	3.96	3.86	3.76	3.70	3.44
15	10.8	7.70	6.48	5.80	5.37	5.07	4.85	4.67	4.54	4.42	4.33	4.25	4.07	3.88	3.79	3.69	3.59	3.52	3.26
16	10.6	7.51	6.30	5.64	5.21	4.91	4.69	4.52	4.38	4.27	4.18	4.10	3.92	3.73	3.64	3.54	3.44	3.37	3.11
17	10.4	7.35	6.16	5.50	5.07	4.78	4.56	4.39	4.25	4.14	4.05	3.97	3.79	3.61	3.51	3.41	3.31	3.25	2.98
18	10.2	7.21	6.03	5.37	4.96	4.66	4.44	4.28	4.14	4.03	3.94	3.86	3.68	3.50	3.40	3.30	3.20	3.14	2.87
19	10.1	7.09	5.92	5.27	4.85	4.56	4.34	4.18	4.04	3.93	3.84	3.76	3.59	3.40	3.31	3.21	3.11	3.04	2.78
20	9.94	6.99	5.82	5.17	4.76	4.47	4.26	4.09	3.96	3.85	3.76	3.68	3.50	3.32	3.22	3.12	3.02	2.96	2.69
25	9.48	6.60	5.46	4.84	4.43	4.15	3.94	3.78	3.64	3.54	3.44	3.37	3.20	3.01	2.92	2.82	2.72	2.65	2.38
30	9.18	6.35	5.24	4.62	4.23	3.95	3.74	3.58	3.45	3.34	3.25	3.18	3.01	2.82	2.73	2.63	2.52	2.46	2.18
40	8.83	6.07	4.98	4.37	3.99	3.71	3.51	3.35	3.22	3.12	3.03	2.95	2.78	2.60	2.50	2.40	2.30	2.23	1.93

Degrees of freedom for denominator

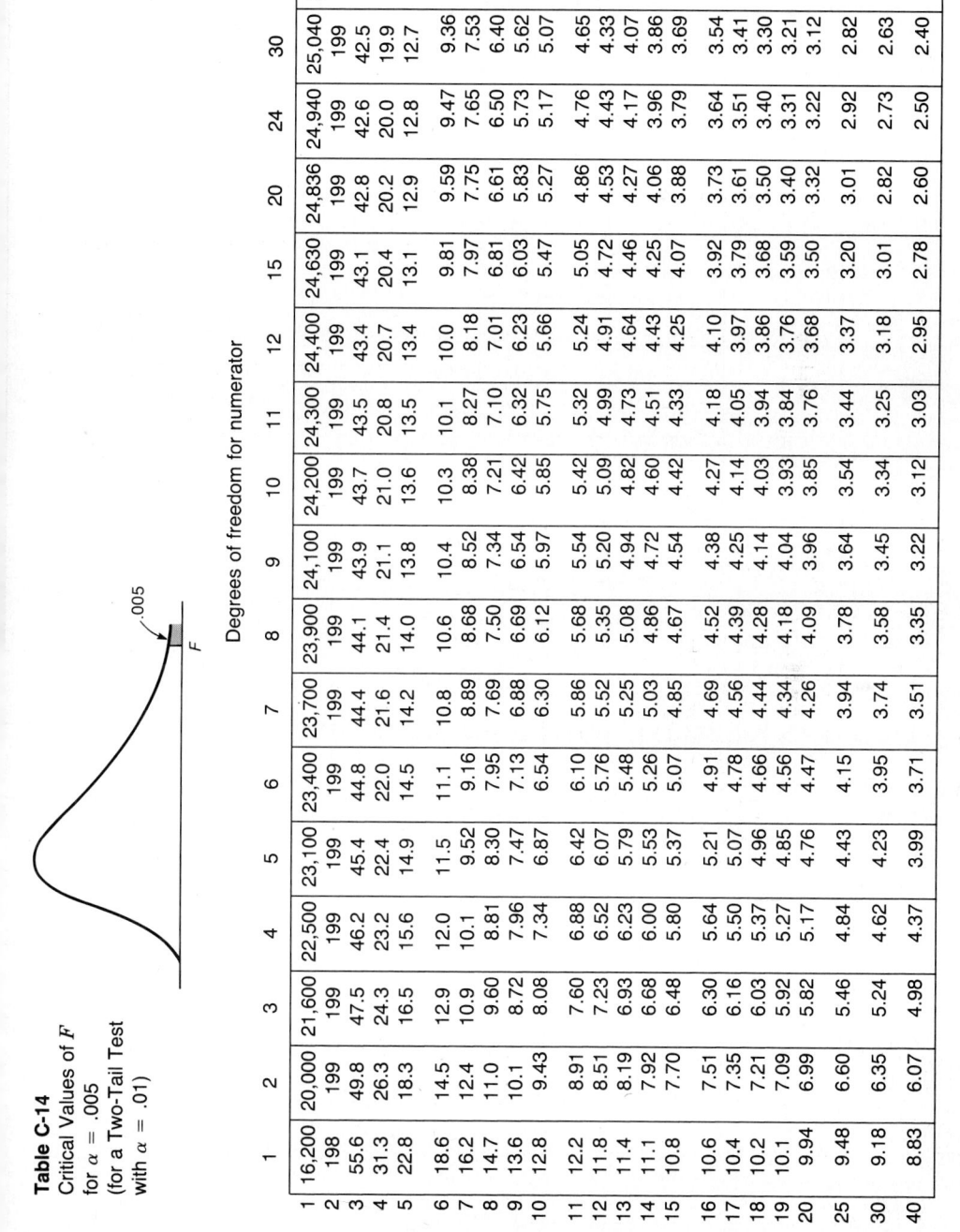

Table C-15
Critical Values of U for a One-Tail Test

n_1	n_2	$\alpha = .05$	$\alpha = .01$
3	2		
	3	9	
4	2		
	3	12	
	4	15	
5	2	10	
	3	14	
	4	18	20
	5	21	24
6	2	12	
	3	16	
	4	21	23
	5	25	28
	6	29	33
7	2	14	
	3	19	21
	4	24	27
	5	29	32
	6	34	38
	7	38	43
8	2	15	
	3	21	24
	4	27	30
	5	32	36
	6	38	42
	7	43	49
	8	49	55
9	1		
	2	17	
	3	23	26
	4	30	33
	5	36	40
	6	42	47
	7	48	54
	8	54	61
	9	60	67
10	1		
	2	19	
	3	26	29
	4	33	37
	5	39	44
	6	46	52
	7	53	59
	8	60	67
	9	66	74
	10	73	81

Table adapted from *Handbook of Statistical Tables*, D. B. Owen, Addison-Wesley Publishing Co., with permission.

Table C-16
Critical Values of U for a Two-Tail Test

n_1	n_2	$\alpha = .05$	$\alpha = .01$
3	2		
	3		
4	2		
	3		
	4	16	
5	2		
	3	15	
	4	19	
	5	23	25
6	2		
	3	17	
	4	22	24
	5	27	29
	6	31	34
7	2		
	3	20	
	4	25	28
	5	30	34
	6	36	39
	7	41	45
8	2	16	
	3	22	
	4	28	31
	5	34	38
	6	40	44
	7	46	50
	8	51	57
9	1		
	2	18	
	3	25	27
	4	32	35
	5	38	42
	6	44	49
	7	51	56
	8	57	63
	9	64	70
10	1		
	2	20	
	3	27	30
	4	35	38
	5	42	46
	6	49	54
	7	56	61
	8	63	69
	9	70	77
	10	77	84

Table adapted from *Handbook of Statistical Tables*, D. B. Owen, Addison-Wesley Publishing Co., with permission.

Answers to Selected Exercises†

These answers are mostly for odd-numbered exercises, although in some cases alternate lettered answers are given such as 14a, 14c, 14e. Both odd and even answers are given for the sample tests.

CHAPTER 1

1-1 Yes   **1-13** (a) 16.4; (c) 40,540; (e) $4\bar{0},000$ **1-14** (a) 1.8 inches; (c) 30°; (e) 5000 people
1-15 $180,000
1-17 (a) 2.2 inches
1-19 (a) .4; (b) .3; (c) They are approximately the same. They differ by only 1 in the last accurate digit.

CHAPTER 2

2-3 (a) 35; (b) 203; (c) 1225; (d) 5; (e) 5; (f) 21; (g) 0; (h) 30; (i) 4⅔; (j) 4⅔; (k) $(\Sigma Y)^2$
2-5 (a) 12; (b) 2; (c) 2; (d) 0; (e) 144; (f) 46; (g) 4.4
2-7 (a) 14.96
2-9 7, 0, 0, or 2, 2, −1, etc.
2-11 $17,385; $16,000; $16,000
2-15 yes, yes
2-17 (b) higher

† Answers may sometimes vary slightly depending on the degree of accuracy and the number of decimal places carried through your solution. Do not be concerned over minor differences between your calculations and those given here.

Answers to selected exercises

2-19 (a) 31; (b) 105; (c) 375; (d) 3
2-21 599
2-23 (a) median; (b) a poll of their subscribers
2-25 (a) *By Definition:*   *By Computational Formula:*

X	$(X-6)$	$(X-6)^2$	X	X^2
4	-2	4	4	16
6	0	0	6	36
8	2	4	8	64
18		8	18	116

$m = 18/3 = 6$
$s = \sqrt{8/2} = 2$

$s = \sqrt{\dfrac{116 - 18^2/3}{2}} = 2$

The definition seems easier in this case.

(b) *By Definition:*   *By Computational Formula:*

X	$X-11.4$	$(X-11.4)^2$	X	X^2
3	-8.4	70.56	3	9
8	-3.4	11.56	8	64
9	-2.4	5.76	9	81
17	5.6	31.36	17	289
20	8.6	73.96	20	400
57		193.20	57	843

$m = 57/5 = 11.4$
$s = 193.20/4 = 6.95$

$s = \sqrt{\dfrac{843 - 57^2/5}{4}} = 6.95$

The computational formula seems easier in this case (and in most cases).

2-27 We expect the mean of a random sample to probably be close to the mean of the population it is drawn from, but we expect the standard deviation of a sample to probably be smaller (that is why we divide by $n-1$ instead of n).

2-29 (a) $m = 34/5 = 6.8$, $s^2 = \dfrac{276 - 34^2/5}{4} = 11.2$, $s = 3.35$; (b) $m = 6.8(10) = 68$, $s = 3.35(10) = 33.5$, $s^2 = 11.2(10)^2 = 1120$; (c) There are none.

2-31 m_Y is 60 more than m_X, $s_Y = s_X$

2-33 (a) $16.75/6 = \$2.79$
(b) $s^2 = \dfrac{59.6875 - 16.75^2/6}{5} = 2.585$, $s = \$1.61$
(c) $m = \dfrac{14.0 + 14.6 + 12 + 14.8 + 15.7 + 14.3}{6} = 14.2\%$
(d) $s^2 = \dfrac{1223.18 - 85.4^2/6}{5} = 1.531$, $s = 1.2\%$

2-35 (a) 13; (b) $13/17 = 76\%$; (c) 16; (d) $16/17 = 94\%$

2-37 Although the averages are the same, the psychotic children's IQs varied more than those of the normal kids. There were both more higher scores and more lower scores among the psychotic kids.

2-39 Pythagorean

Understanding statistics

2-41 In the first experiment a standard deviation of only 22 days makes the difference of 400 days between the means meaningful. However, in the second experiment the standard deviation is about 300 days. The lifetimes of these hearts are much more erratic. We cannot conclude that a difference of 400 days between the means is significant evidence.

2-43 (a) **1980:**

$$m = 12{,}256/6 = 2043$$

$$s^2 = \frac{40{,}586{,}272 - 12{,}256^2/6}{5}$$

$$= 3{,}110{,}270$$

$$s = 1764$$

1981:

$$m = 2468/6 = 411$$

$$s^2 = \frac{1{,}710{,}714 - 2468^2/6}{5}$$

$$= 139{,}109$$

$$s = 373$$

(b) 5.4%, 29.5%, 38.8%, 15.0%, 3.4%, 7.9%
(c) The incidence of measles declined between 1980 and 1981.

2-45 mean, median, mode, 25th percentile: a distribution of temperatures at the South Pole. z-score: any value below the mean. The others cannot be negative.

2-46 (a) $\dfrac{99.1 - 98.6}{.5} = 1$; (c) 0; (e) $\dfrac{98 - 98.6}{.5} = -1.2$

2-47 (a) $32 + 2(1.2) = 34.4$; (c) $32 + (-3)(1.2) = 28.4$; (e) $32 + (-.06)(1.2) = 32.0$

2-49 (a) A:$10 + (-.02)5 = 9.9$, B:$10 + (1.27)5 = 16.4$, C:$10 + (.001)5 = 10.0$, J:$10 + (-2.03)5 = -0.2$, E:$10 + (.48)5 = 12.4$; (b) no; (c) no

(d) L: $\dfrac{10 - 10}{5} = 0$, N: $\dfrac{9 - 10}{5} = -.2$, B: $\dfrac{5 - 10}{5} = -1$, I: $\dfrac{15 - 10}{5} = 1$,

P: $\dfrac{12 - 10}{5} = .4$, S: $\dfrac{11 - 10}{5} = .2$

2-51 true **2-53** (a) $z_{169} = \dfrac{169 - 171}{5} = -.4$, $z_{171} = 0$, $z_{180} = \dfrac{180 - 171}{5} = 1.8$; (b) $171 + (-1)5 = 166$ inches, $171 + (0)5 = 171$ inches, $171 + (.3)5 = 172.5$ inches; (c1) $z_{185} = \dfrac{185 - 171}{5} = 2.8$ is greater than 2;

(c2) $171 + (2)5 = 181$ inches; (c3) $171 + (-2)5 = 161$ inches

2-55 You are very tall relative to the control group. You weigh less than average relative to the control group. Your blood pressure is average relative to the control group.

2-57 (a) 25%; (c) 75%; (e) 10%
2-58 (a) 125,000 families; (c) 375,000 families; (e) 50,000 families
2-59 (a) Not from the table (but $n = 500$); (b) Yes, at $z = 0$ we find $X = 100$; (c) Yes. $P_{50} = 116$; (d) 2%; $50 - 16 = 34$%; (e) 16; (f) $100 + (1.5)16 = 124$; (g) $100 + (-1.2)16 = 80.8$

2-61 Yes **2-63** $z_5 = \dfrac{5 - 3.5}{4/7} = 2.63$

2-65 It's not an average, it's a rate.

Answers to selected exercises

2-67

	1	2	3
(a)	the year 1978	population of the U.S.	number of births
(b)	the year 1983	people over 65 in the U.S.	number of deaths
(c)	the month of July 1981	married couples in the U.S.	number of divorces
(d)	the year one	apples in the Garden of Eden	number of apples eaten

2-69 The base population in 1975 was not the same as in 1976.

2-71

	for	against
ham	.63	.59
spinach	.61	.63
potatoes	.62	.61
cabbage	.64	.60
jello	.70	.58
rolls	.57	.66
bread	.67	.58
milk	.50	.62
coffee	.61	.62
water	.54	.65
cakes	.68	.54
vanilla	.80	.14
chocolate	.51	.75
fruit	.67	.61

CHAPTER 3

3-1 (a) 39; (b) 9; (c) 54%

3-3 (a)

40–44	2
45–49	5
50–54	12
55–59	12
60–64	6
65–69	3

(b) 54.8, 55; (c) 60%

3-7 (a)

$50,000– 74,999	3
75,000– 99,999	3
100,000–124,999	2
125,000–149,999	6
150,000–174,999	5
175,000–199,999	5
200,000–224,999	1
225,000–249,999	1

(b)

$50,000– 89,999	6
90,000–129,999	4
130,000–169,999	8
170,000–209,999	6
210,000–249,999	2

3-9 (a)

0– 4999	44.0
5000– 9999	18.9
10,000–14,999	16.9
15,000–24,999	15.8
25,000 and over	4.5

Understanding statistics

(c) 44.0%, 32.7%; (d) approximately $6500
3-13 (a) 50%; (c) 16%; (e) 98%; (g) 48%; (i) 68%
3-14 (a) 40%; (c) 20%; (e) 90%; (g) 20%; (i) 30%
3-15 (a) 40%; (b) 40%

(c)
2 to 3	15	15%
1 to 2	40	40%
0 to 1	10	10%
−1 to 0	5	5%
−2 to −1	20	20%
−3 to −2	10	10%

3-17
2 to 3	10
1 to 2	30
0 to 1	20
−1 to 0	40

CHAPTER 4

4-5 Both are false
4-7 (a) 8/15; (c) 4/15; (e) 2/15
4-9 (a) 1/2; (c) 1/20; (e) 1/10
4-11 (a) CCC (b1) 1/8; (b2) 1/2; (b3) 1/8
CCW
CWC
WCC
CWW
WCW
WWC
WWW
4-13 (a) 5/36; (b) 1/6
4-15 (a)

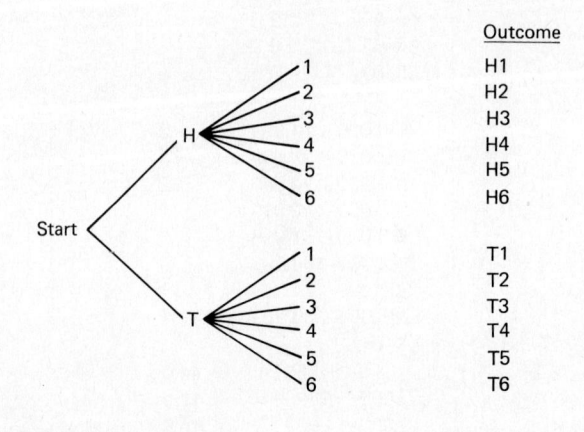

	Outcome
H1	H1
H2	H2
H3	H3
H4	H4
H5	H5
H6	H6
T1	T1
T2	T2
T3	T3
T4	T4
T5	T5
T6	T6

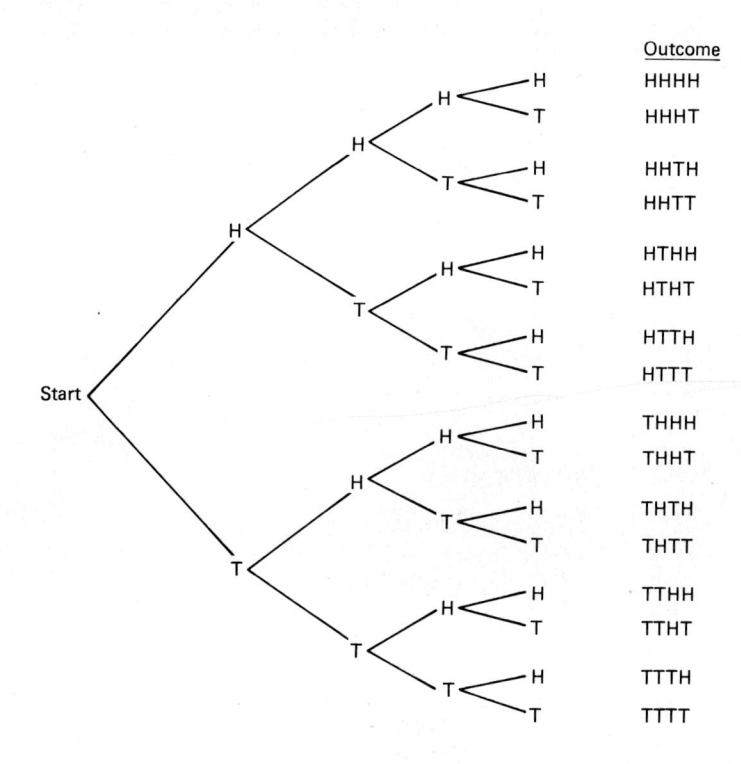

(b)

	Outcome
	HHHH
	HHHT
	HHTH
	HHTT
	HTHH
	HTHT
	HTTH
	HTTT
	THHH
	THHT
	THTH
	THTT
	TTHH
	TTHT
	TTTH
	TTTT

$P(S = 2) = 3/8$
$P(S > 2) = 5/16$
(c) No. $6^4 = 1296$ outcomes!

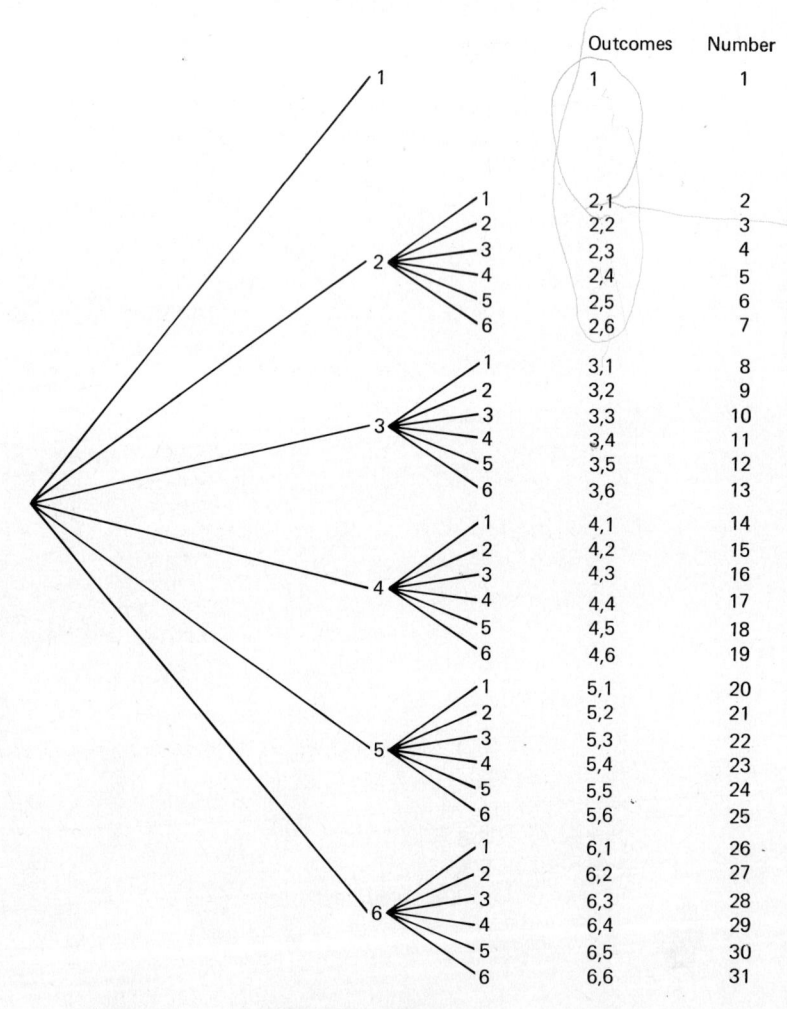

(d)

	Outcomes	Number
	1	1
	2,1	2
	2,2	3
	2,3	4
	2,4	5
	2,5	6
	2,6	7
	3,1	8
	3,2	9
	3,3	10
	3,4	11
	3,5	12
	3,6	13
	4,1	14
	4,2	15
	4,3	16
	4,4	17
	4,5	18
	4,6	19
	5,1	20
	5,2	21
	5,3	22
	5,4	23
	5,5	24
	5,6	25
	6,1	26
	6,2	27
	6,3	28
	6,4	29
	6,5	30
	6,6	31

4-17 (a)

BBB	CBB	LBB
BBC	CBC	LBC
BBL	CBL	LBL
BCB	CCB	LCB
BCC	CCC	LCC
BCL	CCL	LCL
BLB	CLB	LLB
BLC	CLC	LLC
BLL	CLL	LLL

(b) 1/27; (c) 1/9; (d) 7/27; (e) about 14

4-19 (a) .80; (b) .64; (c) .75; (d) 1.00

4-21 No; we expect approximately 10 correct guesses.

4-23 (a) 1/16; (c) 1/16; (e) 1/2

4-25 (a) .16; (b) .68; (c) .50

Answers to selected exercises

SAMPLE TEST FOR CHAPTERS 1 TO 4

(1) a is correct.
(2) random
(3) median, mean, mean, mode
(4) z score, percentile rank, and raw score
(5) a statistic
(6) 1/8, 1/8
(7) 43%, 57%
(8) 62%, 38%
(9) approximately 35
(10) (a) 2.1, 2, both 1 and 2; (b) 6, 2.77, 1.66; (c) $z_6 = 2.35$, and $z_0 = -1.27$; (d) 3
(11) (a) One possible arrangement is

interval	frequency	boundaries
30–39	1	29.5–39.5
40–49	0	39.5–49.5
50–59	2	49.5–59.5
60–69	7	59.5–69.5
70–79	4	69.5–79.5

(b) $PR_{61} = 39$
(12) For one thing, he is not including families with only preschool children.

CHAPTER 5

5-1 The probability of passing varies from course to course.
5-3 (a) 31,824; (c) 7; (e) 1; (g) n
5-4 (a) 4; (c) 16
5-5 1, 21, 210, 1330
5-9 (a)

3	1	1/8
2	3	3/8
1	3	3/8
0	1	1/8

(c) 1/8; (d) 3/8; (e) 7/8; (f) 1/8; (g) 1/2; (h) 7/8
5-11 (a) .17; (b) .001
5-13 (a) .49 (b) .01 (c) .99
5-15 (a) 1/16; (b) 1/216
5-17 (a) approximately one; (b) .26; (c) $P(0 \text{ or } 1)$ since $.74 > .26$
5-19 (a) .21; (b) .11; (c) .00001; (d) .97
5-21 (a) .24; (b) .68; (c) .08
5-23 .63   **5-25** .51   **5-27** .25

CHAPTER 6

6-3 (a) 99.01%; (c) 49.38%; (e) 5.48%; (g) 1.66%
6-5 (a) 2.50%; (b) 2.28%; (c) 31.74%; (d) .0013; (e) .4404; (f) .0668; (g) .0215; (h) .0215
6-7 (a) -2.05; (b) 1.65; (c) 0; (d) ± 1.04; (e) the same
6-9 (a) -2.33; (c) -1.28; (e) 1.65

Understanding statistics

6-11

x	4	7	10	13	16	19	22
z	−3	−2	−1	0	1	2	3

6-13   244
6-15   (a) .0143; (b) .0150; approximately the same
6-17   (a) 73%; (b) 3%; (c) 1%; (d) 50%; (e) 3%; (f) 10%; (g) 90%; (h) 30
6-19   (a) .18; (b) .03; (c) .71; (d) 7 pounds, 10 ounces; (e) 7 pounds, 10 ounces; (f) 5 pounds, 15 ounces
6-21   (a) $35,920; (b) yes; (c) $27,850; (d) $31,240 and $35,320
6-25   The shape of the distribution is not known.
6-27   (a) P_{71}; (b) about the same
6-29   (a) .43; P_{70}; (b) P_{67}

CHAPTER 7

7-3   (a) .6; (b) .4; (c) 8.4 > 5; 5.6 > 5; (d) 8.4; (e) 1.83; (f) .31
7-5   (a)
2, 2
2, 1
2, 0
1, 2
1, 1
1, 0
0, 2
0, 1
0, 0
(c) .001; (d) .17; (e) .52

7-7   (a) .03; (b) .003   7-9   .004   7-11   yes; no
7-13   (a) 0; (b) 0   7-15   .61
7-17   (a) .09; (b) 13; (c) .07, 12
7-19   (a) .9; (b) .01; (c) 7
7-21   (a) .44; (b) 1/900; (c) 1/3600

SAMPLE TEST FOR CHAPTERS 5 TO 7

(1) (a) 7%; (b) 9.6 pounds; (c) 4.3 pounds; (d) .16
(2) yes, if distribution is not normal
(3) The 4 ways an outcome can have one win and 3 losses (WLLL, LWLL, LLWL, LLLW) and the 4 ways there can be 3 wins and one loss
(4) (a) 120; (b) 2 or 3; (c) 1; (d) 1; (e) 117; (f) 231
(5) (a) .5193; (b) .5193; (c) .5222
(6) (a) .00217; (b) .0022 (using $p = .40$); (c) .0039
(7) (a) .0003; (b) yes; (c) no; (d) yes

CHAPTER 8

8-1   (a) $P \neq .40$; (b) $P < .40$; (c) $P > .40$
8-3 H_0: 10% of pet owners own goldfish.
       H_a: The percentage of pet owners who own goldfish is more than 10%.
8-5 H_0: 12% of the students in your school major in mathematics.
       H_a: The percentage of mathematics majors is not 12%.
8-7   This calls for an estimation; there is no hypothesis to test.

Answers to selected exercises

8-9 Mart is correct, Marv is not.
8-11 Type II
8-13 (b) More false positives would occur; that is, more errors of a Type I nature.
8-15 The bigger the expense, the smaller alpha should be.

8-17

spinner is:	number of sixes could be near	conclusion
honest	10	honest—correct
honest	20	biased—Type I error
biased	10	honest—Type II error
biased	20	biased—correct

8-19 $z = \dfrac{49.5 - 60}{6.48} = -1.62$; Danny.

8-21 (a) $p = P$(a genitz is pibled): H_0: 3% of all genitz are pibled. $p = .03$; H_a: Less than 3% of all genitz are pibled. $p < .03$, (one-tail)
(b) $30 - 2.33(5.39) = 17.4$, 17 or less.

8-23 (a) $p = P$(a patron is male): H_a: Less than 75% of the patrons are male, $p < .75$, one-tail; (b) H_0: 75% of the patrons are male, $p = .75$;
(d) $z = \dfrac{59.5 - 75}{4.33} = -3.6$, $\alpha = .0002$.

8-25 H_0: Sex has nothing to do with political affiliation, $p = .42$
H_a: More than 42% of the Republicans are women, $p > .42$, one-tail;
$S_c = 42 + 2.33(4.94) = 53.5$
Decision rule: Reject H_0 if $S > 53$.

8-27 $z = \dfrac{35.5 - 30}{4.58} = 1.20$; $P(S \geq 36) = .12$

8-35 $z = \dfrac{37.5 - 20}{4} = 4$, $\beta = 100\%$

8-37 $z = \dfrac{4.5 - 21}{3.69} = -4.5$, $z = \dfrac{15.5 - 21}{3.69} = -1.49$
$\beta = .0681 - .0000 = .07$

8-41 *Decision rule*: $30 \pm 1.96(3.46) = 23.2$ and 36.8

$p = .4$: $z = \dfrac{36.8 - 20}{3.46} = 4.86$ and $z = \dfrac{23.2 - 20}{3.46} = 0.92$

$\beta = 1 - .82 = .18$, power $= .82$

$p = .5$: $z = \dfrac{36.8 - 25}{3.54} = 3.33$ and $z = \dfrac{23.2 - 25}{3.54} = -.508$

$\beta = 1 - .31 = .69$, power $= .31$

$p = .7$: $z = \dfrac{36.8 - 35}{3.24} = .56$ and $z = \dfrac{23.2 - 35}{3.24} = -3.64$

$\beta = .712 - 0 = .71$, power $= .28$

$p = .8$: $z = \dfrac{36.8 - 40}{2.83} = -1.13$ and $z = \dfrac{23.2 - 40}{2.83} = -5.94$

$\beta = .129 - 0 = .13$, power $= .87$

8-43 (a) $S_c = 25 \pm 1.96(3.54) = 18.1$ and 31.9;
(b) $z = \dfrac{31.5 - 40}{2.83} = -3.00$, $z = \dfrac{18.5 - 40}{2.83} = -7.6$,
$\beta = .0013 - 0 = .0013$
(c) $S_c = 50 \pm 1.96(5) = 40.2$ and 59.8,
$z = \dfrac{59.5 - 80}{4} = -5.1$, $z = \dfrac{40.2 - 80}{4} = -10.0$, $\beta = 0 - 0 = 0$;
(d) as n increases, β decreases.

8-45 $z = \dfrac{13.5 - 22.5}{3.52} = -2.56$, $z = \dfrac{26.5 - 22.5}{3.52} = 1.14$,
$\beta = .8729 - .0052 = .87$

8-47 (a) $S_c = 6.25 + 1.65(2.17) = 9.8$ or more than 9;
(b) $z = \dfrac{9.5 - 10}{2.74} = -.18$, $\beta = .4286$

8-49 (a) $z = \dfrac{8.5 - 10}{2.24} = -.67$, $z = \dfrac{11.5 - 10}{2.24} = .67$
$\alpha = .5028$, 50%
(b) can't tell unless you know p(heads) for a fake coin; (c) 2

8-51 (a) How much will the new technique cost? Will we have to retrain, hire, or fire any employees? Etc.; (c) How expensive is the medicine? How serious is the allergic reaction? What other remedies are available? Etc.

8-55 (a) $p = P$(a rattlesnake has broken fangs)
H_0: The percentage of rattlesnakes with broken fangs is 3%, $p = .03$
H_a: The percentage is less than 3%, $p < .03$ (one-tail test)
(b) $p = P$(a teacher is tardy)
H_0: The percentage of late teachers is 18%, $p = .18$
H_a: The percentage is more than 18%, $p > .18$ (one-tail test)
(c) H_0: Joining the union will have no effect on take-home pay
H_a: Joining the union will change the take-home pay (two-tail test)
(d) H_0: Surprise tests do not affect student grades
H_a: Surprise tests increase student grades (one-tail test)
(e) H_0: Wearing flippers has no effect on the average number of points per game
H_a: Wearing flippers increases the average (one-tail test)

8-57 (a) population: all students at the university, $p = P$(a student is left-handed), H_a: more than 10% are left-handed, $p > .10$
(b) H_0: 10% are left-handed, $p = .10$
(c) one-tail
(d) $S_c = 10 + 1.65(3) = 14.95$. I will reject H_0 if my outcome exceeds 14.95;
(e) Since $16 > 14.95$, she can reject the hypothesis that p is only 10%; at the .05 significance level she has proof that more than 10% are left-handed.

8-59 (a) Population: all cars coming off this assembly line
$p = P$(a car is unfit for delivery); H_a: Less than 20% are defective with new technique, $p < .20$; (b) H_0: 20% are still defective with new method, $p = .20$;
(c) one-tail; (d) $S_c = 16 - 1.65(3.58) = 10.1$; I will reject H_0 if my outcome is less than 10.1; (e) Since $3 < 10.1$, we reject H_0 at the .05 significance level.

Answers to selected exercises

We have evidence that the new technique does decrease the percentage of defective cars.

8-61 (a) Population: all freshmen at Wealth College; $p = P$(a freshman drops out); H_a: Less than 50% will drop out under recent policies, $p < .50$;
(b) H_0: 50% will still drop out, $p = .50$; (c) one-tail;
(d) $S_c = 300 - 1.65(12.2) = 279.8$. I will reject H_0 if my outcome is less than 279.8;
(e) Since $260 < 279.8$, we reject H_0. We have evidence at the .05 significance level that the recent policies have lowered the dropout rate.

8-63 (a) Population: all voters in the politician's district; $p = P$(a voter favors the health care bill), H_a: More than 40% favor the bill, $p > .40$;
(b) H_0: 40% favor the bill, $p = .40$; (c) one-tail;
(d) $S_c = 12 + 2.33(2.68) = 18.3$. I will reject H_0 if my outcome is greater than 18.3; (e) No. Since 14 is not greater than 18.3 we have not proved that the attitude in this district is different from the rest of the nation. We fail to reject H_0; it may be true that $p = .40$.

8-65 (a) Population: all the programming time on WW-TV on Saturday mornings; $p = P$(a particular time is devoted to commercials); H_a: Ralph's claim is wrong, $p \neq .25$; (b) H_0: Ralph is correct, $p = .25$; (c) two-tail; (d) $S_c = 12.5 \pm 1.96(3.06) = 6.5$ and 18.5. I will reject H_0 if my outcome is either less than 6.5 or greater than 18.5; (e) Since 9 is between 6.5 and 18.5, I fail to reject H_0. Ralph may be right.

8-67 (a) Population: all freshmen at this college; $p = P$(a freshman is more interested in being popular than in doing well at school), H_a: The paper's claim is too high, $p < .60$;
(b) H_0: The paper is correct, $p = .60$; (c) one-tail;
(d) $S_c = 60 - 1.65(4.90) = 51.9$; I will reject H_0 if my outcome is less than 51.9; (e) Since 10 is less than 51.9, they have proved that the 60% figure is too high *on their campus*. If it was a school paper talking about their campus, they have evidence that the paper is in error. If it was an article about all freshmen nationwide, they have evidence that the freshmen at their school are different.

8-69 (a) Population: all tadpoles under certain conditions; $p = P$(a tadpole survives to become a frog); H_a: More than 10% will survive, $p > .10$; (b) H_0: 10% will survive, $p = .10$; (c) one-tail; (d) $S_c = 9.8 + 1.65(2.97) = 14.7$. I will reject H_0 if my outcome exceeds 14.7 frogs; (e) If the outcome is 12, we fail to reject H_0; we have not proved that the vitamins have had any effect on survival. If the outcome is 27, then we would reject H_0. We have evidence at the .05 significance level that more than 10% of the tadpoles will survive when given vitamins.

8-71 (a) Population: all possible rolls of the dice; $p = P$(a roll produces doubles);
$$z = \frac{14.5 - 10}{2.89} = 1.56; P(S \geq 15) = .0594;$$ (b) alpha

CHAPTER 9

9-1 If he knows the sample sizes used.
9-3 (a) $dp_c = 0 \pm 2.58(.0484) = \pm.12$; (b) $130/200 - 120/200 = .05$. Fail

to reject H_0. There may be no difference; the two flights could carry the same percentage of business people.

9-5 $dp_c = 0 - 2.33(.152) = -.35$; $d\hat{p} = 6/19 - 18/25 = -.40$. Fail to reject H_0. We have not proved that the aspirins reduce clots.

9-7 $dp_c = 0 \pm 1.96(.0897) = \pm.18$; $d\hat{p} = 42/60 - 29/60 = .22$. Reject H_0. More Catholics than Orthodox Jews oppose birth control.

9-9 $dp_c = 0 + 2.33(.0236) = .055$; $d\hat{p} = \dfrac{432}{500} - \dfrac{401}{500} = .062$; reject H_0; more males eat 3 meals a day.

9-11 $dp_c = 0 - 1.65(.0753) = -.12$; $d\hat{p} = 6/50 - 15/60 = -.13$; reject H_0; the percentage of smokers with lung cancer is higher.

9-13 $dp_c = 0 \pm 1.96(.083) = \pm.16$; $d\hat{p} = 20/60 - 15/60 = .083$; reject H_0; we have not proved that their ESP is different.

9-15 $.50(1457) = 728.5$; $.23(2797) = 643.31$; $dp_c = 0 - 2.33(.0151) = -.035$; $d\hat{p} = .23 - .50 = -.27$; reject H_0; more younger people *report* infirmities.

9-17 $.30(500) = 150$; $.3(500) = 180$; $dp_c = 0 - 2.33(.0297) = -.07$; $d\hat{p} = .30 - .36 = -.06$. Fail to reject H_0. The two groups might do just as well; the answer is 2 quarts.

9-19 $.40(1700) = 680$; $.45(2100) = 945$; $dp_c = 0 - 2.33(.0161) = -.04$; $d\hat{p} = .40 - .45 = -.05$; yes.

9-21 (a) $dp_c = 0 + 2.33(.107) = .25$; $d\hat{p} = 20/30 - 40/70 = .09$; no; (b) $dp_c = 0 + 2.33(.0935) = .22$; $d\hat{p} = 20/60 - 10/40 = .08$; no

9-23 (a) 50%; (b) 25%; (c) $.50 - .25 = .25$; (d) $20(.5) = 10$, $20(.5) = 10$, $50(.25) = 12.5$, and $50(.75) = 37.5$ are all greater than 5; (e) $D = .25 - 1.65(.11) = .07$

CHAPTER 10

10-1 The means should be approximately equal, but the class range and standard deviation would be less than the school range and standard deviation.

10-3 (a) All physicians in some groups; (b, c) $50.76; (d) $28.56; (e) normal; (f) $50.76; (g) $4.04

10-5 (a) $z = \dfrac{39.5 - 50}{7} = -1.5$; $P(x < 40) = .07$; (b) $z = \dfrac{45 - 50}{1.17} = 4.27$; $P(m < 45) = 0$

10-7 (a) $m_c = 470 + 2.33(6) = 484$; (b) yes

10-9 $z = \dfrac{1029.5 - 1000}{30} = .98$; $P(m \geq 1030) = 0.16$

10-11 $m_c = 3.1 + 2.33(.04) = 3.2$; yes.

10-13 Is the number of boys in each school the same? the number of girls? If so, are they both greater than 30?

10-15 $dm_c = 0 \pm 1.96(1.36) = \pm 2.67$; $dm = 82 - 77 = 5$; reject H_0; first method is better.

10-17 $dm_c = 0 - 1.65(.40) = -.66$; $dm = 82.5 - 83.1 = -.60$; fail to reject H_0; there may be no difference.

10-19 $dm_c = 0 \pm 2.58(1.80) = \pm 4.65$; $dm = 120 - 30 = 90$; reject H_0; less habits are bought today.

10-21 $dm_c = 0 \pm 1.96(.274) = \pm.54$; $dm = 2.8 - 3.2 = -.4$; fail to reject H_0; there may be no difference.

Answers to selected exercises

10-23 $dm_c = 0 \pm 1.96(1.84) = \pm 3.6$; $dm = 83 - 78 = 6$; reject H_0; the earlier papers do better.

CHAPTER 11

11-1 Increase α and decrease β.
11-3 $m_c = 3 + 1.73(.27) = 3.5$; $3.8 > 3.5$; reject H_0; the average time for women is more than 3.
11-5 $m_c = 260 - 3.00(2.83) = 251.5$; $250 < 251.5$; reject H_0; the average IQ is less than 260.
11-7 $m_c = 71 \pm 2.06(.6) = 69.8$ and 72.2; 70 lies between 69.8 and 72.2; fail to reject H_0; average could be 71 inches.
11-9 $m_c = 3000 \pm 2.95(125) = 2631$ and 3369; $3472 > 3369$; reject H_0; average is more than 3000 words.
11-11 $\sqrt{\dfrac{108{,}153 - \dfrac{(825)^2}{8}}{7}} = 57.4$; $m_c = 100 \pm 2.36(20.3) = 52$ and 148;
103 is between 52 and 148; fail to reject H_0; mean could be 100.
11-13 This is a *binomial* problem about t scores. $H_0: p = .95$ (95% of the t scores will be less than 1.86); $H_a: p \neq .95$; $S_c = 475 \pm 1.96(4.87) = 465$ and 485; $S = 468$; fail to prove teacher wrong.
11-15 $s_p^2 = \dfrac{34(.2)^2 + 19(.15)^2}{34 + 19} = 0.337$; $s_{dm} = \sqrt{.0337 \left(\dfrac{1}{35} + \dfrac{1}{20}\right)} = .051$,
$dm_c = 0 \pm 2.01(.051) = \pm .14$; $dm = 3.1 - 3.2 = -.1$. *Note:* Since these approximate values are so close, an argument could be made either for or against rejecting H_0. If a more conclusive result is necessary, the experiment should be repeated with more accurate numbers and larger samples.
11-17 $s_p^2 = \dfrac{19(9.4)^2 + 19(11.1)^2}{19 + 19} = 105.8$; $s_{dm} = \sqrt{105.8 \left(\dfrac{1}{20} + \dfrac{1}{20}\right)} = 3.25$; $dm_c = 0 \pm 1.68(3.25) = \pm 5.5$; $dm = 83.3 - 80.7 = 2.6$; fail to reject H_0; we have not proved that the offer of exemptions improves grades.
11-19 $s_p^2 = \dfrac{14(5)^2 + 34(8)^2}{14 + 34} = 52.625$; $s_{dm} = \sqrt{52.625 \left(\dfrac{1}{15} + \dfrac{1}{35}\right)} = 2.24$; $dm_c = 0 \pm 2.01(2.24) = \pm 4.5$; $dm = 29.2 - 24.8 = 4.4$; fail to reject H_0; fail to show a difference in the average age.
11-21 $s_p^2 = \dfrac{49(.05)^2 + 24(.07)^2}{49 + 24} = .00329$; $s_{dm} = \sqrt{.00329 \left(\dfrac{1}{50} + \dfrac{1}{25}\right)} = .014$; $dm_c = 0 - 2.37(.014) = -.033$; $dm = 1.1 - 2.1 = -1.0$; reject H_0; August toads are heavier.
11-23 $s_p^2 = \dfrac{9(10)^2 + 14(10)^2}{9 + 14} = 100$; $s_{dm} = \sqrt{100 \left(\dfrac{1}{10} + \dfrac{1}{15}\right)} = 4.08$; $dm_c = 0 \pm 2.07(4.08) = \pm 8.4$; $dm = 180 - 170 = 10$; reject H_0; male nurses' average is higher.

Understanding statistics

11-25 $s_1^2 = \dfrac{22150 - \dfrac{330^2}{5}}{4} = 92.5$; $s_2^2 = \dfrac{20550 - \dfrac{320}{5}}{4} = 17.5$;

$s_p^2 = \dfrac{4(92.5) + 4(17.5)}{4 + 4} = 55$; $s_{dm} = \sqrt{55\left(\dfrac{1}{5} + \dfrac{1}{5}\right)} = 4.69$;

$dm_c = 0 + 2.90(4.69) = 13.6$; $dm = 330/5 - 320/5 = 2$; reject H_0; Krispy toasters are quicker.

11-27 Pairing the students gives us more information and thus we have more information to work with.

11-29 (a) $s_p^2 = \dfrac{9(4.2)^2 + 9(4.0)^2}{9 + 9} = 16.82$; $s_{dm} = \sqrt{16.82\left(\dfrac{1}{10} + \dfrac{1}{10}\right)} = 1.83$; $dm_c = 0 \pm 2.88(1.83) = \pm 5.27$; $dm = 16.9 - 16.3 = .6$; fail to reject H_0; there may be no difference in mileage. (b) $m_c = 0 \pm 3.25(.123) = \pm .4$; $m = .6$; reject H_0; Flug gives better mileage.

11-31 $m_c = 0 + 1.83(1.58) = 2.9$; $m = 5.8$; reject H_0; the mean weight loss is greater than zero.

11-33 $m_c = 0 + 1.83(1.90) = 3.5$; $m = 10$; reject H_0; pros make more backhand errors.

11-35

d	-3	-4	-2	1	2	-2	-14	1	2	1	$\Sigma d = -18$
d^2	9	16	4	1	4	4	196	1	4	1	$\Sigma d^2 = 240$

$m_d = -1.8$; $s_d = 4.80$; $s_m = 1.52$; $m_c = 0 - 1.83(1.52) = -2.8$; fail to reject H_0; the course may not have made any difference.

SAMPLE TEST FOR CHAPTERS 8 TO 11

(1) Parameter
(2) Rejecting a true null hypothesis; significance level is the probability of making a Type I error.
(3) Examining a statistic, we can never be 100% sure about a parameter.
(4) That the populations are normal, and their variances are approximately equal.
(5) (a) $p = P(\text{red})$; H_0: the wheel is fair, $p = 1/4$; H_a: the wheel is biased, $p \neq 1/4$ (two-tail); (b) $S_c = 15 \pm 1.96(3.35) = 8.4$ and 21.6; I will reject H_0 if my outcome is less than 8.4 or greater than 21.6; (c) Since $S = 12$, we fail to reject H_0; the wheel could be fair.
(6) (a) H_0: the mean parking time is 2 hours 15 minutes; H_a: the mean is greater than 2 hours 15 minutes; (b) $m_c = 2.25 + 1.65(.075) = 2.4$; I will reject H_0 if my outcome exceeds 2.4 hours; (c) Since $2.6 > 2.4$, we reject H_0; the mean exceeds 2.25.
(7) (a) H_0: the mean times are the same; H_a: the mean times are not equal;

(b) $s_1^2 = \dfrac{126.84 - \dfrac{(25)^2}{5}}{4} = .46$; $s_2^2 = \dfrac{96.98 - \dfrac{(22)^2}{5}}{4} = .045$;

$s_p^2 = \dfrac{4(.46) + 4(.045)}{4 + 4} = .2525$; $s_{dm} = \sqrt{.2525\left(\dfrac{1}{5} + \dfrac{1}{5}\right)} = .32$;

Answers to selected exercises

$dm_c = 0 \pm 2.31(.32) = \pm .73$; I will reject H_0 if my outcome is not between $-.73$ and $+.73$; (c) $dm = 25/5 - 22/5 = .6$; we fail to reject H_0; the mean times could be the same.
(8) (a) H_0: the mean times are the same; H_a: the mean of group 1 is longer; one-tail test; (b) $s_{dm} = \sqrt{\dfrac{1}{50} + \dfrac{1}{50}} = .2$; $dm_c = 0 + 1.65(.2) = .33$,

I will reject H_0 if my difference exceeds .33 minutes; (c) $dm = 4.4 - 3.6 = 0.8$; we reject H_0; the movies may have an effect on noise endurance.
(9) (i) to make the groups equivalent and to obtain pairs; (ii) that twins have approximately the same vocabulary skills; (iii) *Solution by Paired t Test:* (a) H_0: there is no difference in the results from the two methods; $\mu_{pop} = 0$; H_a: the results will be different; $\mu_{pop} \neq 0$;

(b)

d	-7	-58	-40	78	10	45	59	0	2	22	111
d^2	49	3364	1600	6084	100	2025	3481	0	4	484	17,191

$$s = \sqrt{\dfrac{17191 - \dfrac{(111)^2}{10}}{9}} = 42.1;\ s_m = 13.3;\ m_c = 0 \pm 2.26(13.3) = \pm 30.1;$$

I will reject H_0 if my outcome is less than -30.1 or greater than 30.1; (c) Since $m = 11.1$, we fail to reject H_0; there may be no difference in learning between these two methods;
Solution by Difference of 2 Means: (a) H_0: no difference between the 2 methods; $\mu_1 = \mu_2$; H_a: $\mu_1 \neq \mu_2$;

(b) $s_1^2 = \dfrac{246223 - \dfrac{(1535)^2}{10}}{9} = 1179;\ s_2^2 = \dfrac{221034 - \dfrac{1468^2}{10}}{9} = 615;$

$s_{pm}^2 = \dfrac{9(1179) + 9(615)}{9 + 9} = 897;\ s_{dm} = \sqrt{897\left(\dfrac{1}{10} + \dfrac{1}{10}\right)} = 13.4;$

$dm_c = 0 \pm 2.10(13.4) = \pm 28.1$;
I will reject H_0 if my difference is less than -28.1 or greater than 28.1; (c) $dm = 1535/10 - 1468/10 = 6.7$; we fail to reject H_0; the two methods may yield the same results.

CHAPTER 12

12-1 Alpha is a probability about H_0, and there is no H_0.
12-3 p lies between $\hat{p} - 1.65\hat{\sigma}$ and $\hat{p} + 1.65\hat{\sigma}$.
12-5 p lies between $.53 \pm 2.58\sqrt{.53(.47)/100}$, between .40 and .66.
12-7 p lies between $.60 \pm 1.96\sqrt{.60(.40)/50}$, between .46 and .74.
12-9 p lies between $.80 \pm 1.65\sqrt{.80(.20)/100}$, between .73 and .87.
12-11 (a) p lies between $.15 \pm 1.96\sqrt{.85(.15)/80}$, between .07 and .23; (b) between 49 and 161.
12-13 (a) p lies between $.32 \pm 1.96\sqrt{.32(.68)/100}$, between .23 and .41; (b) p lies between $.278 \pm 1.96\sqrt{.278(.722)/500}$, between .24 and .32; (c) p lies between $.297 \pm 1.96\sqrt{.297(.703)/1000}$, between .27 and .33;
(e) $\pm 1.96\sqrt{.25/38000} = \pm .005$.

12-15 $(2.58/.05)^2(.25) = 666$; she must time 666 scenes.

12-17
$$\sqrt{\frac{20065 - \frac{(829)^2}{36}}{35}} = 5.28;$$

μ lies between $829/36 \pm 1.96(5.28/\sqrt{36})$, between 21.3 and 24.8.

12-19 μ lies between $14{,}000 \pm 1.96(1000/\sqrt{100})$, between \$13,800 and \$14,200.

12-21 μ lies between $23.1 \pm 1.96(4.1/\sqrt{1000})$, between 22.8 and 23.4.

12-23 μ lies between $12.3 \pm 1.96\left(\frac{5}{\sqrt{64}}\right)$, between 11.1 and 13.5.

12-25 (a) 3300; (b) 3500; (c) 400; (d) 7 months

12-27
$$s = \sqrt{\frac{165 - \frac{(37)^2}{10}}{9}} = 1.77;$$

μ lies between $3.7 \pm 3.25(1.77/\sqrt{10})$, between 1.9 and 5.5.

12-29 μ lies between $29.7 \pm 2.88(3.2/\sqrt{19})$, between 27.6 and 31.8.

12-31 μ lies between $123.1 \pm 1.75(8/\sqrt{16})$, between 119.6 and 126.6.

12-33 probably not

12-35 $s_p^2 = \dfrac{18(1.4)^2 + 12(2.1)^2}{18 + 12} = 2.94;$

$\mu_1 - \mu_2$ lies between $(5.1 - 3.8) \pm 2.75\sqrt{2.94\left(\dfrac{1}{19} + \dfrac{1}{13}\right)}$, between $-.4$ and 3.0.

12-37 $\mu_1 - \mu_2$ lies between $(17.5 - 14.7) \pm 1.96\sqrt{(1.1)^2/50 + (1.3)^2/50}$, between 2.3 and 3.3.

12-39 $p_1 - p_2$ lies between $\left(\dfrac{20}{80} - \dfrac{24}{120}\right) \pm 2.58\sqrt{\dfrac{.25(.75)}{80} + \dfrac{.20(.80)}{120}}$, between $-.11$ and .21.

CHAPTER 13

13-1 (a) You cannot predetermine both the row totals and also the column totals; (b) Attitude toward the new highway and age are independent. Since 38.67 exceeds 5.99 we reject H_0; more young people are for the highway, more older people are undecided. (c) In the last cell $E = \dfrac{4(20)}{50} < 5$. Her friend could either (1) increase the sample size, n, (2) combine the last two columns into "not for," (3) discard the few undecideds.

13-3 (a) H_0: use of word processor and grades are independent; H_a: use of word processor and grades are dependent; (b) $2 \times 2 = 4$; (c) $X_c^2 = 9.49$.

Answers to selected exercises

(d)

	always	sometimes	never	
A, B	20.33	19.67	20	60
C, D	20.33	19.67	20	60
F	20.33	19.67	20	60
	61	59	60	$n = 180$

(e) $X^2 = 59.57$; (f) Since 59.57 exceeds 9.49, we reject H_0; the increased use of the word processor is related to higher grades.

13-5

	*	#	†	
history	61.30	52.85	52.85	167
physical science	71.58	61.71	61.71	195
social science	91.04	78.48	78.48	248
computer science	66.08	56.96	56.96	180
	290	250	250	$n = 790$

Since $X^2 = 116.39$ exceeds $X_c^2 = 12.59$, we reject H_0; symbols and type of manuscript are dependent.

13-7

46.5	7.5	6
77.5	12.5	10

No, since $X^2 = .40 < X_c^2 = 5.99$, we fail to reject H_0.

13-9

9.95	5.05
14.59	7.41
25.2	12.8
13.26	6.74

Since $X^2 = 22.61$ exceeds $X_c^2 = 7.82$ we reject H_0; a relationship does exist.

13-11

20	20	20
14.7	14.7	14.7
18	18	18
12	12	12
11.3	11.3	11.3
9	9	9
9.3	9.3	9.3
5.7	5.7	5.7

Since $X^2 = 26.3$ does not exceed $X_c^2 = 29.14$ we fail to reject H_0; we fail to prove that pole position affects final position.

13-13

11	9	20
9	11	20
20	20	$n = 40$

Without correction factor, $X^2 = \dfrac{(11 \cdot 11 - 9 \cdot 9)^2 (40)}{20^4} = .40$; with correction factor, $X^2 = \dfrac{(|121 - 81| - 20)^2 (40)}{20^4} = .10$; since $X_c^2 = 3.84$ we fail to show that there is any difference resulting from the two approaches.

13-15 (a) $z = .89$; (b) $z = .89$; (c) $X = .89$; (d) $X_c = 2.58$; (e) they are all equal; (f) they are mathematically equivalent.

13-17 | 16 | 8 | 32 | 24 | Since $X^2 = 11.7$ exceeds $X_c^2 = 7.82$ we reject his theory; there are more drinkers and fewer abstainers than he theorized.

13-19 observed: | 100 | 600 | 300 | expected: | 250 | 500 | 250 |
Since $X^2 = 120$ exceeds $X_c^2 = 5.99$ we reject H_0; we do not have a binomial distribution with $p = .50$ in this population.

13-21 Combining the last 3 cells so that $E > 5$ and using $\alpha = .01$, we have $X^2 = 182.3$ bigger than $X_c^2 = 11.34$; hence we reject the theory that the bugs are distributed randomly; they tend to cluster together.

13-23 (a) $p_1 - p_2$ lies between $\left(\dfrac{30}{100} - \dfrac{40}{200}\right) \pm 1.96\sqrt{\dfrac{.3(.7)}{100} + \dfrac{.4(.6)}{200}}$, between $-.01$ and $.21$, that is, p_1 could equal p_2;
(b) $\hat{p} = (30 + 40)/(100 + 200) = .23$; $d\hat{p}_c = 0 \pm 1.96\sqrt{.23(.77)/100 + .23(.77)/200} = \pm.101$; since $dp = .30 - .20 = .10$ lies between $-.101$ and $.101$ we fail to reject H_0, p_1 could equal p_2; note, however, $.10$ and $.101$ are extremely close; if the results were very important we may wish to repeat the experiment (with larger samples if possible);

(c)

30	40	70
70	160	230
100	200	$n = 300$

$X^2 = \dfrac{[30(160) - 40(70)]^2 (300)}{100(200)(70)(230)} = 3.73$ is less than $X_c^2 = 3.84$; so we fail to reject H_0.

13-25 (a) $E = 10{,}654/12 = 887.8$; since $X^2 = 114.14$ exceeds $X_c^2 = 24.72$, we reject H_0; homicide is more likely in December and July, less likely in February and March; (b) $\hat{p} = (1042 + 1024)/(10654 + 10654) = .10$; $d\hat{p}_c = 0 \pm 2.33\sqrt{.10(.90)(1/10654 + 1/10654)} = \pm.0096$; $d\hat{p} = 1042/10654 - 1024/10654 = .002$; we fail to reject H_0; there is not a significant difference; (c) $m_1 = 744/[10(28) + 3] = 2.6$ exceeds; $m_2 = 789/[10(31)] = 2.5$; true.

13-27 (a)

i, vi	5.9
ii, v	35.3
iii, iv	88.7

(b) No, since $X^2 = 20.31$ exceeds $X_c^2 = 11.07$.

Answers to selected exercises

13-29 $E = 235/3 = 78.3$; $X^2 = 1.6$ does not exceed $X_c^2 = 5.99$; we fail to show any difference between the three pills.

SAMPLE TEST FOR CHAPTERS 12 AND 13

(1) It would extend from $-\infty$ to $+\infty$.
(2) (a) p lies between $30/200 \pm 1.96\sqrt{.15(.85)/200}$, between .10 and .20; (b) if the association was representative of the profession, and not biased by sex.
(3) $n \geq (1.96/.03)^2(.25) = 1.067.1$; i.e., $n \geq 1068$.
(4) (a) The response time of all emergency calls received between 8 A.M. and 4 P.M. by the Emergency Medical Service in New York City; was this an average day or was 88 calls extremely heavy or extremely light? were all calls reported? etc.; (b) μ lies between $28.8 \pm 1.96(15/\sqrt{88})$ between 25.7 and 31.9 minutes.
(5) $(p_1 - p_2)$ lies between $(36/50 - 18/50) \pm 1.96\sqrt{.72(.28)/50 + .36(.64)/50}$, between 18% and 54%; bias against women; the fact that women have entered marginal positions only recently and therefore have less time in such jobs, etc.

(6)

29.7	391.2	39.1	460
151.7	1998.4	199.8	2350
23.6	310.4	31.0	365
205	2700	270	$n = 3175$

Since $X^2 = 77.99$ exceeds $X_c^2 = 9.49$, we reject H_0; there are more short father–short son and tall father–tall son pairs than expected from a random distribution; the heights of father and son are dependent variables.

(7)

20	60	80
30	90	120
50	150	$n = 200$

Yes, since $X^2 = \dfrac{[20(90) - 60(30)]^2(200)}{50(150)(80)(120)} = 0$

CHAPTER 14

14-1 r close to -1 indicates that those with high grades in Spanish I received low grades in Spanish II; r close to $+1$ indicates that those with high grades in Spanish I received high grades in Spanish II; r close to zero indicates that there is very little connection between grades in Spanish I and grades in Spanish II.

14-3 (a)

X	Y	X^2	Y^2	XY
1	1	1	1	1
2	1	4	1	2
3	2	9	4	6
4	3	16	9	12
10	7	30	15	21

Understanding statistics

(b) 4; (c) 10; (d) 7; (e) 21; (f) 30; (g) 15; (h) 100; (i) 49; (j) $4(21) - 10(7) = 14$; (k) $\sqrt{4(30) - 10^2} \sqrt{4(15) - 7^2} = 14.83$; (l) $14/14.83 = .94$

14-5 (a)

X	Y	X^2	Y^2	XY
2	−1	4	1	−2
2	3	4	9	6
0	−2	0	4	0
3	4	9	16	12
4	4	16	16	16
1	0	1	0	0
12	8	34	46	32

(b) 6; (c) 12; (d) 8; (e) 32; (f) 34; (g) 46; (h) 144; (i) 64; (j) $6(32) - 12(8) = 96$; (k) $\sqrt{6(34) - 144} \sqrt{6(64) - 64} = 113$; (l) $96/113 = .85$.

14-7 r is near -1; $r = -.98$

14-9 $r = .9795$

14-11 (a) $r = .96$; (b) $r = .99$

14-13 $r = \dfrac{5(17274) - 348(248)}{\sqrt{5(24258) - 348^2} \sqrt{5(12306) - 248^2}} = .95$

14-17 (a) H_0: the population has zero correlation; H_a: the population has nonzero correlation; (b) two; (c) $r_c = \pm.87$; (d) fail to reject H_0 since $r = -.24$ lies between $-.87$ and $+.87$; there is not enough evidence to support nonzero correlation at the .01 significance level.

14-19 (a) H_0: the population has zero correlation; H_a: the population has nonzero correlation; (b) two; (c) $r_c = \pm.96$; (d) Since $r = .99$ exceeds .96 we reject H_0; there is evidence of nonzero (positive) correlation.

14-21 (a) H_0: the population has zero correlation; H_a: the population has positive correlation; (b) one; (c) $r_c = .98$; (d) Since $r = .99$ exceeds .98 we reject H_0; we have evidence of positive correlation.

14-23 $r = .90$ exceeds $r_c = .34$; reject H_0; there is positive correlation.

14-25 (a) $r = \dfrac{7(1845) - 215(59)}{\sqrt{7(7475) - 215^2} \sqrt{7(507) - 59^2}} = .36$; $r_c = .67$; fail to show positive correlation;

(b) $r = \dfrac{6(3950) - 50(480)}{\sqrt{6(426) - 50^2} \sqrt{6(39400) - 480^2}} = -.52$; $r_c = -.73$; fail to show negative correlation;

(c) $r = \dfrac{6(522) - 52(62)}{\sqrt{6(458) - 52^2} \sqrt{6(840) - 62^2}} = -.40$; $r_c = .81$; fail to show nonzero correlation

14-27 (a) 83; (b) 83 is the predicted average score

14-29 (a) $r = -.61$; (b) $b = -2.64$; (c) $Y = 74.4 - 2.64(X - 8.9)$; (d) $Y = 68.9$

14-31 $Y = 50 - 1.4(48 - 36) = 33.2$

14-33 (b) $r = \dfrac{8(-21{,}071) - (-546)(372.5)}{\sqrt{8(39{,}992) - (-546)^2} \sqrt{8(24{,}311.25) - 372.5^2}} = .998$;

Answers to selected exercises

(c) $Y = \dfrac{372.5}{8} + \left(\dfrac{34{,}817}{21{,}820}\right)\left(X - \dfrac{-546}{8}\right) = 46.6 + 1.60\,(X + 68.25)$;

(d) $Y = 46.6 + 1.60(8.25) = 59.8$

14-35 $1 - (.3)^2 = 91\%$

14-37 $\sqrt{(.4)^2(4)} = .8$

CHAPTER 15

15-1 Are two variances equal?

15-3 (a) $H_a: \sigma^2 > 5^2$

$X^2 = \dfrac{30(8)^2}{5^2} = 76.8$ exceeds $X_c^2 = 43.77$; reject H_0; evidently the standard deviation is more than 5; (b) $H_a: \mu \neq 125$; $m_c = 125 \pm 1.96(8)/\sqrt{31} = 122.2$ and 127.8; $m = 150$; reject H_0; evidently the mean distance is more than 125 feet.

15-5 (a) $H_a: \sigma^2 \neq 10^2$; $s^2 = \dfrac{232{,}931 - \dfrac{1525}{10}}{9} = 40.94$; $X^2 = \dfrac{9(40.94)}{10^2} = 3.685$; $X_c^2 = 2.70$ and 19.02; it is possible that $\sigma = 10$; we do not have strong enough evidence to reject that hypothesis; (b) $ND_L(\mu = 152.5, \sigma = 10)$;

$z_{162} = \dfrac{162 - 152.5}{10} = .95$; $P(L \geq 162) = 1 - .83 = .17$

15-7 $H_0: \sigma = 10$; $H_a: \alpha < 10$; $\alpha = .05$; degrees of freedom $= 7$; $X_c^2 = 2.17$; $X^2 = (n-1)s^2/\sigma^2 = 7(64)/100 = 4.48$. The sample value of s is not low enough to establish that σ is less than 10.

15-9 $H_0: \sigma = 1$; $H_a: \sigma > 1$; $\alpha = .05$; degrees of freedom $= 24$; $X_c^2 = 36.42$; $X^2 = (24)9/1 = 216$. Since $X^2 > X_c^2$ we have evidence that σ is more than 1.

15-11 $H_0: \sigma_1^2 = \sigma_2^2$; $H_a: \sigma_1^2 \neq \sigma_2^2$.
(a) Degrees of freedom (numerator) $= 34$, degrees of freedom (denominator) $= 19$; $F_c = 2.39$; $F = s_1^2/s_2^2 = .2^2/.15^2 = .04/.0225 = 1.78$. Fail to reject H_0. It is not unreasonable to assume variances are equal.
(b) Degrees of freedom (numerator) $= 14$, degrees of freedom (denominator) $= 14$; $F_c = 2.95$; $F = .81/.49 = 1.65$. Fail to reject H_0. It is not unreasonable to assume $\sigma_1^2 = \sigma_2^2$.
(c) Degrees of freedom (numerator) $= 19$, degrees of freedom (denominator) $= 19$; $F_c = 2.51$; $F = 11.10^2/9.48^2 = 1.37$. Fail to reject H_0. It is not unreasonable to claim $\sigma_1 = \sigma_2$.
(d) Degrees of freedom (numerator) $= 9$, degrees of freedom (denominator) $= 9$; $F_c = 4.03$; $F = 3.2^2/3.0^2 = 1.14$. Fail to reject H_0. It is not unreasonable to claim that $\sigma_1^2 = \sigma_2^2$.

15-13 Given $\alpha = .01$, $s_1 = 40$, $n_1 = 20$, $s_2 = 100$, $n_2 = 20$. $H_0: \sigma_1^2 = \sigma_2^2$; $H_a: \sigma_2^2 > \sigma_1^2$; $F = 100^2/40^2 = 6.25$; $F_c = 3.00$. Since $F > F_c$ we reject H_0. The action in the oat plants is more erratic.

15-15 Given $\alpha = .01$, $s_1 = 1.3$, $n_1 = 10$, $s_2 = 2.4$, $n_2 = 10$. $H_0: \sigma_1^2 = \sigma_2^2$; $H_a: \sigma_1^2 \neq \sigma_2^2$; $F_c = 6.54$; $F = 2.4^2/1.3^2 = 3.41$. Fail to reject H_0. It is not unreasonable to assume the pills are approximately equal in variability of effect.

15-17 $s_w^2 = \dfrac{2.12^2 + 3.3^2 + 3.5^2}{3} = 9.21$, d.o.f. $= 4 + 4 + 4 = 12$;

m	49	44	48.2	$\Sigma m = 141.2$
m^2	2401	1936	2323.24	$\Sigma m^2 = 6660.24$

$s_m^2 = \dfrac{6660.24 - (141.2)^2/3}{2} = 7.21$

$s_A^2 = 5(7.21) = 36.1$, d.o.f. $= 3 - 1 = 2$; $F = 36.1/9.21 = 3.92$ does not exceed $F_c = 3.98$; we fail to reject H_0; the three population means could be equal.

15-19 $s_w^2 = \dfrac{s_1^2 + s_2^2 + s_3^2}{N} = \dfrac{129.2 + 254.2 + 38.3}{3} = 140.6$; $s_A^2 = ns_m^2 = 5(172,186) = 860,930$; $F = \dfrac{s_A^2}{s_w^2} = 6125$; clearly larger than F_c for $\alpha = .01$.

15-21 H_a: The mean acidity is not the same in all 3 grape varieties; $s_w^2 = \dfrac{20^2 + 23^2 + 18^2}{3} = \dfrac{1253}{3} = 417.67$; $s_m^2 = 281.3$; $s_A^2 = 15(28.3) = 4220$; $F = \dfrac{4220}{417.67} = 10.10$; $F_c = 3.23$; this indicates that the grape varieties differ in acidity.

CHAPTER 16

16-1 Test 2 needs normal distributions and knowledge of the variances. Test 3 needs matched pairs. Test 2 is more powerful than test 1, and test 3 is more powerful than test 2. You would use the most powerful test that you could in any given situation.

16-3 (a) $S_c = 18(.5) \pm 1.96\sqrt{18(.5)(.5)} = 4.8$ and 13.2; $S = 13$ plus signs. We fail to show that the variations in weights are not random. They may be random.

(b) $s = \sqrt{\dfrac{1904 - 24^2/20}{19}} = 9.93$; $m_c = 0 \pm 2.09\ (9.93/\sqrt{20}) = \pm 4.64$; $m = 24/20 = 1.2$. Fail to show variations in weight are not random. They may be random.

16-5 $H_a: p \neq .5$, $n = 65$; $\sigma = \sqrt{\dfrac{.25}{65}} = .062$; $\hat{p}_c = .50 \pm 1.96(.062) = .38$ and $.62$; $\hat{p} = \dfrac{40}{65} = .62$; our sample value equals the critical value; this is probably strong enough to indicate that H_0 is not correct; further tests could be done to see if the pattern holds.

16-7 $H_a: p > .5$, $\sigma = \sqrt{\dfrac{.25}{28}} = .168$; $\hat{p}_c = .50 \pm 1.65(.168) = .22$ and $.78$; $\hat{p} = \dfrac{17}{28} = .61$; fail to show that Adam's median is bigger than 8.5.

Answers to selected exercises

16-9 $H_a: p > .5$ (*more* than *half* the time, no traffic stopped);
$\sigma = \sqrt{\frac{.25}{40}} = .079$; $\hat{p}_c = .50 + 2.33(.079) = .68$; $\hat{p} = \frac{30}{40} = .75$;
indicates that median number of times that traffic stopped is less than 1.

16-13 $\mu_R = 2(13)(12)/(13 + 12) + 1 = 13.48$;
$\sigma_R = \sqrt{\frac{2(13)(12)[2(13)(12) - 13 - 12]}{[(13 + 12)^2(13 + 12 - 1)]}} = 2.44$;
$R_c = 13.48 \pm 1.96(2.44) = 8.7$ and 18.3;
since $R = 12$ we fail to reject H_0; his wins and losses may be randomly distributed.

16-15 H_a: fast and slow people do not arrive randomly; $n_1 = 14$, $n_2 = 11$;
$\mu_R = \frac{2(14)(11)}{14 + 11} + 1 = 13.32$; $\sigma_R = \sqrt{\frac{308(283)}{25^2(24)}} = 2.41$;
$R_c = 13.32 \pm 1.96(2.41) = 8.6$ and 18.0;
$R = 10$; fail to show people do not arrive randomly by speed.

16-17 (a) TFFFF
FTFFF
FFTFF
FFFTF
FFFFT

$\frac{2 + 3 + 3 + 3 + 2}{5} = 2.6$; (b) $\frac{2(1)(4)}{1 + 4} + 1 = 2.6$

(c)
2	4
3	9
3	9
3	9
2	4
13	35

$\sigma = \sqrt{\frac{35 - \frac{13^2}{5}}{5}} = .49$; $\sigma = \sqrt{\frac{2(1)(4)[2(1)(4) - 1 - 4]}{(1 + 4)^2(1 + 4 - 1)}} = .49$

16-19 $U = 4 + 4 + 4 = 12$ (one-tail test); $U_c = 12$; we fail to reject H_0; we fail to prove that tutoring does help.

16-21 two-tail test; $U_1 = 3(7 + ½) + 2(5 + ⅔) + 2(4 + ½) + 2(½) = 47.5$; $U_2 = 9(8) - 47.5 = 24.5$; $U_c = 70$; fail to reject H_0; the quality could be the same.

16-23 $n_1 = 110$, $n_2 = 90$, $N = 200$; mean $= \frac{110(90)}{12} = 825$

T	T^3	$T^3 - T$
40	64,000	63,960
40	64,000	63,960
40	64,000	63,960
40	64,000	63,960
40	64,000	63,960
		319,800

Understanding statistics

$$C = \frac{110(90)}{12(200)(199)}(319{,}800) = 6629.0;$$

$$\sigma = \sqrt{\frac{110(90)(201)}{12} - 6629.0} = 399.0$$

16-25

x	143	141	132	128	118	118	107	101	100	93	89
z	142	140	135	130	120	120	102	97	95	60	60

$n_1 = 11$, $n_2 = 11$, $N = 22$; $\mu = \frac{11(11)}{2} = 60.5$; $\sigma = \sqrt{60.5(23)} = 37.3$; $U_c = 60.5 \pm 1.96(37.3) = -12.6$ and 133.6; $U_1 = 11 + 10 + 8 + 7 + 2(5) + 5 + 4 + 4 + 2 + 2 = 63$; $U_2 = 121 - 63 = 58$; fail to reject H_0; neither country's agents are significantly superior.

16-27

T	T^3	$T^3 - T$
10	1,000	990
30	27,000	26,970
30	27,000	26,970
30	27,000	26,970
		81,900

$$C = \frac{50(50)}{12(100)(99)}(81900) = 1723.5; \mu = \frac{50(50)}{12} = 208.3;$$

$\sigma = \sqrt{208.3(101) - 1723.5} = 139.0$; $U_c = 208.3 + 1.65(139.0) = 437.6$; $U = 25(45 + 5/2) + 15(30 + 15/2) + 10(10 + 20/2) = 1942.5$; reject H_0; the new type of gasohol gives better mileage.

SAMPLE TEST FOR CHAPTERS 14 TO 16

(1) (a), (c), and (d) positively; (b) and (e) negatively

(2) r cannot exceed 1.

(3) (a) $r = \dfrac{6(277{,}482.8) - 6814(314.8)}{\sqrt{6(28{,}608{,}788) - 6814^2}\sqrt{6(17{,}161.5) - 314.8^2}} = -.69;$

(b) $b = \dfrac{-480{,}150.4}{125{,}222{,}132} = -.00383$; (c) $Y = 52.47 - .00383(X - 1135.7)$

(4) $s^2 = \dfrac{1006.0246 - \dfrac{(100.3)^2}{10}}{9} = .00173$; $s = .042 < .10$;

$X^2 = \dfrac{9(.00173)}{(.01)^2} = 155.7 > 16.92 = X_c^2$; we reject H_0; σ is greater than .01.

(5) $H_0: \sigma_1 = \sigma_2$; $H_a: \sigma_1 \neq \sigma_2$ (two-tail)

$s_1^2 = \dfrac{2{,}126{,}672 - \dfrac{4606^2}{10}}{9} = 572.0$, d.o.f. $= 9$

Answers to selected exercises

$$s_2^2 = \frac{1{,}995{,}822 - \frac{4456^2}{10}}{9} = 1136.5, \text{ d.o.f.} = 9$$

$F = \frac{1136.5}{572.0} = 1.99$, does not exceed $F_c = 4.03$; we fail to reject H_0; the variances may be the same.

(6) $s_w^2 = \frac{68.9 + 122.9 + 39.6 + 122.9}{4} = 88.6$, d.o.f. $= 3 + 3 + 3 + 3 = 12$; $m_1 = 110.75$, $m_2 = 111.25$, $m_3 = 138.75$, $m_4 = 146.25$;

$$s_m^2 = \frac{65{,}282.75 - \frac{507^2}{4}}{3} = 340.17$$

$s_A^2 = 4(340.17) = 1360.68$, d.o.f. $= 4 - 1 = 3$

$F = \frac{1360.68}{88.6} = 15.36$ exceeds $F_c = 3.49$; reject H_0; the averages are not all the same;

(7) $n = 37 + 28 = 65$; $\mu = 65(.5) = 32.5$; $\sigma = \sqrt{65(.5)(.5)} = 4.03$
$S_c = 32.5 + 1.65(4.03) = 39.2$; $S = 37$ does not exceed 39.2; we fail to reject H_0; we have not proved that Channel 0 is more popular.

ANSWERS TO EXERCISES IN APPENDIX A

1 (a) 23.73; (b) 4.97; (c) .046; (d) 1.202; (e) 9.628; (f) .0012; (g) 21.25; (h) .5; (i) .02; (j) 0
2 (a) −2.33; (b) .081, .41, .6, 4.51, 4.7; (c) −.273, .273, .41
3 (a) true; (b) true; (c) false; (d) true; (e) 5, 6; (f) 4, 5, 6; (g) 0, 1, 2, 3, 4; (h) 0, 1, 2, 3, 4, 5; (i) 3, 4; (j) 2, 3, 4, 5
4 (a) 5%; (b) .3%; (c) .37; (d) .032; (e) .375, 37.5%; (f) .263, 26.3%; (g) 11.5; (h) 8; (i) 30%; (j) 25%
5 (a) −15; (b) +15.35; (c) −16.5; (d) −2; (e) −5
6 (a) 7.446; (b) −41; (c) 11.167; (d) 6.92 and −.92; (e) true
7 (a) +16; (b) .343; (c) 1/16; (d) .3087
8 (a) 1.550; (b) 2.315; (c) 4.45

ANSWERS TO EXERCISES IN APPENDIX B

1 (a) independent; (b) independent; (c) dependent; (d) independent
3 A and B are dependent events.
5 (a) no; (b) yes; (c) yes; (d) no
7 (a) .20; (b) no
9 1/1,000,000
11 (a) 1/16; (b) 1/169; (c) 2/169; (d) 9/169; (e) 1/17; (f) 4/663; (g) 8/663; (h) 11/221
13 (a) 1/2; (b) 3/8   15 2/663   17 .08

Index

Alternative hypothesis, 112
Analysis of variance, 251, 258, 262
Approximate number, 4, 5
Arithmetic review, 292
Average, 11

Bar graph, 37
Best-fitting line, 243
Binomial coefficient, 69
Binomial distribution, 67
    approximation by normal, 98
Binomial probability table, 79, 304
Binomial variable, 68
Boundaries, histogram, 38

Calculators, 4
Central limit theorem, 156
Chi-square distribution, 208
Chi-square table, 317
Chi-square test, 204
Class survey, 9
Confidence intervals, 188
Contingency table, 204, 213, 215

Correction factor (2 × 2 contingency table), 215
Correlation, 229
Correlation coefficients, 229

Data, 3
Decision rules, 115
Degrees of freedom, 172, 208, 215
Dependent event, 296
Descriptive statistics, 2
Deviations from the mean, 17
Differences:
    between means, 163, 165, 199
    between proportions, 198
Distribution, 3
Distribution-free tests, 269

Equally likely outcomes, 56
Estimate, 141
Estimation, 141, 188
Event, 56
Exact numbers, 5
Expected results, 205

INDEX

F ratio, 255
F ratio test, 258
F tables, 320–323
False positive, 77, 193
Frequency table, 36

Goodness of fit, 215
Graphs, 36, 43
Grouped data, 16

Histogram, 37, 59
Hypothesis testing, 111, 130

Independence, 204
Independent events, 294
Independent variables, 207
Interval, histogram, 39
Interval estimate, 188

Linear correlation, 231
Linear relationship, 229

Mann-Whitney U test, 277
Matched pairs, 182, 270
Mean, 11
Means, comparison of, 163
Measure of central tendency, 11
Measure of variability, 16
Median, 11
Median test, 274
Mode, 11
Motivated hypothesis, 113
Multiplication rule, 70
Mutually exclusive events, 298

Nonparametric test, 269
Normal curve, 83, 85
   table, 310
   use of, 86
Normal distribution, 82–97
   table, see Normal curve table
Null hypothesis, 112

Observed results, 206
Octothorpe, 219
One-tail tests, 113

Paired differences, 179
Parameter, 12
Pascal's triangle, 72, 73, 303
Percentile rank, 26, 43
Pooled estimate, 145
Population, 4
Power, 128
Prediction, 242
Probability, 55, 294
Random order, 56
Random sample, 4
Randomness, test for, 274
Range, 16
Rank order, 269
Rates, 31
Raw score, 3, 24, 88
Regression line, 243
Replacement, 297
Robust test, 270
Rounding off, 4
Runs test, 274

Sample, 4
   large, 156, 194
   small, 171, 196, 200
Sample mean, 155
Sample means, theoretical
   distribution of, 155
Sample size, 191
Scatter, 16
Scattergram, 230
Screening test, 77, 193
Sign test, 270
Significance level, 119
Spread, 16
Standard deviation, 16, 158
Statistical errors, 115
Statistical hypotheses, 111
Statistical inference, 2
Statistics, 12
Student's t distribution, 171

t curve, 172
t scores, 172
t table, 316
t test, 174
Tree diagram, 63
Two-sample test, 141, 176
Two-tail test, 113
Type I error, 116
Type II error, 116

U table, 324, 325

Variability, 16
Variable, 67
Variance, 18, 250
Variances, comparison of, 251, 255
Vital statistics, 31

z scores, 23, 24, 43

Frequently Used Formulas *(Continued from front cover)*

Chapter 11

One-Sample Tests
1. Degrees of freedom $= n - 1$
2. $\mu_m = \mu_{pop}$
3. $s_m = \dfrac{s}{\sqrt{n}}$
4. $m_c = \mu_m + t_c s_m$
5. Experimental outcome, $m = \dfrac{\Sigma X}{n}$

Two-Sample Tests
1. Degrees of freedom $= n_1 + n_2 - 2$
2. $\mu_{dm} = \mu_1 - \mu_2$
   (if H_0 states that $\mu_1 = \mu_2$, then $\mu_1 - \mu_2 = 0$)
3. $s_{dm} = \sqrt{\dfrac{s_1^2}{n_1} + \dfrac{s_2^2}{n_2}}$
4. $dm_c = \mu_{dm} + t_c s_{dm}$
5. Experimental outcome, $dm = m_1 - m_2$

Chapter 12

Distribution of Sample Proportions
1. $\mu = p$
2. $\sigma = \sqrt{\dfrac{pq}{n}}$
3. $\hat{\sigma} = \sqrt{\dfrac{\hat{p}\hat{q}}{n}}$
4. p is between $\hat{p} - z_c \hat{\sigma}$ and $\hat{p} + z_c \hat{\sigma}$

Distribution of Sample Means

5. $s_m = \dfrac{s}{\sqrt{n}}$
6. μ is between $m - z_c s_m$ and $m + z_c s_m$ (large samples)
7. μ is between $m - t_c s_m$ and $m + t_c s_m$ (small samples)

Differences between Two Proportions

8. $\hat{\sigma} = \sqrt{\dfrac{\hat{p}_1 \hat{q}_1}{n_1} + \dfrac{\hat{p}_2 \hat{q}_2}{n_2}}$
9. $p_1 - p_2$ is between $d\hat{p} - z_c \hat{\sigma}$ and $d\hat{p} + z_c \hat{\sigma}$
10. $s_p^2 = \dfrac{(n_1 - 1)s_1^2 + (n_2 - 1)s_2^2}{n_1 + n_2 - 2}$
11. $s_{dm} = \sqrt{s_p^2 \left(\dfrac{1}{n_1} + \dfrac{1}{n_2}\right)}$
12. $\mu_1 - \mu_2$ is between $dm - z_c s_{dm}$ and $dm + z_c s_{dm}$ (large samples)
13. $\mu_1 - \mu_2$ is between $dm - t_c s_{dm}$ and $dm + t_c s_{dm}$ (small samples)